CW00739147

# MODERN CONSTRUCTION HANDBOOK

ANDREW WATTS

MODERN CONSTRUCTION SERIES

# CONTENTS

## AIMS OF THIS BOOK

The Modern Construction Handbook is a textbook for students and young practitioners of architecture, as well as students of structural and environmental engineering who wish to broaden their study by examining the architect's point of view of building practice and available technologies. It shows the principles of the main construction methods used today and illustrates this through drawings and typical generic details that can form a starting point for a particular design.

Each of the book's six chapters examines a particular aspect of construction from Materials to Walls, Roofs, Structure, Environment and Fittings. Each double page explains specific elements which are accompanied by drawn and annotated generic details. Throughout the book, built examples by high profile designers are used to illustrate generally accepted principles. The construction techniques described can be applied internationally.

## NARRATIVE STRUCTURE

This book is structured so that it reads in two ways: 1) as a continuous 'narrative' read in sequence which may serve as a textbook for a teaching course and 2) as a subject-orientated handbook whereby the user may focus on a specific topic for reference purposes. The chapters on Materials, Walls and Roofs are introduced first to explain the interrelationships of constructional logic rather than simply to follow a narrative of the sequence of construction.

The Materials chapter focuses on the physical properties of materials and compares such topics as strength, stiffness, thermal expansion and related properties. In addition, the production process of each material is discussed in order to appreciate the constraints imposed upon design by materials. An additional introductory essay on embodied energy discusses the importance of the judicious use of materials.

## WALLS CHAPTER

The Walls chapter sets out generic wall types in terms of the principal material used and distinguishing between loadbearing and non-loadbearing facades. In addition to explaining how each wall type is constructed, the text describes recent technical developments that allow the reader to follow trends in their use. One introductory essay discusses wider trends of the increased use of thermal insulation, rainscreens and internally drained and ventilated framing. A second essay describes the development of 'thin' and 'layered' facades

## ROOFS CHAPTER

The Roofs chapter separates drainage, finishes and substrates into separate components. This approach allows the reader to appreciate that these constituent parts can be assembled in a wide range of combinations of substrate and finish. The chapter begins with an essay focusing on trends of the increased use of roofs a 'non accessible' facades and as usable external spaces.

# INTRODUCTION

## STRUCTURE CHAPTER

The Structure chapter introduces the most commonly used generic structural 'elements' and shows their interface with walls and roofs through illustrations of typical structural details. Where most textbooks on structure focus on structural action and calculation, this book focuses on joints, junctions and interface with the external envelope. In other words, it derives from the architect's perspective. The reader clearly sees how structure is not an isolated element; rather, it is designed in close conjunction with the external envelope.

## ENVIRONMENT CHAPTER

The Environment chapter picks up themes of Structure, Walls and Roofs and shows how they can be used to modify environmental conditions through a set of active controls and passive controls. The chapter starts with a discussion of the changing relationship between structure, cladding and energy. Support services such as sanitation, drainage and the design of sanitary spaces are discussed, as these form a significant part of the architect's traditional contribution in this area.

## FITTINGS CHAPTER

The Fittings chapter shows how stairs, lifts, partitions, floors, ceilings and doors are detailed. This chapter focuses on issues of the prefabrication and site-based techniques and the influence that each has on the completed construction. Stairs, lifts and escalators are treated as 'fittings' in a wider sense of the word to describe those elements of construction which fit into the structure and serviced space within an external envelope.

## QUALIFYING COMMENTS

The building techniques discussed and the built examples shown are designed to last for an extended period with a relatively high performance. Consequently, buildings for exhibitions and for temporary use are excluded. In addressing an international readership, references to national legislation, building regulations, codes of practice and national standards have specifically not been included. This book explains the principles of accepted building techniques currently in use. Building codes throughout the world are undergoing increased harmonisation because of increased economic and intellectual globalisation. Building components and assemblies from many different countries are often used in a single building. Since building codes are written to protect users of buildings by providing for their health and safety, good construction practice will always uphold these codes as well as assist their advancement. The components, assemblies and details shown in this book describe many of the building techniques used by the building industry today, but this book does not necessarily endorse or justify their use since techniques in building are in a continual state of change and development. Values given for the physical properties of materials are average figures. They are to be used for general comparison purposes and not for structural design.

# 1

## MATERIALS

## An introduction to embodied energy

The energy consumed in the manufacturing of construction material, its transport, its installation and ultimate renovation or demolition is almost always in the form of non-renewable fossil fuel which produces carbon dioxide ($CO_2$) emissions as a result of these processes. Atmospheric emission levels are widely regarded as one of the most significant considerations associated with the production of building material and, particularly, the emission of $CO_2$ as the chief contributor towards climate change.

The energy embodied in a building material can be considered to be the sum of the energy used in the manufacture of its components, transportation, assembly on site, together with its ultimate demolition and associated recycling. This embodied energy must be seen in relation to the overall amount of energy expended by a building during its lifetime. The total energy consumption of a building can be considered to be the sum of its embodied energy and its operational energy consumed in using the building during its lifetime. Embodied energy currently accounts for only around 10% of the lifetime energy expenditure of a building, based on a 50 year period. However, levels of the operational energy component are set to reduce significantly. This will be done by improving thermal insulation and by a mixture of passive and active controls that control ventilation heat loss and solar heat gain. This would make the embodied energy component in a building even more significant.

Since reductions in the operational energy component are currently being researched, the relationship between embodied energy and operational energy is not discussed here. Instead, this text focuses on how typical forms of construction might best be combined to create buildings of low embodied energy, regardless of the individual levels of embodied energy in the materials used.

## Embodied energy values

The figures for embodied energy used in this chapter are those quoted by The Institution of Structural Engineers of the United Kingdom in their publication Building for a Sustainable Future: Construction without Depletion, which lists the total embodied energy for seventy-one materials including transportation but excluding demolition or recycling. These figures are consistent with an earlier UK Steel Construction Institute's publication 'A Comparative Environmental Life Cycle Assessment of Modern Office Buildings' by K J Eaton and A Amato. Precision in measuring embodied energy is difficult when the amount of energy used in the production, transportation to site and installation of the material varies between locations.

## Levels of embodied energy in building materials

Current research has focused on levels of embodied energy in individual materials used in buildings rather than in specific types of construction, which use a combination of materials. The body of research reveals that natural materials such as stone have low embodied energy. Reinforced concrete, bricks, concrete blocks, timber elements and other wood products such as laminated timber and plywood have higher EE values. Timber, excluding transportation, has a very low embodied energy, but much of the timber used in highly industrialised countries is imported, which forms a relatively high proportion of the total embodied energy level of this material. The embodied energy of steel is higher still, with aluminium having by far the most embodied energy of the common building materials. These findings have led to an assumption that natural materials are more sustainable.

## Comparisons of embodied energy in typical forms of building construction

In order to compare levels of embodied energy, two worked examples have been taken for comparison. The first example is a set of options for the structural frame of an office building. Similar spans and loadings are used for a wide range of buildings in the UK, from offices to educational buildings and hospitals. The structural frames investigated use either steel or reinforced concrete. Timber was not considered as it is not economical for large spans with high loadings. The second example comprises a set of options for wall cladding in different materials. The two sets of examples could be used as part of the same building. Either a steel or concrete frame could be used with any of the cladding options.

The examples were intended to be used to find out how materials for both structure and external wall might best be used both separately and together to achieve low levels of embodied energy in their construction. Examples of roofs were not used since they would have broadened the scope of investigation too widely into comparisons of construction methods for pitched and flat roofs in various configurations.

The effect of thermal insulation on levels of embodied energy in the wall cladding options was investigated as a separate exercise. Since most forms of thermal insulation can be used with most forms of building construction, a comparison was made between the use of a material with very high levels of embodied energy (expanded polystyrene) and one with very low levels of embodied energy (mineral fibre insulation). Taking the 3x3 metre bay with a 1x1 metre window used in the cladding options, it was found that expanded polystyrene accounted on average for 15% of the overall embodied energy of the wall panel. Mineral fibre insulation accounted for on average less than 1% of the overall embodied energy. Since the type of insulation used with a particular form of construction has such a large effect on the level of embodied energy, it was decided to omit thermal insulation from the following worked examples.

## Example 1: options for an office-building frame

The level of embodied energy in a 6000 mm wide x 16000 mm long bay with intermediate columns was calculated for each option. Four options were examined: in situ concrete flat slab with concrete columns; prestressed precast concrete hollowcore (wide plank) deck on a steel frame; light gauge steel floor cassette on a steel frame; composite concrete steel deck on a steel frame. The structural options are based on alternative designs for a typical office building. For each option the weight of structural components was multiplied by the embodied energy (GJ/m$^2$). Embodied energy levels for each material were then added together to find the overall value for the complete structural assembly. Fire protection was added to the steel options to make them as closely comparable as possible. For each option, the total embodied energy was derived from the summation of the calculated embodied energy for each component. In all options an alternative has been calculated using multi-cycle steel (steel that will be recycled at the end of the building's life) and recycled steel reinforcement bars (recycled from an Electric Arc Furnace as is common practice) from figures published in The Steel Construction Institute's 'A Comparative Environmental Life Cycle Assessment of Modern Office Buildings'. Crushed concrete can replace up to 20% of aggregates in structural concrete, but the embodied energy would not change substantially as aggregates have a low embodied energy, therefore this option was not considered. The columns in all options have been omitted, as their contribution is deemed insubstantial.

The results are as follows:
1a) Insitu concrete flat slab (virgin steel reinforcement) 1.73 GJ/m$^2$
1b) Insitu concrete flat slab (recycled steel reinforcement) 1.41 GJ/m$^2$
2a) Precast concrete hollowcore floor / steel beams (virgin steel) 1.27 GJ/m$^2$
2b) Precast concrete hollowcore floor / steel beams (multi-cycle steel) 1.05 GJ/m$^2$
3a) Light gauge steel floor cassette / steel beams (virgin steel) 1.46 GJ/m$^2$
3b) Light gauge steel floor cassette / steel beams (multi-cycle steel) 1.20 GJ/m$^2$
4a) Composite concrete steel deck / steel beams (virgin steel) 2.28 GJ/m$^2$
4b) Composite concrete steel deck / steel beams (multi-cycle steel) 1.78 GJ/m$^2$

It is clear from this information that the construction technique with the highest embodied energy is the composite concrete steel deck. The lowest figure was for the prestressed precast concrete hollowcore floor supported off a steel frame, closely followed by the light gauge steel floor cassette supported off a steel frame. The embodied energy for the composite concrete steel deck solution is 80% greater than the prestressed precast hollowcore solution and 56% greater than the light gauge steel cassette solution. The four options were re-examined using multi-cycle structural steel and recycled reinforcement bars. The results remained similar with the exception that the embodied energy for the light gauge steel cassette solution was only very marginally greater than the precast concrete hollowcore solution. In conclusion, the options with the greatest degree of material efficiency and prefabrication resulted in solutions with the least embodied energy. The present popularity of preassembled modular construction appears to be in tune with sustainable construction.

## Example 2: options for wall cladding

The level of embodied energy (EE) in a 3000 x 3000 mm (10ft x 10 ft) bay of a facade was calculated for each option. Since thermal insulation has been omitted from the calculations, the U-value of each construction is not considered here. For each cladding panel option, it was assumed that each panel would be supported by a floor slab/perimeter beam both top and bottom. Components within each panel were sized accordingly. Like the structural options, levels of embodied energy were calculated based on the weight of each material used in the assembly multiplied by the EE value for each. The results are as follows:

| | GJ/m² |
|---|---|
| 1) Timber framed wall+double glazed window | 0.48 |
| 2) Timber framed curtain walling | 0.82 |
| 3) Reinforced concrete panel+dg window | 0.70 |
| 4) Facing brick cavity wall+double glazed window | 2.33 |
| 5) Facing brick cavity wall with steel shelf angle | 2.54 |
| 6) Aluminium framed curtain walling | 2.48 |
| 7) Steel framed curtain walling | 1.26 |
| 8) Bolt fixed glazing | 0.80 |

The wall configuration with the lowest embodied energy is the timber framed wall. The highest are the aluminium framed curtain walling and the facing brick cavity wall. When steel is used for framing solution, the embodied energy of the curtain walling drops to approximately half the level of aluminium.

Among the options for opaque walls with a 1 metre square window, levels of embodied energy are relatively similar between timber and reinforced concrete but the brick cavity wall has almost 5 times the embodied energy of the timber framed wall. Among the glazed curtain wall options, levels of embodied energy are higher for steel than timber but bolt fixed glazing provides the lowest solution. With the obvious exception of aluminium and brick cavity, the levels of embodied energy are relatively similar for a given configuration of opaque and glazed areas.

The EE level of facing brick is nine times that of concrete block and twice that of common brick. Bolt fixed glazing minimises the use of materials, using glass as panels which are unrestrained along their edges. The use of aluminium framing results in there being three times as much embodied energy in the frame as in the glass. The use of steel framing is much better, using 60% more embodied energy compared to the bolt fixed glazed units. Nevertheless, from a performance point of view, there are problems to overcome concerning thermal breaks in steel.

## What are the best mixes between high and low embodied energy materials in typical forms of construction?

In the examples, the combination of materials consuming the least EE tends to be the combinations using the materials most efficiently. A building constructed from a high embodied energy material, such as a steel framed building clad in glazed panels, uses far less material than an equivalent design in reinforced concrete and has a lower level of embodied energy. What is surprising is the comparison between different types of glazed walling and cavity brick construction. An uninsulated cavity wall with regularly spaced windows and a shelf support angle has three times as much embodied energy as double glazed bolt fixed glazing, but both forms of construction have a similar U-value.

The exercise suggests that low levels of embodied energy can be achieved using mixed methods of lightweight and heavyweight construction in a single building. From the above examples, the best combination might be a timber clad precast concrete solution or a bolt fixed glazed facade with a light gauge steel cassette solution. The worst combination would be a brick clad composite concrete steel deck solution but this building combination has been extensively constructed in recent years.

This puts greater emphasis on the need for an efficient use of material where high embodied energy materials are used. Aluminium seems to be a very extravagant material in terms of energy. This is mainly because when used in building components, aluminium is required to have high rigidity as well as high strength. The material is used for the precision of its extrusions and castings rather than for any real qualities of lightness. This is not the case where aluminium is used in other industries. In aircraft design, aluminium is used for its strength and lightness. Rigidity is much less important, as part of the aircraft can deflect without significantly impairing performance.

## Recycling and sustainability

In attempting to reduce the embodied energy level in the construction of a building, the levels of EE in different materials might suggest that prestressed concrete and timber are preferable to steel and aluminium. However, these criteria must be seen within the overall context of recycling non-renewable resources and a sustainable approach to the use of renewable resources.

Bricks and blocks can be re-used, but are not used in many parts of the world, where the ready supply of timber, for example, can provide a more appropriate material. Steel and aluminium, with much higher embodied energies, are more easily recycled. Once manufactured, these metals are in a fairly 'closed' cycle of re-use. Aluminium is one of the easiest materials to recycle and at reasonable cost. Energy savings are made by recycling the material.

The conversion of scrap aluminium back to high-grade metal requires only about 5% of the energy needed to make the same amount of metal from bauxite.

The manufacture of building materials has a direct impact on our environment. Removing raw materials from their natural environment can cause long-term damage to the environment. The extraction of ores to make steel and aluminium leaves holes in the ground. Mining areas can be re-instated. Trees cut down as a result of timber production can and should be replanted and replaced. The cost of this work will have to be added into future calculations of embodied energy. Care should be taken to establish the source of timber. An area of tropical forest corresponding to the size of the United Kingdom is being destroyed or seriously degraded every year. Therefore, timber should be selected from an audited and sustainable forest or from plantations already established on degraded land. Planning the life of the building as appropriate in environmental terms will become a higher priority. This might lead to the preassembly of shop-constructed components which can be readily adapted to suit changing needs using a kit of parts or can be dismantled and reassembled as the need arises. This might lead to a scenario where modern high embodied energy buildings will be demolished and recycled rather than maintained, but masonry buildings of heritage may be allowed to be remain without fear of being recycled!

In conclusion, a well conceived light steel structure with bolt fixed glazed facades or a prestressed precast concrete structure with a timber facade may provide an optimised solution for low embodied energy and, hence, provide a sustainable construction. Both solutions rely on a high degree of specialist prefabrication and preassembly. This is good news and bad news for traditional construction. The focus should be on prefabrication and efficient use of carefully selected material. It appears that prefabrication and preassembly can provide the means to a future in construction that is flexible, adaptable and sustainable.

## Embodied energy values used GJ/tonne

|  | GJ/tonne |
|---|---|
| 1) Aluminium alloy | 200 |
| 2) Synthetic rubber | 150 |
| 3) Structural steel | 26.8 |
| 4) Steel used in windows | 31 |
| 5) Float glass | 15 |
| 6) Softwood | 13 |
| 7) Plasterboard | 2.7 |
| 8) Facing bricks | 11.7 |
| 9) Mortar | 0.84 |
| 10) In situ concrete structure above ground | 1.09 |
| 11) Steel reinforcement | 26.8 |
| 12) Plywood | 17 |
| 13) Concrete block | 1.31 |
| 14) Wall insulation | 35 |
| 15) Plastic | 150 |

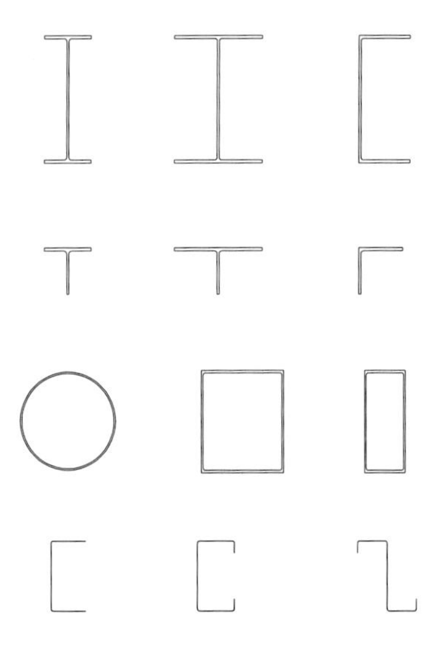

20E

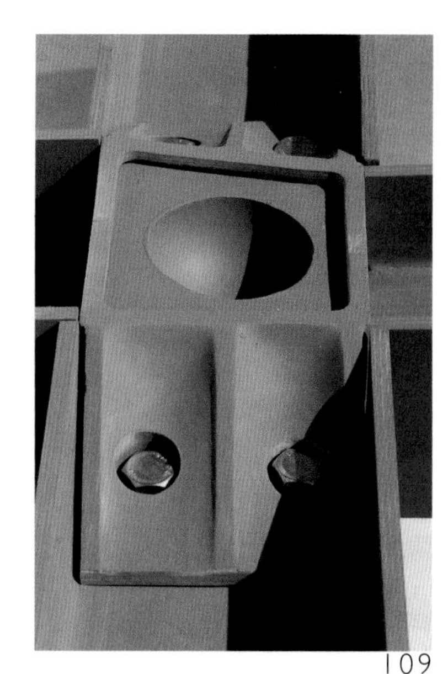

109

Steel is an iron-based metal alloyed with small amounts of other elements, the most important being carbon. The three main forms of steel used in the building industry are sections, sheets and castings. Steel sections are currently formed using a rolling process. It can be extruded to form complex sections, but this currently has only limited applications due to the high pressure needed to extrude steel. Aluminium is a much softer material, making it easier to extrude. Even when aluminium is alloyed with other materials such as bronze, the extrudable size of sections drops dramatically. Extrusions in steel cannot exceed shapes that fit into a circle approximately 150mm (6") in diameter. This is too small for structural sections, but their smooth appearance makes them suitable for components such as stiffeners in curtain walling (to provide a fin that is visually more refined than an I-section or a tee). Currently, it is still far easier to roll steel sections than to extrude them.

Historically, cast iron and wrought iron were the forerunners of steel. Cast iron, a brittle material with high compressive strength, came into general use as a building material at the end of the 18th century, while wrought iron was developed some 50 years later. Wrought iron is a more ductile material and has greater tensile strength, making it less susceptible to shock damage. (The Eiffel Tower, in Paris, was one of the last large structures to be constructed in wrought iron). By the end of the 19th century, both materials had been superseded by steel. Steel was first produced around 1740, but was not available in large quantities until Bessemer invented his converter in 1856. This device introduced a method of blasting air into the furnace (hence blast furnace) to burn away the impurities that inhibited the extraction of a purer iron. By 1840, standard shapes in wrought iron, mainly rolled flat sections, tees and angles were available which could be fabricated into structural components which are then assembled by riveting them together. By 1880, the rolling of steel I-sections had become widespread, leading the way to this material eventually replacing wrought iron as a material of choice.

## Production process of raw material

There are several steps in the manufacture of steel. First, iron is refined from ores containing iron oxide. The iron oxide is heated in a blast furnace until it is molten, using carbon as a reducing agent. The molten material is poured into moulds to produce pig iron. It is then re-heated to remove impurities, including carbon, to make cast iron that has a carbon content of 2.4 per cent to 4 per cent. Steel is produced by reducing the carbon content to approximately 0.2 per cent, with materials such as manganese and silicon added to halt the oxidation process and stabilize the carbon content. It can be poured when molten to make castings or formed into ingots to be rolled into sheets or sections.

## Properties and data

The main properties of structural carbon steels are as follows:

Density: Mild steel = 7850 kg/m$^3$ (490 lb/ft$^3$)
Design strength:
Approximate range 275 N/mm$^2$ to 800 N/mm$^2$
(5.7 x 10$^6$ to 1.6 x 10$^7$ lbf/ft$^2$)
Young's modulus = 205 kN/mm$^2$ (4.2 x 10$^9$ lbf/ft$^2$)
Coefficient of Thermal Expansion
= 12 x 10$^{-6}$ K$^{-1}$ (6.7 x 10$^{-6}$ °F$^{-1}$)
Thermal conductivity = 45 W/m°C
(26 BTU/hr.ft.°F)
Specific heat capacity= 480 J/kg°C (0.11 BTU/lb°F)

For comparison with other materials, steels have the following general properties:

· High strength in both tension and compression.
· High stiffness. High rigidity in both tension and compression.
· Its appearance is smooth in sheet form; rougher of texture in rolled sections and castings, even with paint applied.

14A

· Lighter than an equivalent structural member in reinforced concrete.
· High ductility, deforming long before it fails.
· High impact resistance.
· High heat conductor.
· High electrical conductor.
· Thermal expansion approximately half that of aluminium.
· Susceptible to continuous rusting, excluding weathering steels.
· Low fire resistance.

101D

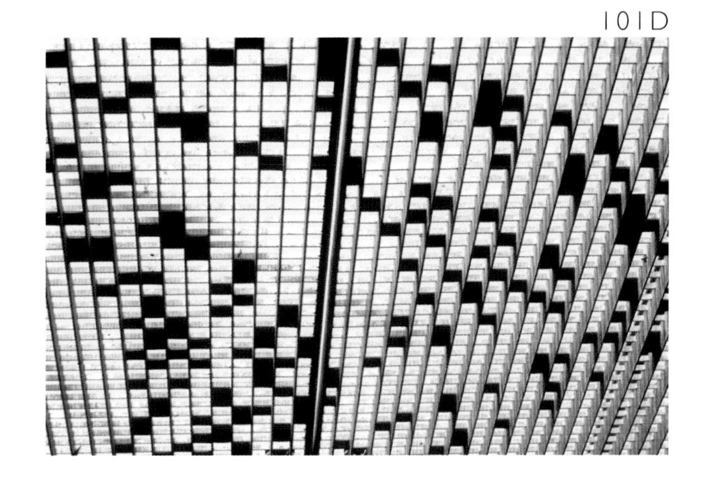

## Material selection

Hot rolled structural mild steels are made in three types called 'grades' increasing in design strength from 275N/mm$^2$ to 400N/mm$^2$ (5.7 x 10$^6$ lbf/ft$^2$ to 8.3 x 10$^6$ lbf/ft$^2$), varying slightly in different regions of the world. High strength steels can reach design strengths of 800N/mm$^2$ (17.6 x 10$^6$ lbf/ft$^2$). Since the Young's Modulus is constant for all these types, the strength of material increases but the stiffness remains constant. Steel also increases in cost with higher levels of strength, both in the cost of the raw material and in the working of the material. In addition, as strength increases in the material, welding becomes more difficult and consequently more specialized. In some high strength steels, which undergo

27B

73D

55D

heating and quenching during their manufacture, the effects of welding could potentially undo the work of manufacture if sufficient care is not taken. Standard rolled sections are manufactured in the low to medium strength grades but higher strength steels are made mostly in the form of plate, due to lower demand for their use. Consequently, compound shapes for structural components, such as beams and columns, must be specially fabricated.

Cold worked mild steels are used for much smaller scale structural components such as lightweight structural framing in metal framed housing and low-rise commercial buildings, used mainly in the USA, and drywall partitioning. Cold-formed steel sections are made from structural carbon steel to form sheets or strips approximately 1.5mm (1/16in) thick. Complex sections are formed by folding and pressing, rather than rolling which is the case with hot formed sections.

## Working with the material

Sections and sheets can be curved to small radii. Bolting and welding are the most common methods of joining sections, sheets and castings. Steel can also be sawn and drilled. An essential characteristic of steelwork is that it will continue to rust if a surface protection is not provided. When drilling or cutting the material, the newly exposed surface requires protection, which is particularly important if the material has been factory coated prior to drilling and cutting. The economic protection is galvanising, a zinc coating that is corrosion resistant, applied to the steel in a hot dip bath or as a flame spray. Galvanising occurs after fabrication of steel components to cover all the welding and drilling. This process can cause distortion of smaller steel components, so may not suit all types of fabrication. The appearance of galvanising when new is a mottled shiny grey, turning to a dull grey with weathering as the zinc oxidizes. Its visual appearance is often not suitable for exposed structural steelwork or cladding in buildings, where paint coatings are more common. Flame sprayed aluminium can be used as an alternative to galvanising. Paint can be applied by hand on site or in a factory as part of a proprietary finish. Care must be taken to ensure that touching up on site of visible components is done in controlled conditions that ensure the finish both matches and blends into the surrounding coating.

When used as primary structure in a building, steelwork requires fire protection. This can be done by either encasing the material in concrete, by enclosing it in a fire resistant board, or by coating it in intumescent paint. A spray-applied coating that yields a very rough, fibrous surface appearance is often used where the steel frame is concealed behind finish materials.

## Coatings

Many factory applied proprietary systems are available for coating steel; the most common types are thick organic coatings and powder coating.

63B

63A

PVDF (polyvinylidene di-fluoride, also called PVF2 in Europe), is sometimes used, and is discussed further in the section on aluminium. Organic coatings provide high levels of protection against corrosion but have a distinctive orange peel texture. They are applied to steel coil, from which sheet is cut, during manufacture. These finishes have methods of touching up surfaces that become exposed or are damaged during installation, but colour matching remains an important consideration in successful retouching.

## Recycling

Steel can be recycled at reasonable cost, and requires much less energy than the original production process.

71A

13C

13B

77D

Stainless steel is an alloy of steel which contains between around 11 per cent to around 25 per cent chromium, together with nickel in some types, giving it properties that are distinct from carbon steels, the main one being a high resistance to corrosion without the need for an additional coating. Since the material is considerably more expensive than carbon steels, stainless steel is most commonly used in small building components and in cladding panels where durability is a prime concern.

## Properties and data

Density = 7850 to 8000 $kg/m^3$
(490 to 500 $lb/ft^3$)
Young's Modulus :
In the longitudinal direction =
190-200$kN/mm^2$
($3.9 \times 10^9$ to $4.1 \times 10^9$ $lbf/ft^2$)
In the transverse direction =
195-205 $kN/mm^2$
($4.0 \times 10^9$ to $4.2 \times 10^9$ $lbf/ft^2$)
Coefficient of Thermal Expansion
$=13 \times 10^{-6}$ to $17 \times 10^{-6}$ $K^{-1}$
($7.2 \times 10^{-6}$ to $9.4 \times 10^{-6}$ $°F^{-1}$)

72B

77C

Stainless steel has the following general properties:

· Highly resistant to corrosion and usually requires no further coatings.

· Higher fire resistance than carbon steels.

· A risk of bimetallic corrosion at the junction of stainless steel and carbon steel when they are used together.

Separation at junction is usually required, such as a nylon or neoprene spacer.

## Material Selection

Although the material develops a thin oxide layer that protects it from further corrosion, different grades of stainless steel are available to suit the severity of exposure from polluted urban to maritime to rural environments. A limited range of standard sections is available and usually in small sizes only. The need for a high degree of fabrication of members can make construction time slower than that for carbon steel applications. For example, plate is folded to form angles and tubes, and hollow sections are formed by bending and seam welding. As with carbon steels, the high strengths types, which have been heat-treated, are more difficult to weld, as the process can undo the heat strengthening. Different finishes are available which are achieved by using a variety of rolling techniques from smooth to textured, in an appearance from matt to polished. In addition, the sheet can be coloured as part of the manufacturing process.

## Working with the Material

The fabrication of stainless steel follows the traditional pattern of fabrication for carbon steel members except that more use is made of pressing and bending to form suitable shapes. Fabrication of stainless steel should be kept entirely separate from that of carbon steel to ensure that the processes of cutting and grinding do not cause impregnation of carbon steel particles onto the stainless surface, which can lead to rusting. Fabricated elements should seek to eliminate standing seams or edges where water can collect, in order to avoid crevassing corrosion. Stainless steel has high ductility which gives the material excellent resistance to impact loading.

77B

8L

104D

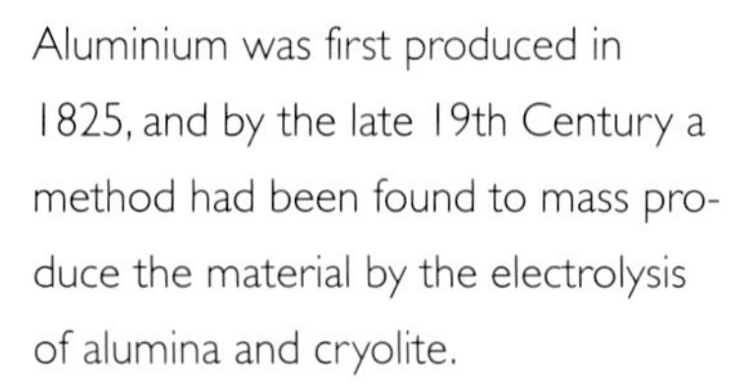

Aluminium was first produced in 1825, and by the late 19th Century a method had been found to mass produce the material by the electrolysis of alumina and cryolite.

## Production process of raw material

Aluminium is made from bauxite, which is essentially an hydrated alumina, or aluminium oxide. Mined bauxite is treated chemically to remove impurities and obtain alumina, which is aluminium oxide. This is then reduced to aluminium by electrolysis. Because aluminium has a very high melting point (2450°C) it cannot be electrolysed on its own, and so it is dissolved in molten cryolite. A high electric current is passed through the alumina-cryolite mixture at around 1000°C, and the molten aluminium is tapped off. Aluminium alloys are either formed directly, followed by continuous casting, or are cast into solid ingots. The metal is then cast into ingots which form the basis for producing aluminium alloys. Pure aluminium is too soft for structural use and is therefore combined with other metals to form alloys to increase its strength and hardness, though reducing its ductility. Magnesium, silicon and manganese are the most common additives. Aluminium alloys make strong, lightweight structural components. In common with steel, aluminium is a material that can be extruded, rolled and cast into complex shapes: plates, sheets, extrusions and castings.

## Properties and data

The main properties of aluminium alloys are as follows:

Density = 2700 kg/$m^3$ (169 lb/$ft^3$)
Design strength
Heat treated = 270 N/$mm^2$ ($5.6 \times 10^6$ lbf/$ft^2$) for extrusions and 235 N/$mm^2$ ($4.9 \times 10^6$ lbf/$ft^2$) for plate
Fully softened = 105 N/$mm^2$ ($2.1 \times 10^6$ lbf/$ft^2$) for plate
Young's Modulus = 70 kN/$mm^2$ ($1.4 \times 10^9$ lbf/$ft^2$)
Coefficient of Thermal Expansion = $23 \times 10^{-6}$ $K^{-1}$ ($12.8 \times 10^{-6}$ $°F^{-1}$)
Thermal conductivity = 200 W/m°C (116 BTU/hr.ft.°F)
Specific heat capacity = 880 J/kg°C (0.21 BTU/lb °F)

For comparison with other materials, aluminium alloys have the following general properties:

· Lightness, weighing about a third that of steel.
· High tensile strength, similar to that of steel.
· High impact resistance. (compared to steel)
· High corrosion resistance, but aluminium requires protective coating in very polluted or severe atmospheric conditions.
· Coatings are not applied solely for appearance.
· High heat conduction.
· High electrical conduction.
· Poor stiffness.
· Low resistance to soft impact, but absorbs impact energy which localises damage. (Whereas a soft, or low level impact, such as a kick, would not damage a steel panel, it will dent one in aluminium.
· A high impact, such as a car reversing into a panel, would cause a large steel panel to buckle across its entire height and length, but one made of aluminium will again dent only around the impact area).
· Thermal expansion approximately twice that of steel.
· Poor fire resistance.

101C

69A

69B

## Material selection

Pure aluminium and its alloys are in two broad groups: the non heat-treated alloys, also called fully softened alloys, whose strength is produced from being cold worked, and the heat-treated alloys whose strength is produced by heat treatment. The non heat-treated types are generally not as strong, but have better corrosive resistance.

Structural use of aluminium alloys is limited by two significant disadvantages: they are more expensive to manufacture than steel and they deform more easily under load. Aluminium alloys are more elastic than steel, restricting their use to components and assemblies where this is not a constraint. Since the Young's Modulus of aluminium is one third that of steel, buckling is an important issue in its structural use. The potential of this material as a full structural material in buildings is beginning to be recognised. The Media Centre at Lord's Cricket Ground, London, England, is a recent notable example because of the full structural use of aluminium in a large-scale building frame.

## Working with the material

On exposure to the atmosphere, aluminium forms a protective coating of aluminium oxide. Under adverse conditions, the oxide film can break down locally, but it usually reforms to a greater thickness preventing further attack. Aluminium can be exposed to the weather, in non-polluted environments and away from maritime conditions, without the need for additional treatment provided the surface is maintained. Over time, it loses its initial bright appearance and assumes a dull grey sheen. Aluminium should be cleaned regularly to avoid pits forming in the material surface. However, one way to avoid this is to anodise or paint the material with a proprietary coating.

Aluminium is susceptible to electrolytic corrosion in contact with certain materials such as copper. Therefore direct contact with copper and copper-rich alloys, such as brass and bronze is avoided and the material should be used in a way that water does not flow onto it from copper. However, water flowing from aluminium to copper or lead is not harmful. There is no corrosive action between aluminium and zinc or zinc coatings and galvanised surfaces. Some timber preservatives contain compounds harmful to aluminium. Untreated tim-

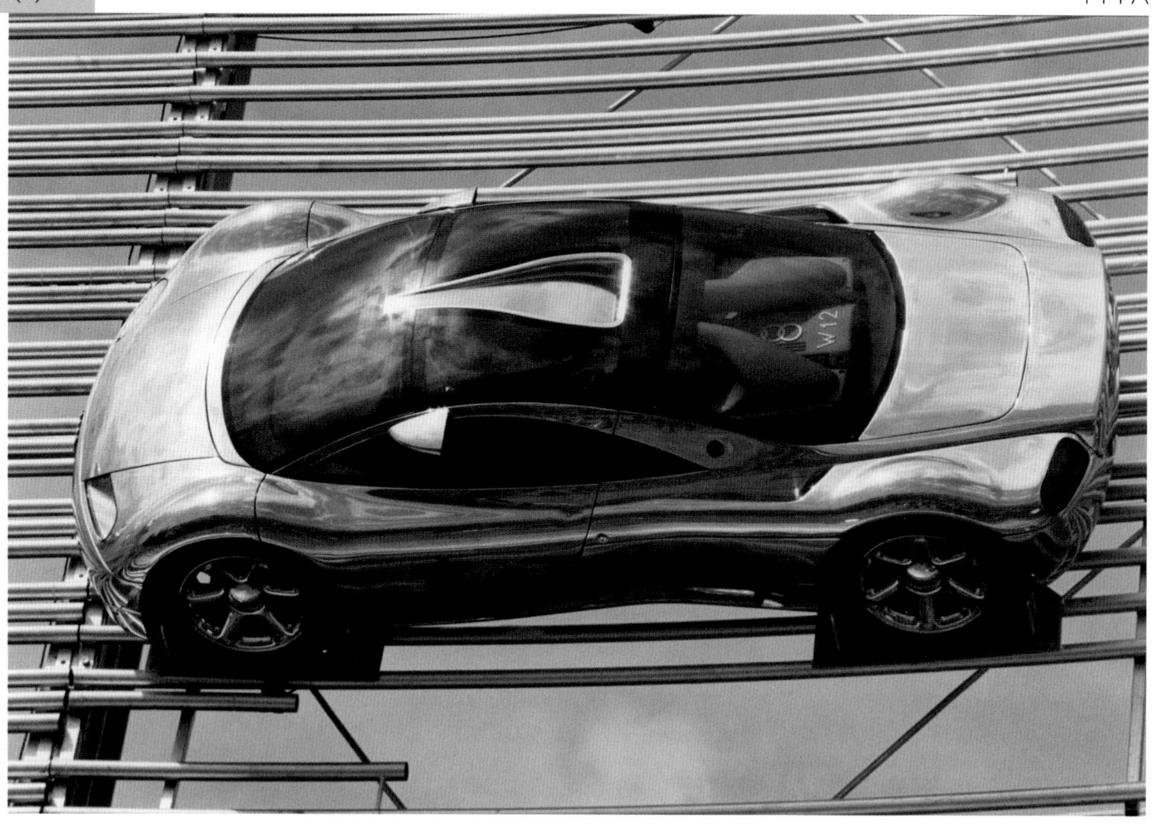


ber affects the material to a much lesser extent.

Aluminium can be cut and drilled, riveted, bolted, screwed and glued. The material can also be welded. However, welding is usually done using the fully softened alloys, since this process can undo the work of the heat treatment in the other alloy types. Since the design strength of the fully softened alloys is half that of the heat-treated types, the section sizes used in welded aluminium structures can often be similar to that of comparable steel structures, but with considerably less weight. The design strength of the heat-treated alloys, which is similar to the bottom end of the design strength of steel, can be exploited in extrusions, which require no welding in their manufacture. Extrusions can be used to form complex profiles, such as those needed in window sections or walkway decking, and be much lighter than an equivalent member in steel. The material can also be cast to form complex shapes that are more economic in large quantities than an equivalent fabricated component.

## Anodising

Anodising produces a fine translucent film over the surface of aluminium. The anodising process results in the replacement, by electrochemical means, of the metal's naturally formed oxide film by a dense chemically resistant artificial film many times the thickness of its natural equivalent. This film is extremely hard, gives added protection against abrasion, and reduces the adhesion of dirt particles. Anodising is carried out by immersing the aluminium in an electrolyte and applying an electrical current, creating an oxide layer integral with the underlying metal. The anodic film is porous and must be sealed. This is done by immersing the anodised aluminium in boiling water or steam. The anodised coating can be dyed; the sealing then assists its colour-fastness. Anodising should be carried out after welding. The process of welding would otherwise break down the anodising process at heat-affected locations. Broken-down anodising could result in weld impurities that would impair its structural effectiveness.

Aluminium's natural finish, often referred to as mill finish, can be worked to produce a polished, ground or brush-grained finish. Etching gives a matt and non-directional finish with no direct reflections. Anodising generally follows these processes, which increases durability and enhances long-term appearance. Brightening is not suitable for architectural alloys which are only 99.5% aluminium because the brightening is not uniform. Chemical brightening on other alloys dissolves and flattens surface irregularities found in extruded or sheet aluminium surfaces, and produces a mirror finish with a very high reflectivity. It can be anodised without dulling the surface.

102A

## Coatings

Aluminium can be coated in a wide range of colours through the use of proprietary processes. Plastic coatings provide a durable paint surface; polyester powder coating is one of the most common finishes. Plastic coatings are dip-coated, sprayed or electrophoretically deposited underwater. The electrostatically applied finish ensures that an even coat is built-up on the metal. These paints fade and lose their shine with time, though the change is slow and even.

PVDF (polyvinylidene di-fluoride), also called PVF2 in Europe, and powder coatings are most commonly used. PVDF is a spray-applied finish, which is highly resistant to fading in sunlight, making it very suitable for external use where colour stability is an important consideration such as in wall cladding. Powder coating is applied in an electrolytic process that provides a softer, and less expensive coating than PVDF. It is not as resistant to fading in sunlight, but is a harder finish and less expensive, making it suitable for both an economic external finish and excellent for internal use. All these finishes have methods of touching up surfaces that become exposed or are damaged during installation or use, but colour matching remains an important consideration.

## Recycling

Aluminium is one of the easiest and cheapest materials to recycle. The conversion of scrap back to high-grade metal requires only about 5% of the energy needed to make the same amount of metal from bauxite.

111B

101A

## Production process of raw material

The manufacture of float glass is the first stage of production. Float glass is made by pouring molten glass onto a bath of molten tin. The glass floats on top and is drawn off as it solidifies. It is available in thicknesses ranging from 2mm to 25mm (1/8in to 1in). Most float glass has a green tint caused by small amounts of iron oxide in the glass. Adding different oxides to the mix during the manufacturing stage can alter the tint of the glass.

## Properties and data

Density: Float glass = 2520 kg/m$^3$ (158 lb/ft$^3$)
Tensile strength: Float glass = 35-55N/mm$^2$ (7.4 x 10$^7$ lbf/ft$^2$)
Compressive strength:
Float glass = 3800–4670 N/mm$^2$ (1.0 x 10$^7$ to 1.2 x 10$^7$ lbf/ft$^2$)
Young's Modulus:
Float glass = 70 kN/mm$^2$ (1.4 x 10$^9$ lbf/ft$^2$)
Coefficient of Thermal Expansion = 23 x 10$^{-6}$ K$^{-1}$ (12.8 x 10$^{-6}$ F$^{-1}$)
Thermal conductivity = 0.7–1.1 W/m°C (0.4 BTU/hr.ft.°F)
Specific heat capacity = 820–995 J/kg°C (0.19 - 0.24 BTU/lb °F)

Approximate and easily available maximum glass sheet sizes
1) Float glass
Maximum size 3180 x 6000mm (125in x 235in) for thicknesses from 2mm to over 25mm
2) Clear toughened glass
Maximum size 4200 x 2000mm (165in x 80in) for thicknesses from 6mm to12mm for use in double glazed units.
3) Laminated glass
Maximum size 2700x4500mm (125in x 165in) for thicknesses from 10mm to 12mm (3/16in x 3/8in) for use in double glazed units.
4) Rough cast wired glass
Maximum size 3700 x 1840mm (145in x 72in) for thickness 7mm (1/4in)
5) Polished wired glass
Maximum size 3300x1830mm (130in x 72in), thickness 6mm (1/4in)
6) Body tinted float glass
Maximum size 2540 x 4600mm (100in x 180in) for thicknesses from 6mm to12mm

Glass block sizes for external walls:
190 x 190 x 100mm thick (metric)
(8in x 8in x 4in thick (imperial)
146 x 146mm (nominal 6in x 6in)
197 x 197mm (nominal 8in x 8in)
197 x 95mm (nominal 8in x 4in)
Typical thickness 98mm (nominal 4in x 8in)

Glass block sizes for internal partitions
120 x 120 x 40mm (nominal 6 x 6 x 1 3/4in)
and
200 x 200 x 50mm (nominal 8 x 8 x 2in).

45C

8G

The characteristics common to different glass types are:
· Variable tensile strength.
· It is prone to fracture resulting from tiny cracks or imperfections.
· Variable impact resistance.
· Non-corrosive.
· Non-combustible.
· High heat conduction.
· Low thermal expansion.

## Material selection

Heat soaked glass is made by re-heating float glass then cooling it quickly, which puts the surface of the glass into compression and removes impurities such as nickel sulphide. When broken, fully toughened glass disintegrates into tiny, comparatively harmless, pieces. Its strength, measured in terms of impact resistance, is up to five times that of float glass. The toughening process can produce minor distortions in the glass, usually caused by roller marks. Heat soaked glass is a partially toughened glass for use where full toughening is not required or as part of a laminated glass construction.

Laminated glass is made by bonding two or more sheets of glass together with a film of plastic in between. The film can be clear or translucent. When broken the glass stays together due to this bonding between the layers. This makes it very useful for glazed roofs. Laminated glass breaks into larger, sharper fragments than toughened glass. Combining several laminates together in one sheet can make anti-vandal and even bullet resistant glass.

Wired glass is made by sandwiching a steel wire mesh between two layers of glass, which are then rolled flat. The wire holds the glass together for a period of time during a fire and so prevents the passage of smoke from one side to the other. The roughcast product can be polished to provide a more transparent finish. Wired glasses cannot be toughened and are not regarded as safety products.

Fire resistant glass is formed by glass sandwich panels containing an intumescent layer. In a fire the layer of gel, or salts, reacts to the rise in temperature to provide a degree of insulation against radiant heat.

Variable or switchable transmission glass is a new form of treated glass. Though expensive, it aims to reduce internal heat loss from inside as well as to reduce solar gain. It is able to change its own thermal and light transmission perfor-

24B

111C

24A

mance by means of an electrical signal. In many applications, it turns from transparent to an opaque white when an electric signal is introduced. Applications include glazed partitions in office buildings.

Glass blocks can be solid or hollow. Solid blocks are used as paving for floors. The hollow type is used for walls and consists of two half-bricks fused together to give a smooth appearance on both faces.

Glass can be mounted in double- and triple-glazed units to provide greater thermal insulation and sound insulation than is achieved by an equivalent single sheet of glass. An insulated unit can be a mix of float, laminated or other glasses. To improve thermal performance, the air gap between the layers can be evacuated to create a vacuum or be replaced with a low conductivity gas such as argon. The maximum size of double glazed units is determined by the maximum sizes of glass types used. However, the size of unit is usually determined by windloading rather than maximum glass sizes.

## Working with the material

Float, toughened and laminated glass can be curved. Flat glass is heated and moulded to shape in either one or two directions. Float glass can be cut, drilled, screwed and glued. It can also be bolted using proprietary systems. Toughened glass cannot be cut, drilled or surface worked after manufacture but laminated glass can be drilled with specialist equipment.

## Surface and body treatments

Float, toughened and laminated glass can be further treated to allow varying levels of light transmission and thermal insulation. Sometimes this is done during the manufacturing process. These treatments are body tinting, screen-printing, sand blasting and acid etching, coatings, including low-E and fritting, and curving.

Body tinted glass is produced by small additions of metal oxides to the glass, reducing solar gain. A limited range of tints is available, including shades of green, grey, bronze and blue.

Fritted glass is made by printing ceramic designs onto float glass, which is then toughened. This

10D

10C

process involves fusing coloured frit (powdered glass), through a stencil onto the surface of the glass, providing a permanent durable finish. This treatment can be used to help reduce solar gain. Surface printing with a high level of detail can be achieved by screen-printed dots, lines or meshes.

Sand blasting and acid etching are surface treatments which produce a uniform, matt translucent finish. The microscopically pitted surface has a tendency to retain dirt and grease, making it difficult to clean.

A low-emissivity coating (low-E) is applied to glass to improve its thermal insulation. The coating is a microscopically thin layer of metal which allows maximum daylight and short-wave heat to enter the building but reduces heat loss by reflecting long wave radiation trying to escape at night. The coating is hardly visible.

## Recycling

Glass is one of the easiest materials to recycle and is economically viable. Enormous energy savings in glass manufacture can be made by recycling the material.

2A

26E

26G

## Historical context

Concrete had been used for over 2000 years before huge advances were made during the 19th century with the development of reinforced concrete, a technique that involved incorporating metal rods to compensate for the inherent weakness of the material in tension. François Hennibique constructed in 1892 what is thought to be the first building with a reinforced concrete frame. Ernest Ransome patented a similar system in the US in 1895. By the beginning of the 20th century, in Europe and in North America, iron rods and wires were being used as reinforcement in patented floor systems and structural frames. Modern reinforced concrete is essentially a development of these systems.

## Properties and data

The main properties of concrete are as follows:

Density:
Concrete with dense aggregate:
2240 to 2400 kg/m$^3$ (140 to 150 lb/ft$^3$)
Concrete with lightweight aggregate:
320 to 2000 kg/m$^3$ (20 to 125 lb/ft$^3$)
Design strength = 35 N/mm$^2$
Young's Modulus:
= $1.5 \times 10^4$ to $3.0 \times 10^4$ N/mm$^2$
($3.1 \times 10^8$ to $6.2 \times 10^8$ lbf/ft$^2$)
Coefficient of Thermal Expansion
= $12.0 \times 10^{-6}$ K$^{-1}$ to $7.0 \times 10^{-6}$ K$^{-1}$
($6.7 \times 10^{-6}$ to $3.9 \times 10^{-6}$ F$^{-1}$)
reducing with age.
Thermal conductivity:
Concrete with dense aggregate:
= 1.0 W/m°C
(0.58 BTU/hr.ft.°F)
Concrete with lightweight aggregate: = 0.5 W/m°C
(0.29 BTU/hr.ft.°F)
Specific heat = 840 J/kg°C
(0.20 BTU/lb °F)

Concrete has the following properties:

- Easily moulded.
- High strength in compression.
- High acoustic insulation for both air-borne and structure-borne sound.
- High fire resistance, but appropriate cover to steel reinforcement is needed.
- Most shrinkage of the material occurs, as creep, during the first year after casting.
- Moisture movement occurs but is less significant than timber.
- Slightly permeable to water.
- It will not set properly if the air temperature approaches freezing point. If the air temperature is too high, it will set too quickly, causing cracking.

## Material selection

Concrete is a dense material, composed of cement and aggregate mixed with water. It sets to form a hard, brittle material, strong in compression but weak in tension. Portland cement is the most widely used binding agent for concrete and consists of lime and clay mixed together at high temperature, which is then crushed to form a fine powder.

The type and relative proportions of cement and aggregate will vary according to use and desired appearance. The mix should make economic use of the cement. Varying the composition of the constituents produces different strengths of concrete. The most important factors are the water to cement ratio and the proportion of cement to aggregate. A typical mix is the 1:2:4 (cement: fine aggregate: coarse aggregate). The amount of water used in the mix affects both workability and strength. Less water increases strength but reduces the workability, making it more difficult for the concrete to flow around the reinforcement when it is being poured in place. The reduced workability is improved by using additives, such as plasticisers, to the mix. The coarse aggregates, consisting of small stones, form most of the mass, while spaces between these stones are filled by fine aggregate and cement which bind the mixture together. It is important when pouring concrete, that the mix-

26C

26D

ture is properly vibrated to ensure that the coarse aggregate is fully surrounded by the finer material.

Concrete shrinks as it dries, and can take more than a year to reach its final size, though it continues to shrink indefinitely by tiny amounts. The rate of shrinkage is considerable in the period immediately following the pouring, but slows down by the end of the first year. After approximately 28 days, concrete approaches its design strength. It is important that the mixture should not dry out too quickly to enable the chemical reactions between the constituents to take place. Controlling the drying process is known as 'curing'. Having dried out, concrete subsequently absorbs water, but any expansion is always much less than the original shrinkage during the curing process. Plasticisers can be used to vary the rate of drying, which reduces construction time. Plasticisers also increase the workability of the wet concrete.

In reinforced concrete, steel and concrete are combined to take advantage of the compressive strength of concrete and the tensile strength of the steel used. Two types are used; mild steel reinforcement is used to form complex elements and has a yield-strength of around 250N/mm$^2$, and high yield reinforcement is used elsewhere with a yield-strength of around 460N/mm$^2$. As the concrete sets it shrinks, gripping the steel bars thus producing a monolithic structural material for use as frames, walls and floors. These can be cast-in-place at the site, or off-site as precast elements. With either method, the cost of formwork, or mould, can represent up to half of the cost, and therefore the efficient use and re-use of formwork is essential.

Cast-in-place reinforced concrete is usually delivered to site ready mixed and is poured either by pumping or by crane bucket. Cast-in-place reinforced concrete is made by setting steel reinforcing rods, often in the form of a cage, between formwork, made from plywood, steel or timber boards. Wet concrete is poured and compacted. The reinforcement must have an appropriate cover of concrete to protect it in the event of fire. The cover also protects the steel from corrosion.

A series of bolts holds the sides of the formwork apart and shoring on the outside of the formwork supports the weight of the wet concrete. The formwork is removed, or struck, after 3 to 4 days and the holes left behind are filled with grout, which is a mixture of cement and water. Since the holes remain very apparent, they are arranged at regular centres in order to enhance their appearance.

The technique of manufacturing precast concrete, developed in the 1950s, has the advantage of a

1D

95

reduced construction time, as no curing, and much reduced formwork, is needed on site. Concrete is cast in moulds in a factory, and then delivered to site. Precast techniques have two main advantages over cast-in-place construction. The first is that quality can be easier to control in a workshop, and the second is that precast elements can be assembled rapidly on site. Cast-in-place concrete needs time to gain strength before it can support another element and this slows down construction dramatically.

Precast concrete is usually more expensive than cast-in-place due to additional transport costs. To be economic, and comparable in cost with cast-in-place construction, the number of different components should be kept to a minimum, because fewer moulds, prototypes and trial panels would be used. On larger projects, it may be cost effective to manufacture precast components on site in a temporary facility. This avoids the need to transport components, though such conditions may not be ideal for high quality work. Precast concrete systems are manufactured as proprietary systems for both structural frames and wall cladding panels. There is a significant move towards re-using the formwork from one application to the next.

Components are cast in moulds which are made from glass reinforced polyester, steel or concrete. Accurate manufacture is difficult with plywood and timber due to the thermal movement of these materials, making the mould sizes unreliable. Panels are cast either face-up or face-down. In face-down casting the inside surface of the mould forms the component finish. In face-up casting the surface of the mould forms the back of the component so that structural ribs can be formed, leaving the panel face flat. An alternative method, which avoids complex formwork, is tilt up construction. This is a partially precast method where a wall is cast flat on the ground at the site, directly adjacent to the floor slab. Once it has cured sufficiently, it is then lifted up into the vertical position and bolted directly in place.

Prestressed concrete is a precasting technique that is typically used in floor structures. It allows increased clear spans to be used with thinner slabs resulting in lighter structures. The reduced cracking and reduced deflections in this technique are a distinct advantage over normal reinforced concrete. In common with other precasting techniques, this rapid construction method allows construction times to be reduced.

In prestressing, high strength steel wires, rods or cables are passed through a small diameter tube set into a precast component, typically a beam or a tee-section used as a structural deck. The wires, set in the bottom half of the beam, are tensioned at the ends, resulting in the beam or tee arching up in a slight camber. The tubes are then filled with grout. When the prestressed component is set in place, the imposed loads flatten the pre-camber and reduce the deflection experienced by the component.

When an ordinary reinforced concrete beam is loaded, compression is created in the top half of the beam and tension in the bottom part. The compressive stresses in the concrete created by the tension wires must be overcome before the beam bends. However, when a prestressed beam or similar component is loaded, the concrete is in compression throughout its depth, allowing it to be shallower in depth than a reinforced concrete equivalent. Prestressed concrete is more suitable for large spans than ordinary reinforced concrete because components can be shallower and as a consequence be lighter, reducing the dead load. Prestressing increases resistance to shear forces compared to a reinforced concrete beam, making it possible to achieve a more slender section than using cast-in-place techniques.

Where these components are too large to be prestressed, or where the construction method dictates that they are to be built on site, the technique of post-tensioning can provide a similar method of reducing the depth of beams, decks or related assemblies such as masonry arches. Although this technique is used mainly in bridge-building, post tensioning allows individual large-scale components, such as frames and floors, to use prestressing techniques with a site-based construction method. Post tensioning is more suited to large assemblies such as arches and long-span floor decks than to smaller-scale assemblies where the time needed on site to effect the tensioning would slow down the speed of construction.

10E

65

Lightweight concrete is used primarily for toppings in profiled steel/concrete composite construction. It is also used for non-loadbearing components such as precast wall panels. It is not used for high strength applications, but is suitable for most structural applications where weight is an important consideration. The material is typically made from crushed pumice or clinker, giving it better properties of heat and sound insulation than cladding panels made from other materials.

Ferro-cement is a concrete-based material which is typically used to make yacht hulls. This material is beginning to be used as a structural material in buildings, and is suited to complex shapes with a high quality smooth finish. A recent example is the curved roof trusses of the Menil Museum in Houston, Texas, USA. Ferro-cement consists of a cement mortar-based mix with a high degree of steel reinforcement. It has good tensile strength in thin sections. Complex shapes can be formed by applying the wet mix onto a steel mesh by hand without the need for formwork.

## Working with the material

Concrete is compacted by vibration when poured in order to remove air voids in the mix and to achieve an even distribution of the material in the formwork. Too little compaction of the wet material can result in air pockets being left in the mix, weakening the concrete. Too much compaction brings the fine aggregate to the surface, making the surface crumbly, and causing surface staining. A vibrating instrument, inserted into the wet concrete, is used to ensure an even consistency. The reinforcement is given an adequate cover of concrete to protect it in the event of fire. The cover also reduces the possibility of water reaching the steel which can cause it to rust. In the manufacture of precast components the mould is typically compacted, either by placing it on a vibrating table or by applying a surface mounted vibrator.

## Finishes

Concrete adopts the texture of the formwork. Steel formwork leaves a smooth appearance while softwood boards leave an imprint ranging from fairly smooth plywood to rougher finishes of planks. Boarded formwork produces a pattern of joints across the concrete face. Other finishes include tamping, where a board is moved in a tapping action across the surface of the concrete to form a directional texture; trowelling, where concrete is smoothed with a hand tool; and power floating, where a poured slab is smoothed with a machine to provide a finish that avoids the need for an additional layer of smooth screed.

The fine aggregate and cement determine the colour of the concrete. Changing the attributes of the fine aggregate has a dramatic effect on the appearance of the concrete, whereas large aggregate has little effect unless retardants are used and the large aggregate is exposed. Colour additives can also be added to the mix. Unfortunately, slight variations in the proportions of additives have a dramatic impact on the appearance of the concrete, making it hard to achieve consistent coloration between batches.

Paint provides a thin decorative layer, but is prone to flaking and requires re-coating at regular intervals. Colour stains are an alternative as they are absorbed by the top surface of the concrete forming a permanent coloured finish. Sand blasting and acid etching, processes more commonly associated with glass, can also be used. Bush hammering the material exposes the large aggregate and provides a rough texture to the surface. Conversely, polishing provides a smooth shiny surface. However, both of these surface treatments are very labour intensive. Unpolished concrete has a dusty surface and it may be appropriate to seal the surface.

## Recycling

Concrete can be recycled by crushing the material and using it as an aggregate in new concrete. Although the use of recycled concrete is new, it has been successfully used in new reinforced concrete structures.

25

The principles of using materials are common to stone, brick and block. The relationship between the masonry unit and the mortar is of particular importance.

Mortars for loadbearing masonry and cladding panels use the same materials and follow the same principles in stone, brick and concrete blockwork. The compatibility of masonry and mortar is an essential factor. It is usual to use the weakest mortar that will adequately sustain the load, because increasing strength can result in too much rigidity which would cause cracking.

The strength is varied by altering the proportions of the binders cement and lime. Stronger mixes have more cement, while lime provides flexibility which allows the brickwork to move without cracking. Different mixes of mortar are used depending on whether the masonry is used as cladding or as loadbearing masonry. Mortar mixes have to achieve a balance between strength and flexibility. The comparatively low permeability of lime gives greater resistance to rain penetration. Lime also makes a mortar lighter in colour than a cement-based one. Mortar mixes are a balance between the needs of loads, structural movement, water permeability and preferred colour.

In terms of appearance, it must be noted that the mortar accounts for between 10 and 20 per cent of the masonry, and consequently makes an important contribution to the colour of the surface. The colour of the mortar can be controlled by using pigment additives. In addition, the way that the joint is made has an important effect on the flatness or amount of shadow perceived.

In stonework, crushed stone is often added to the mortar mix as fines instead of sand so that the joints match the stone as closely as possible. The mix depends on the type of stone and its intended use. As a result the mortar is the best compromise between load, structural movement, water permeability and preferred colour. Mortar joints can also be reinforced with steel mesh.

## Recycling

Natural stone can be recycled if the mortar is soft enough to be removed, particularly if a lime putty mortar has been used. Otherwise, stone and brick are used for structural fill.

## Masonry block

Masonry blocks are made from concrete in a range of sizes and strengths. Some types are designed to withstand large compressive forces and are therefore made with a high-density concrete. They can withstand forces of up to around 20N/mm$^2$ (4.2 x 10$^5$ lbf/ft$^2$). Other types are designed to provide limited amounts of thermal insulation and are made with aerated concrete or with insulation bonded to one side. Most types of block are manufactured to course with brick but the size of the block will depend on its weight, since the block must be lifted by hand. Various thicknesses of block are available for different applications.

## Material selection

Types used are dense aggregate blocks, light aggregate blocks and aerated blocks. Blocks with dense aggregates are more commonly used for loadbearing walls. Those with lightweight aggregates and aerated mixes are used for non-loadbearing walls and partitions. All these types are made in solid, cellular and hollow form. Cellular types allow the voids to be filled with concrete and reinforcement to improve structural performance. The main advantage of block is its low cost and the speed with which walls can be built. Although

50

some blocks, called fairfaced blocks, are made as a facing material, it is more usual to conceal blocks, or face them with another material such as render.

## Working with the material

Block can be easily cut and drilled. Metal fixings in the form of brackets and dowels are used for walls where the blockwork does not provide sufficient stability. These fixings are used extensively in blockwork cladding which is supported by a structural frame.

SIZES

Common metric block size:
Length 390mm x Height 190mm (15 5/8in x 7 5/8in)
Overall dimensions
(Length 400mm x Height 200mm with 10mm joint) (16in x 8in with 3/8in joint)
A range of thicknesses is made.
Common UK block size to course with UK brickwork:
Length 440mm x Height 215mm
(Length 450mm x Height 225mm with 10mm joint)
Blockwork has the following general properties:

- Heavy.
- High compressive strength.
- Very low tensile strength.
- High resistance to weathering
- High impact resistance.
- High fire resistance.
- Susceptible to thermal and moisture movement.
- Not waterproof.
- Low seismic resistance.

## Properties and data

The main properties of masonry blockwork are as follows:
Density:
Dense aggregates = Over 1500 $kg/m^3$ (94 $lb/ft^3$)
Lightweight aggregates =1000 to 1500 $kg/m^3$ (62 to 94 $lb/ft^3$)
Aerated = 500 to 1500 $kg/m^3$ (31 to 94 $lb/ft^3$)
Compressive strength
Dense concrete blockwork = 10$N/mm^2$ to 20 $kN/mm^2$
($209 \times 10^3$ to $418 \times 10^3$ $lbf/ft^2$)
Aerated concrete blockwork = 3.5$N/mm^2$ to 7.0 $N/mm^2$
($73 \times 10^3$ to $146 \times 10^3$ $lbf/ft^2$)
Young's Modulus :
Dense concrete blockwork = 5.0$N/mm^2$ to 25.0 $kN/mm^2$
($1.1 \times 10^5$ to $5.2 \times 10^5$ $lbf/ft^2$)
Aerated concrete blockwork = 2.0$N/mm^2$ to 8.0 $kN/mm^2$
($4.2 \times 10^4$ to $1.8 \times 10^5$ $lbf/ft^2$)
Coefficient of Thermal Expansion:
Dense concrete blockwork = $6 \times 10^{-6}$ to $12 \times 10^{-6}$ $K^{-1}$
($3.4 \times 10^{-6}$ to $6.8 \times 10^{-6}$ $°F^{-1}$) at 5% moisture content
Aerated concrete blockwork = $8 \times 10^{-6}$ $K^{-1}$ ($4.5 \times 10^{-6}$ $°F^{-1}$) at 5% moisture content
Thermal conductivity:
Dense concrete blockwork = 1.2 W/m°C (0.69 BTU/hr.ft.°F)
Aerated concrete blockwork = 0.3 W/m°C (0.17 BTU/hr.ft.°F)
Specific heat capacity = 840 J/kg°C (0.2 BTU/lb °F) for dense concrete blockwork

## Production process of raw material

Stone is cut or hewn from large blocks which have been cut, blasted or split from the bedrock. Being a natural material, the appearance and durability of stone, even from the same block, can vary enormously. To control quality, it is sometimes best to select cut stone at the quarry. Since no two cuts produce the same appearance, when defining the required quality it is advisable to define a limited band between the most veined and least veined stone acceptable. As stone is a naturally occurring material, the properties are not as controlled as man-made materials such as steel. Stone from a particular quarry is usually tested before sale in order that its physical and mechanical properties are known. When stone is used in structural applications, the material undergoes rigorous testing to determine its performance in the intended application.

Natural stone has the following properties common to most types:

· Durable
· A heavy material, weighing as much as reinforced concrete.
· High compressive strength.
· Low tensile strength.
· Finish can be adversely affected by weathering due to exfoliation as a result of a freeze/thaw cycle, pollution, salts, etc.
· Low moisture movement.
· Brittle, but high impact resistance improving with thickness.
· High fire resistance.
· Low seismic resistance

## Material selection

Natural stone is a brittle material that is strong in compression but is weak in tension. It is used mostly for wall facings and pavings, although the high unit cost means that it is rarely used in a traditional loadbearing capacity. Most stone has the strength and durability of block and brickwork. The most widely used types are granite, limestone, sandstone, marble and slate. Igneous rocks, such as granite, are formed directly from molten magma. Sedimentary rocks, such as limestone and sandstone, are made up from the eroded elements of earlier rocks laid down in beds near the earth's surface, and are often composed of loose material bound together by cement-like materials. Metamorphic rocks, such as slate and marble, are igneous or sedimentary rocks which have undergone a chemical transformation due to high temperature and pressure.

Granite, within the building industry, refers to coarse-grained igneous rocks. It has a wide range of colours, and is extremely hardwearing. Most granites are grey or pink, with mixtures of white/grey and pink/grey depending upon their geographical source.

Limestone is made up from rock material bound together by calcium carbonate, in the form of the mineral calcite. Many limestones contain a proportion of the mineral dolomite. The colour is generally light, ranging from near white through to brown and grey. Chemical impurities can cause a darkening of the colour. Limestones vary in texture and can range from a sand-textured and coarse material to one that is so fine-grained as to lack visible particles. Limestones such as Portland Stone (Indiana Limestone in the US) are strong and durable.

Most sandstones consist mainly of quartz grains cemented together by mineral solutions. Calcareous, dolomitic, ferruginous and siliceous cements are common. Small amounts of other minerals, often iron compounds, give the stone its colour. Sandstones vary in colour from dull crimson to pink or green/brown mixtures to blue/grey. Sandstones vary enormously in durability from soft,

## Properties and data

The main properties of stone are as follows:

Density:
Natural stone: 2200 to 2600 $kg/m^3$ (137 to 162 $lb/ft^3$) wet
Cast stone: 2100 $kg/m^3$ (131 $lb/ft^3$)
Characteristic compressive strength:
Granite = 30 $N/mm^2$ to 75 $N/mm^2$ ($6.3 \times 10^5$ to $1.5 \times 10^6$ $lbf/ft^2$)
Sandstone = 10 $N/mm^2$ to 30 $N/mm^2$ ($2.0 \times 10^5$ to $6.2 \times 10^5$ $lbf/ft^2$)
Coefficient of Thermal Expansion = $7.9 \times 10^{-6}$ $K^{-1}$ ($4.4 \times 10^{-6}$ $°F^{-1}$)
Thermal conductivity = 1.0 W/m°C to 1.3 W/m°C
(0.58 to 0.75 BTU/hr.ft.°F)

easy to work types with low strength and high porosity, to relatively durable types with strengths approaching those of granites, with lower porosity.

Marble is a metamorphic rock formed by the recrystallisation of limestone or dolomite through a combination of heat and pressure. The crystalline structure is seen in a fractured surface, which gives it a sparkling appearance. During metamorphism, impurities in the original limestone, such as different minerals, are incorporated into the rock and appear as bands or as discrete inclusions scattered through the calcite mix. No true marble shows fossils. Veined marbles are the result of minerals deposited from solutions penetrating cracks and fissures. Some marbles contain fragments of earlier crushed rocks. A wide range of colours and textures may be found. The presence of iron gives rise to shades of yellow, brown and red as a result of oxidation.

Slate is a crystalline rock produced by dynamic metamorphism of clays and shales, causing it to be orientated along a single grain or 'slaty cleavage' which allows the rock to be split into sheets. It is used mainly for roof slates and for durable surface finishes such as floors. The colour of slate varies from grey to green to black to red.

## Working with the material

Stone is shaped either by cutting or sculpting, which is a slow and difficult process. Metal fixings in the form of brackets and dowels are used where the mortar joints do not provide sufficient stability for lintels and copings. These fixings are used extensively in stone cladding panels where the material is supported by a structural frame.

Stone has a very different appearance when polished, though not all types benefit from the process. Depending upon type and application, stone can be finished to different levels of sheen from a reflective polished finish to a matt honed finish. Polishing does not change the structure or weathering of stone and is typically applied where stone is used as a flooring material. Additional finishes include etching and needle gunning, but stone used externally usually has no additional treatment after being cut to size.

## Reconstituted stone

Reconstituted stone, also referred to as cast or 'reconstructed' stone, is made from cement and crushed stone that is cast in a mould. It is used either as a structural material or as a facing to a concrete component. Steel reinforcement is used if the casting has a structural function.

Reconstituted stone has the following properties:

- Very durable.
- A heavy material, weighing as much as reinforced concrete.
- High tensile and compressive strength.
- High impact resistance.
- High fire resistance.
- Susceptible to shrinkage like concrete.

8R

8Q

101G

80C

73D

26A

## Production process of raw material

Bricks are made by cutting or moulding clay. They are then baked in a kiln to form a hard, brittle unit. A very wide range of colours and textures is available, from the precisely dimensioned, evenly coloured types which are wire cut from a clay extrusion to the less regular handmade bricks which are individually formed in moulds. Extruded wirecut bricks are sometimes made with holes running through them to reduce their weight or to allow reinforcing rods to link them together. Hand-made bricks often have an uneven appearance which provides a rich visual texture as a completed wall.

Most bricks in the UK are made in a single standard size: 215mm long x 102.5mm wide x 65mm high, (7 5/8" l. x 3 5/8" w. x 2 3/8" h.) though metric size bricks are also available. When constructing brickwork a 10mm joint is used throughout, resulting in vertical courses 75mm high and 225mm long. Two bricks laid side by side with a 10mm joint are equal to a brick length. The modular nature and size of bricks makes brickwork a very flexible medium. The weight and size of one brick allow it to be lifted with one hand. The modular nature of brickwork imposes a strict discipline on the detailing of openings and corners if expensive specially shaped bricks, called 'specials' are to be avoided. Bricks are economically and easily transported in large or small quantities.
Brickwork has the following general properties:

· A heavy and durable material.
· High compressive strength.
· Very low tensile strength.
· High resistance to weathering.
· High thermal mass.
· High acoustic mass.
· High impact resistance.
· High fire resistance.
· Susceptible to thermal and moisture movement.
· Low seismic resistance

## Material selection

The most widely used types are common, facing, engineering and calcium silicate bricks. Common bricks are the weakest type, while engineering bricks are the strongest. Facing bricks are usually used on the external face, while cheaper bricks can be used within the wall. Engineering bricks are used for their high strength and are almost impervious to water, and so are often being used below ground in addition to structural applications where brick is used as a primary loadbearing element. Calcium silicate bricks are made from a mixture of sand and lime which are compressed under steam pressure in a mould. They are low to medium strength, and their water absorption is comparable to that of clay bricks. They are mainly used in internal walls where their light appearance can be exploited. Brickwork is strong in compression, but is not used to resist tensile forces.

## Working with the material

Brick can be easily cut and drilled. Metal fixings, in the form of brackets and dowels, are used for walls where the brickwork would not be sufficiently stable by itself. These fixings are used extensively in brick cladding which is supported by a structural frame. Bricks are made in a vast range of colours from reds to blues. Colour is defined by the type of clay, combined with the way it is fired and any pigments which may be added. Calcium silicate bricks are white.

As new brickwork and mortar dries out after rain, traces of salt deposits are sometimes left on the surface. These appear in the form of white stains known as efflorescence. It can be easily removed with a brush and water. Where moisture continues to penetrate the brickwork this efflorescence will continue to manifest itself.

## Properties and data

The main properties of brick are as follows:

Density:
Average brickwork = 1700 kg/m$^3$ (106 lb/ft$^3$)
Design strength = 5.0 N/mm$^2$ to 25.0 N/mm$^2$
(1.0 x 10$^5$ to 5.0 x 10$^5$ lbf/ft$^2$)
Coefficient of Thermal Expansion
= 5.0 x 10$^{-6}$ K$^{-1}$ to 8.0 x 10$^{-6}$ K$^{-1}$
(2.8 x 10$^{-6}$ to 4.5 x 10$^{-6}$ °F$^{-1}$)
Thermal conductivity: = 1.3 W/m°C at 5% moisture content
(0.75 BTU/hr.ft.°F)
Specific heat capacity = 800 J/kg°C (0.19 BTU/lb °F)

114B

## Production process of raw material

PLASTICS are resinous polymer-based materials divided into two groups. These are thermoplastics, which melt at high temperature, and thermosetting plastics, which set hard and do not melt on further reheating. They are used mainly in cladding systems but composites are beginning to be used as fully structural materials in relatively modest applications such as footbridges. The most commonly used types, under the generic names used in the building industry rather than their polymer-based names, are as follows:

· Polycarbonate
· Acrylic sheet
· PVC-U

COMPOSITES comprise two or more materials combined together where the properties of each constituent can complement the others. Although reinforced concrete, plywood and other more traditional materials are sometimes referred to as composites, the term is generally used in the building industry to refer to polymer-based composites. These materials have a polymer resin reinforced with thin fibres, usually glass fibres or carbon fibres. Glass-fibre-reinforced polyester (GRP) was first used during the Second World War for radar covers and was later applied in a GRP boat for the US Navy in 1947. Carbon-fibre was developed by the Royal Aircraft Establishment in Farnborough, England, during the 1960's and was applied in composites for use in compressor blades in jet engines later in that decade. Carbon-fibre-reinforced polymer is much stronger and stiffer than GRP but remains extremely expensive in relation to most metals or other plastics and has yet to find a significant use in building construction.

## Polycarbonate

Polycarbonate is used in building construction largely as a substitute for glass, especially where moulded shapes are required which would be too difficult or too expensive to make in glass. The material was first made in the 1950's as part of the research into polyesters and was marketed as Lexan sheet in the early 1960's. Polycarbonate is made by polymerisation, where a polymer is melted and extruded into strands which are chopped to produce polycarbonate granules. The granules can then be extruded or moulded to form single sheet, twin-wall sheet or complex shapes.

114A

## Properties and data

Density : 1200-1260kg/$m^3$
(75 to 78 lb/$ft^3$)
Tensile strength = 56N/$mm^2$ to 75N/$mm^2$ ($1.2 \times 10^6$ to $1.6 \times 10^6$ lbf/$ft^2$)
Compressive strength = 100N/$mm^2$ to 120N/$mm^2$
($2.1 \times 10^6$ to $2.5 \times 10^6$ lbf/$ft^2$)
Young's Modulus = 2.3 - 2.8kN/$mm^2$ ($4.8 \times 10^7$ to $5.8 \times 10^7$ lbf/$ft^2$)
Coefficient of Thermal Expansion :
60-75 $\times 10^{-6}$ $K^{-1}$
($33.5 \times 10^{-6}$ to $42 \times 10^{-6}$ $°F^{-1}$)
Thermal conductivity :
0.18 - 0.22 W/m°C (0.1 to 0.13 BTU/hr.ft.°F)
Specific heat capacity = 1200 - 1300 J/kg°C (0.29 – 0.31 BTU/lb °F)

Polycarbonate has the following general properties:
- A strong material with low stiffness.
- High transparency can be obtained.
- A tough but ductile material.
- High impact resistance.
- Flame resistant, tending to melt rather than ignite, but is still combustible.
- Poor scratch resistance without silicone coatings.
- Recyclable.

## Material selection

Polycarbonate is a thermoplastic used for its high strength, ductility, lightness and transparency. It is fire retardant, and can be easily moulded to complex shapes. The use of polycarbonate is limited by its combustibility. Polycarbonate is an extruded materal used as sheets and sections. Polycarbonate sheet is available in thicknesses from 2mm to 25mm (1/16in to 1in). Its impact resistance is higher than that of toughened or laminated glass. The two main disadvantages of polycarbonate over glass are that it is less durable, scratching easily which makes the surface dull with time and its greater combustibility. Polycarbonate also has greater thermal expansion than glass.

Twin-walled sheet is an extrusion of two layers separated by parallel fins, giving the material greater rigidity for use as a board. The air gap between the two layers provides a degree of thermal insulation. The maximum sheet size is approximately 2 x 6 metres. It can be sawn, cut and drilled. It expands 20% more than glass. For example, a 1.5 metre (5ft) wide sheet will expand up to 3mm (1/8in). The material has a density of 1200 kg/$m^3$, which is half that of glass, although in practice the real weight saving is about one third.

Polycarbonate can be coated with acrylic to prevent yellowing, and other coatings are used to enhance abrasion resistance. It provides an average of 85% light transmission for a sheet 5 to 6mm thick.

## Acrylic Sheet

Acrylic sheet, developed as Perspex in the 1930's, is a plastic based on polymers of acrylic acid. The most common is polymethyl methacrylate, or PMMA, made by polymerising an MMA monomer with a catalyst to form a powder which can then be extruded, moulded or cast.

## Properties and data

Density : 1150 -2000kg/m$^3$
(72 to 125 lb/ft$^3$)
Tensile strength = 38N/mm$^2$ to 80N/mm$^2$
(7.9 x 10$^5$ to 1.7 x 10$^6$ lbf/ft$^2$)
Compressive strength = 45N/mm$^2$ to 80N/mm$^2$
(9.4 x 10$^5$ to 1.7 x 10$^6$ lbf/ft$^2$)
Young's Modulus = 1.8 - 3.4kN/mm$^2$
(3.8 x 10$^7$ to 7.1 x 10$^7$ lbf/ft$^2$)
Coefficient of Thermal Expansion
60 x 10$^{-6}$ to 70 x 10$^{-6}$ K$^{-1}$
(33.5 x 10$^{-6}$ to 39 x 10$^{-6}$ °F$^{-1}$)
Thermal conductivity : 0.2 W/m°C
(0.11 BTU/hr.ft.°F)
Specific heat capacity
= 1280 - 1500 J/kg°C (0.30 – 0.36 BTU/lb °F)

Acrylic sheet has the following general properties:
· High transparency and optical clarity.
· Weathers well with high resistance
to yellowing.
· Hard but brittle.
· Poor scratch resistance.
· Easily recycled.
· Combustible.

## PVC-U

The term PVC-U, or UPVC, denotes the unplasticised form of polyvinyl chloride, or PVC. This rigid form of PVC is used for a wide range of building components from guttering and ground drainage pipes to window frames. The material can be easily extruded or moulded to complex shapes and is available in a range of colours. The low thermal conductivity and flexibility of the material combined with its ability to be extruded makes it very suitable for window frames where the material is, in effect, its own thermal break, minimising the risk of condensation on the face of the frame inside the building.

## Properties and data

Density : 1400kg/m$^3$ (87 lb/ft$^3$)
Young's Modulus = 0.1 to 4.0 kN/mm$^2$
(2.1 x 10$^6$ to 8.3 x 10$^7$ lbf/ft$^2$)
Coefficient of Thermal Expansion
= 70 x 10$^{-6}$ K$^{-1}$ (39 x 10$^{-6}$ °F$^{-1}$)
Thermal conductivity = 0.3W/m°C (0.17 BTU/hr.ft.°F)
Specific heat capacity = 1300 J/kg°C (0.30 BTU/lb °F)

PVC-U has the following general
properties:
· Available in a range of colours.
· Weathers well but is susceptible to fading, particularly with brighter colours.
· Tough but flexible.
· Recyclable.
· Combustible.

## Glass Reinforced Polyester (GRP)

This material was first commercially available in the 1930's from the Owens-Corning Fiberglas Co in the USA, but has only slowly been introduced into building. Its main use is in specially fabricated wall cladding panels. Glass reinforced polyester, or GRP, is made from a combination of glass fibre mat and polyester resin. It is a thermosetting composite which has high tensile, shear and compressive strength combined with lightness and resistance to corrosion. However, like aluminium, it deflects considerably under high loads and requires stiffening.

## Properties and data

Density : 1600-1950 kg/m$^3$
(100 to 120 lb/ft$^3$)
Tensile strength = 300 N/mm$^2$ to 1100 N/mm$^2$
(6.3 x 10$^6$ to 2.3 x 10$^7$ lbf/ft$^2$)
Compressive strength =360N/mm$^2$ to 880N/mm$^2$
(7.5 x 10$^6$ to 1.8 x 10$^7$ lbf/ft$^2$)
Young's Modulus : = 35 - 45 kN/mm$^2$
(7.3 x 10$^8$ to 9.4 x 10$^8$ lbf/ft$^2$)
Coefficient of Thermal Expansion :
8.5-25x10$^{-6}$ K$^{-1}$
(4.8 to 14.0 x 10$^{-6}$ °F$^{-1}$)
Thermal conductivity: 0.4 - 1.2W/m°C
(0.23 to 0.7 BTU/hr.ft.°F)
Specific heat capacity = 100 - 1400 J/kg°C
(0.02 – 0.34 BTU/lb °F)

GRP has the following general properties:
· Strong but light.
· High stiffness compared with plastics.
· High impact resistance.

## Working with the material

GRP is made by embedding glass fibres, usually as a woven cloth, into a polyester resin which are then hardened by a catalyst. Glass fibre is a flexible sheet material formed from fibres drawn from molten glass, and has a tensile strength ten times that of steel. Polyester resin, the other component of GRP, becomes a solid material when a chemical catalyst is added. The material is made either by craft-based open moulding methods to make panels or by pultrusion to make continuous sections in the manner of extrusions in other materials. Injection moulding is sometimes used but this is usually limited to small components. The craft-based methods use either hand lay-up technique, where glass-fibre or glass-fibre cloth is laid in an open mould and coated with resin and catalyst, or the spray technique where a mixture of fibres and resins is sprayed onto the mould. Air bubbles are removed by hand-rolling or by a suction method. The face of the mould is coated with a releasing agent to facilitate removal when set.

The manufacture of GRP panels is a craft-based workshop activity rather than a machine-based industrial technique. It is formed in moulds but without high temperatures or expensive equipment. This is in contrast to the emerging technique of pultrusion where machinery is used to draw a mixture of resin and chopped fibres through a die to produce a material with constant cross-section. A wide range of sections is now available from I-sections to channels, tubes and planks for use in footbridges. Sections can be bolted together using techniques broadly similar to those used in steel construction.

111E

111F

111G

46F

46D

46E

## Properties and data

Density:
Pine softwood = 3.9 $kg/m^3$ (0.243 $lb/ft^3$)
Mahogany type hardwood = 7.5 kg/ $m^3$ (0.47 $lb/ft^3$)
Design strength:
Pine softwood = 5.3 $N/mm^2$ ($1.1 \times 10^5$ $lbf/ft^2$)
Mahogany type hardwood = 12.5 $N/mm^2$ ($2.6 \times 10^5$ $lbf/ft^2$)
Young's Modulus = 10 $kN/mm^2$ ($2.1 \times 10^8$ $lbf/ft^2$)
Coefficient of thermal expansion:
Pine softwood = $34 \times 10^{-6}$ $K^{-1}$ across grain ($19 \times 10^{-6}$ $F^{-1}$)
$3.5 \times 10^{-6}$ $K^{-1}$ along grain ($1.9 \times 10^{-6}$ $F^{-1}$)
Mahogany type hardwood = $40 \times 10^{-6}$ $K^{-1}$ across grain ($22.2 \times 10^{-6}$ $F^{-1}$)
$4.0 \times 10^{-6}$ $K^{-1}$ along grain ($2.2 \times 10^{-6}$ $F^{-1}$)
Thermal conductivity:
Pine softwood = 0.14 W/m°C across grain ($8.1 \times 10^{-2}$ BTU/hr.ft.°F)
Mahogany type hardwood = 0.21 W/m°C across grain ($12.1 \times 10^{-2}$ BTU/hr.ft.°F)
Specific heat = 3.0 J/kg°C ($7.2 \times 10^{-4}$ BTU/lb°F)

Standard sheet sizes:

| | |
|---|---|
| Plywood | 1220 x 2440mm (4'x8') |
| | 1525 x 3660mm |
| Thicknesses 4mm to 25mm (1/4" to 1") | |
| Chipboard | 1200 x 2400 mm |
| | 1200 x 4800mm |
| | Thicknesses 4mm to 25mm (1/4" to 1") |
| MDF | 1220 x 1525mm |
| | 2440 x 3050mm |
| | Thicknesses 4mm to 25mm (1/4" to 1") |

## Timber has the following general properties:

· Fibrous and elastic, making it strong in tension and compression. Like metals, it performs better in tension than in buckling.
· Undergoes varying degrees of moisture movement.
· Straight grain is stronger than an irregular grain and is easier to work. An irregular grain gives a rich, textured appearance, usually resulting from knots which weaken the timber. Knots are actually where the branches were located and leave a distinct pattern in the cut timber.
· Timber is prone to rot, particularly where it cannot be adequately ventilated or is subject to continuous cycles of wetting and drying.
· Theoretically, wood will last forever if it stays either completely dry or completely wet. Wooden piles rot only in the area of fluctuation in the water table or tide. The causes of decay are a mixture of trapped air and water, where fresh air cannot ventilate and dry the wet material. This results in the growth of fungi, causing dry rot, or

46A

46B

46C

insect attack. This is prevented by impregnating the outer zones of timber with preservative chemicals, but these can harm the natural environment if the chemicals are allowed to leach into the ground.

## Material selection

Commercial timber is classified as either softwood, from conifers, or hardwood, from broad-leaf trees. Softwood is used for most structural timber, as it is easily worked due to its softness and straightness of grain, though oak is common for exposed structural timber frames. Typical types are cedar, Douglas fir, western hemlock, pine, redwood (scots pine), spruce, whitewood and yew. Most softwood comes from the coniferous forests of the northern hemisphere. Hardwoods have high strength and durability, a rich grain and varied colour. They are more expensive than softwoods which make them too expensive for most current structural use, but suitable for joinery and finishes. Typical varieties of hardwood are ash, beech, birch, cedar, iroko, mahogany, maple, oak, teak and walnut. Hardwood occurs in most parts of the world but is obtained mainly from both northern temperate forests and tropical rainforests. Since hardwoods constitute a large proportion of the rainforests, and take much longer to grow than softwoods, the world's supply is depleting at an alarming rate. As a result, the use of certain tropical hardwoods has come under considerable scrutiny in recent years.

Laminated timber is constructed from planks glued together to form sections which are larger than could be achieved with natural timber. Laminated timber works well in both tension and compression. As the natural growth defects of timber reduce strength, individual boards are positioned so as to reduce the cumulative effect on the strength of the overall member. The strength of laminated timber therefore approaches that of defect-free solid timber.

70A

70B

56C

79A

56B

56A

90C

Plywood sheet was invented in the USA in the 1930's, and is made by bonding together veneers peeled from logs. Both softwoods and hardwoods are used. They are layered so that the grain of each veneer is set at right angles to the one either side, providing strength in both directions as well as minimising thermal movement. Good quality plywood has very little thermal movement. Plywood is graded according to its resistance to moisture penetration which is determined mainly by the glue used. The most common types are interior grade, exterior grade and marine grade. The face veneers vary from a rough finish, such as Douglas fir to a smooth one like birch. The different veneers have little influence on either moisture penetration or strength. Because plywood performs well in shear, it is used as sheathing to timber wall panels. Plywood also has high impact resistance. It can be bent to small radii, sometimes using a steam treatment.

Recent developments in plywood have produced blockboard and laminboard which are made by applying veneers to a core made from solid timber core blocks also known as a stave-core The core consists of strips of solid wood ranging from 7 to 30mm (1/4in to 1 3/16in) thick. Laminboard, a heavier material than blockboard, has a core of solid strips up to 7mm (1/4in) thick, laminated together. The grain of the face veneers is set at right angles to that of the core strips. These boards are not suitable for forming curves. The advantage of these panels is that they are lighter and cheaper than plywood. Laminboard is useful where weight is important, but is unstable in wet areas because the endgrain to the blocks is exposed which can deteriorate rapidly.

Chipboard, medium density fibreboard (MDF), and hardboard are the most common particleboards. All are made from mixing wood particles, normally wood waste, with an additive to form a sheet material that is cured between heated plates. Chipboard was developed primarily for the furniture industry in the 1960's. However, it is also used as a flooring material in wood construction as a cheaper alternative to plywood or timber boards. Chipboard is not as strong, or as rigid, as an equivalent plywood sheet, since the wood fibres are shorter. Like plywoods, it resists shear forces but is less capable of resisting impact damage and is harder to fix with screws and nails. Also, it cannot be curved except in a pre-formed process. Chipboard is also more prone to creep

68A

68C

68B

under prolonged loading than plywood and timber. MDF has smooth faces and a uniform cross section. This gives it a great advantage over other boards of having smooth exposed edges that require no lipping or trim when cut. For this reason it is often used for built-in furniture and interior panel systems.

## Working with the material

Before the 20th century most timber joints were made to work in compression, so the high tensile strength of the material was not used at junctions. However, the development of modern glues and metal connectors has resulted in a wide range of contemporary tension and compression joints. The most common joints in timber that perform well in tension and compression are the bolted metal connector, split ring connector and the nail-plate connector.

The bolted connector is a thin plate of galvanised steel with projecting teeth. The connector is set between the timber sections being joined, and is embedded into both sections as the bolt fixing is tightened. The split ring connector works in a similar way. Both fixings transmit shear forces at the joint across the bolted connection.

The nail-plate, also called a gang-nail plate, is a galvanised thin metal plate which is pressed to form a series of nail-like projections on one side. The plate is pressed into the timber sections being joined to form a patch connection on either side. This is done under factory-controlled conditions, and is most commonly used in prefabricated floor and roof trusses.

Glued connections in timber are glue-welded joints that are as strong as the wood itself. Widely used types are urea-formaldehyde, phenol-formaldehyde and resorcinol. Glued connections provide structural continuity at the joint. They do not break down when exposed to the weather or changes in temperature.

In joinery, particularly for built-in items, simple jointing techniques that allow parts to be made quickly and efficiently tend to be used. Most joints are designed to be manufactured easily using machinery, which has led to the demise of traditional techniques such as dovetailing. However, the availability of adhesives has enabled simple mitred and butt joints to be used.

68D

68E

68F

44G

## Properties and data

Density:
Copper = 8900 kg/m$^3$ (560 lb/ft$^3$)
Zinc = 7140 kg/m$^3$ (445 lb/ft$^3$)
Lead = 11,340 kg/m$^3$ (705 lb/ft$^3$)

Tensile strength:
Copper = 216–355 N/mm$^2$
($4.5 \times 10^5$ to $1.0 \times 10^6$ lbf/ft$^2$)
Zinc = 139–216 N/mm$^2$
($3.8 \times 10^5$ to $5.9 \times 10^5$ lbf/ft$^2$)
Lead = 15–18 N/mm$^2$
($3.1 \times 10^5$ to $3.8 \times 10^5$ lbf/ft$^2$)

Young's Modulus:
Copper = 117–132 kN/mm$^2$
($2.4 \times 10^9$ to $2.8 \times 10^9$ lbf/ft$^2$)
Zinc = 110 kN/mm$^2$ (pure)
($2.3 \times 10^9$ lbf/ft$^2$)
Lead = 15–18 kN/mm$^2$
($3.1 \times 10^8$ to $3.8 \times 10^8$ lbf/ft$^2$)

Coefficient of Thermal Expansion:
Copper = $17 \times 10^{-6}$ K$^{-1}$ across grain
($9.5 \times 10^{-6}$ °F$^{-1}$)
Zinc = 23 to 40 $\times 10^{-6}$ K$^{-1}$ (across grain)
($12.8 \times 10^{-6}$ to $22.5 \times 10^{-6}$ °F$^{-1}$)
Lead 29.5 $\times 10^{-6}$ K$^{-1}$ (across grain)
($16.5 \times 10^{-6}$ °F$^{-1}$)

Thermal conductivity:
Copper: = 300 W/mK
(173 BTU/hr.ft.°F)
Zinc: = 113 W/mK
(65.4 BTU/hr.ft.°F)
Lead: = 35 W/mK
(20.2 BTU/hr.ft.°F)
Specific heat:
Copper: = 390 J/kg°C
(0.093 BTU/lb °F)
Zinc: = 385 J/kg°C
(0.092 BTU/lb °F)
Lead: = 388 J/kg°C
(0.093 BTU/lb °F)

## Material selection

Apart from steel and aluminium, copper, zinc and lead are the most widely used metals in the building industry. These metals are produced as sheets of up to approximately one metre (3ft 3in) wide, sections and castings. All three metals have excellent weather-resisting properties, however, they are susceptible to attack by pollutants such as acids. Their use is restricted to non-structural metals and alloys, because none has the strength or rigidity of steel, or the combination of strength and lightness of aluminium. They are generally used as cladding materials for walls and roofs. Copper is also used extensively for water supply pipework and in electrical wiring.

## Working with these materials

Copper is strong in tension, tough and ductile, but is not as malleable as lead. The material has a shiny red/yellow colour when new, slowly developing a protective sulphate layer on its surface when exposed to the atmosphere. This patina has a characteristic green colour which has a fairly consistent colour and texture. Copper is available with a pre-weathered, or pre-patinated, finish which is chemically induced. This finish varies slightly from naturally weathered cladding. Well maintained copper cladding, which has oxidised atmospherically, will last from 30 to 50 years.

There is a variety of copper types available, each of which is suitable for a particular task. For instance, deoxidised copper is suitable for welding, while fire-refined tough-pitch copper, with its tougher resistance to corrosion, is used for cladding. Otherwise, both have similar properties. Copper forms a high proportion of metal in

alloys such as bronze, which is primarily a mixture of copper and tin. Brass is primarily a mixture of copper and zinc, and aluminium bronze is primarily a mixture of copper and aluminium. Copper can be cut, drilled, nailed, welded and soldered with hand power tools, making it versatile for site-intensive work. When used as cladding, joints between sheets are made by folding the edges together. This method takes advantage of the fact that copper can be bent along an edge fairly easily, but is rigid enough to remain folded.

Zinc sheet for cladding is made from either commercial zinc or from an alloy of zinc with small amounts of copper and titanium added. The properties of the two types are similar, but zinc alloy has better tensile strength and resistance to creep, which is long-term plastic deformation under load. Zinc is a durable material, although it is more brittle than copper. It is manufactured as a white coloured metal, but when exposed to the atmosphere a carbonate is slowly formed which forms a protective coating that is grey-white in colour. The material has a linear thermal expansion that is similar to lead and higher than copper. Zinc can be cut, drilled, nailed, welded and soldered with hand power tools, making it reasonably versatile for use on site. Its rigidity makes it well suited to standing seam joints. Well maintained zinc cladding can last for between 30 and 50 years. Zinc is liable to attack from copper alloys, so that rainwater running off copper should be avoided. Apart from its use in alloys, zinc is used as a protective coating to steel, applied through a process of galvanising and sherardising

Lead is an extremely durable, ductile and malleable material, making it extremely useful for roof coverings and flashings in traditional roof construction. However, while its relative softness allows lead to be formed into complex shapes, its lack of rigidity means that a supporting material must be provided beneath it. Timber boards are most commonly used for this. Lead has low resistance to creep.

On exposure to the atmosphere, a protective coating of lead carbonate is slowly formed on its surface. This gives weathered lead a dull grey appearance. Lead is a poisonous material, leading to increased awareness of the dangers of water run off from lead cladding reaching the water supply. Lead can be cut, drilled, nailed, welded in a process called leadburning, and soldered with hand power tools. Due to its lack of rigidity, lead sheet is most commonly jointed by dressing it over rounded timber battens or rolls. Where standing seams are used, they often incorporate a steel angle to keep the line of the joint straight and vertical. Lead is alloyed with tin to form solders for jointing and sealing.

## Electrolytic action

When different metals are near each other, rainwater running from one to the other can cause corrosion by electrolysis. Run off from copper and zinc will attack cast iron, mild steel, galvanised steel and aluminium. In addition, copper will attack zinc. Lead is much more resilient and does not attack other metals with the exception of aluminium when used in marine environments. When roofs or walls are made from either copper or zinc, typically other metals are not usually used in adjacent components, such as gutters and flashings, where water run off is likely to occur.

Plaster provides a smooth, continuous finish to a wall or ceiling. Its soft, fibrous nature makes it easy to work and repair. Plaster is often used as a wet mix to a surface by trowel as well as manufactured into lining boards that are fixed on site. The method used generally depends on the nature of the background surface where boards are preferred for use over uneven surfaces.

Renders perform a similar function to plaster on the external surfaces of walls and soffits. Their characteristics are similar to plaster. They can also be used in a board form, fixed to a layer of rigid insulation.

Plasters and renders are used not only for their appearance and suitability for the background, but also for reasons of cost, speed of construction and availability of materials. They are applied either in one, two or three coats.

## Properties of plasters and renders

These materials have the following properties:

- High resistance to impact damage.
- Easily drilled, filled and repaired.
- Moderate to low sound absorption.
- Low resistance to perpetual damp.
- High fire resistance.

## Tiling

Tiling provides a hard, impervious finish that is resistant to water penetration and surface damage and is easy to clean. When the building fabric has dried out, tiles are fixed to the background surface with cement or proprietary adhesive. Joints are grouted.

## Floor tiles

There are two types of tiles for flooring; these are ceramic and quarry tiles. Ceramic tiles are manufactured from refined clays, while quarry tiles are made by extruding or pressing natural clays. Whereas the former are available in a wide range of colours, the latter are produced only in their natural colours of reds, browns and blues.

Floor tiles are made in many sizes and thicknesses. They are laid either on a bed of sand-cement mix or adhesive. Each bay of tiles is edged with a 6mm (1/4in) wide movement joint, which can be filled with a flexible seal, such as polysulphide. Maximum bay size is usually six metres x six metres (20 x 20 ft). The bay size depends on whether the area is reached by sunlight and/or moisture. Metal edging strips, or a rigid sealant such as epoxy-polysulphide, is used to protect bay edges in large areas of tiling.

## Interior wall tiles

Glazed tiles are used for walls, and are made by applying glaze to a fired tile, then firing it a second time. The most common glaze used is enamel, available in a wide range of colours. There is almost no moisture movement or thermal expansion of most glazed tiles.

Tiles are fixed to a background of cement render or gypsum plaster once it has dried out; typically the drying out takes up to four weeks. Mostly proprietary adhesives are used to fix the tiles, although occasionally sand-cement mortars are used. The bedding depth for adhesive is 3mm to 6mm (1/8in to 1/4in). For a full set mortar bed, it is between 6mm and 12mm (1/4in to 1/2in). The joints between tiles are filled with a resilient material that is compatible with the tile to seal the gap without shrinking. Large areas of tiling are divided into bays 3m to 5m (10ft to 16ft) in either direction to compensate for movement in the substrate beneath. Joints are aligned with any movement joints

44C

44D

in the structure. Movement joints are 5mm to 10mm (3/16in to 3/8in).

Many smaller tiles have spacer lugs that assure a narrow, uniform joint. Plain edge tiles are spaced at least 2mm (1/16in) apart to allow for movement, although wider joints are used for their visual effect. Jointing mortars or grouts are generally cement-based but an ever-expanding range of products is becoming available.

## Exterior wall tiles

Glazed wall tiles for external use must be frost proof, and adequately resist airborne pollutants, scratching and fading caused by UV light. Tiles are fixed with an exterior quality adhesive. The bedding and grouting must be weather- and water-resistant. Bedding depth is 3mm to 6mm (1/8in to 1/4in). Movement joints must coincide with those of the background wall and must not bridge them.

Grouting can either be impervious, sealing the external wall completely, or be porous, to allow the wall behind to dry out. In common with internal tiles, joint widths should be a minimum of 2mm (1/16in).

44E

# 2

## WALLS

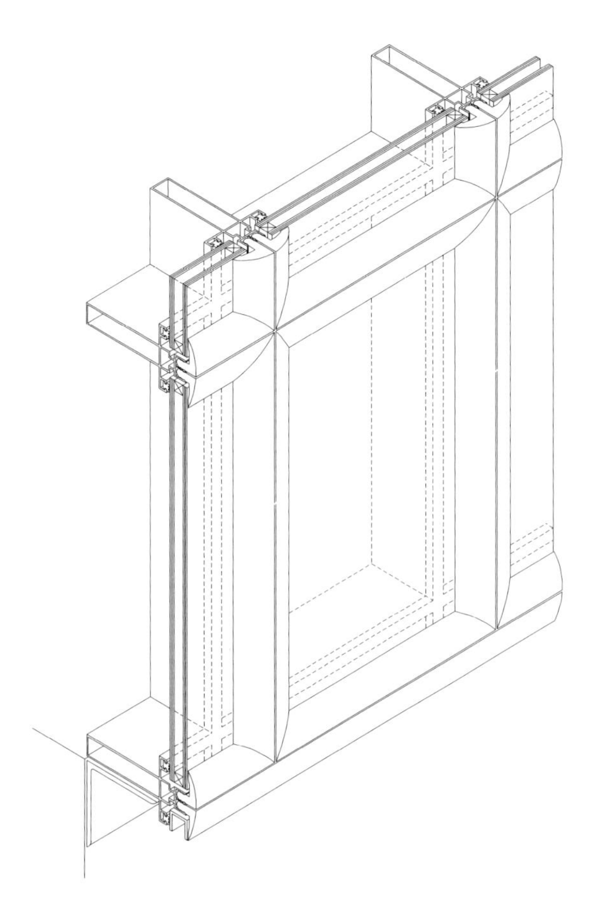

Structural gasket glazing: falling from use.

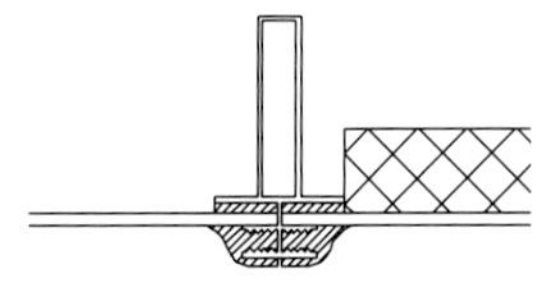

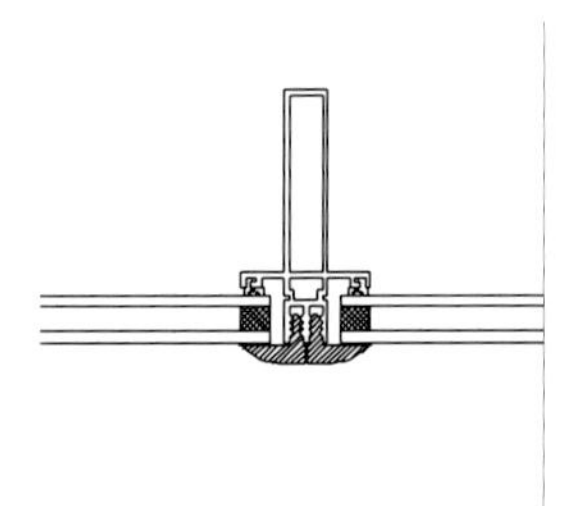

37B

'Cladding' is a word used to describe non-loadbearing external walls in buildings. This clearly distinguishes cladding from load bearing external wall systems, which in addition to keeping out the effects of the weather, form an integral part of the building structure. Almost all wall types used today in buildings are non-loadbearing. Consequently, most forms of wall construction fall into the category of 'cladding'. Cladding systems are fixed back to the main structure of a building, but do not contribute to its structural integrity. In other words, cladding does not form an integral part of a 'monolithic' or 'loadbearing' construction, where the outer surface of the structure performs an additional role as a weather-tight skin. However, even monolithic structures are usually clothed in an additional layer of cladding to improve their environmental or weatherproofing performance. As a result, almost all buildings are 'clad' with an additional skin to provide a substantially better performance than would be the case with a single material such as loadbearing brick or reinforced concrete structures. There has been a recent renewal of interest in loadbearing wall design for building types other than housing, where this form of construction still dominates.

The most profound change in walls over the past 15 years has been in the area of energy conservation. This has led to a huge shift in design criteria, primarily for cladding systems. During the late 1970's, the increased use of thermal insulation brought with it an increased risk of condensation occurring in cladding assemblies. This condensation can occur both inside the panel and on the outer surface of framing members. The avoidance of condensation has been a major development since that date. The increased use of natural ventilation, both as an air supply and as a means of cooling a building at night has had an effect on cladding design, particularly in the integration of opening lights and louvered slots in panels. The arrival of the 'deep plan' building, with distances of up to 18 metres (60 ft) between external walls has led to a need for greater levels of daylight entering a building. This, in turn, has increased the need for both solar shading and glare control. In addition, the cost of photovoltaic cells, which can generate electricity when exposed to sunlight, has reduced considerably over the past 15 years. As a result, they are increasingly used in large-scale applications for buildings. These energy-led changes in wall design have generated

37A

a range of technical developments, which have changed the emphasis of facade design. The expression of 'structure' in buildings is giving way to an expression of 'energy-concerns'. This is largely because the effect of additional layers of external insulation, blinds, shading and so on often renders the structure almost invisible. This section explores these changing design criteria, examines the responses to these issues and concludes with their effect on facade design.

## Increased use of thermal insulation and the reduction of condensation risk

The use of thermal insulation has increased dramatically during the past 15 years. It is common to design an external wall to achieve a U-value of 0.25 $W/m^2K$ (0.144 $BTU/h.ft^2.^{\circ}F$). This compares with a typical level of 0.6 $W/m^2K$ (0.34 $BTU/h.ft^2.^{\circ}F$) in the early 1980's.

The accompanying risk of condensation has led to considerable change in cladding design. 'Interstitial condensation' can occur within a wall assembly because of damp air that penetrates an assembly where it may condense and cause damage. In addition to dew point calculations to assess risk, vapour barriers are added to walls to stop the passage of damp air towards colder surfaces within the building envelope. Walls are particularly vulnerable at the edges, where joints are formed. Consequently, junctions at these points are designed to avoid the penetration of condensation by lapping or sealing edges. Most commonly, interstitial condensation is avoided by placing a vapour barrier on the warm (in winter) side of the insulation. An alternative measure, where appropriate, is to ventilate the construction in order to allow air to move freely and draw away any damp air.

Condensation can also occur on the face of a material inside the building. This condition occurs where there is continuity in a material from the outside to the inside of a building that allows a direct passage of heat or cold through the external envelope. When this 'thermal bridge' occurs, condensation from water vapour within the building can form on the face of the material. Condensation will drip down the face of the material, which will both inconvenience building users and damage the construction. Thermal bridges are avoided by providing a separating component in a low conductivity material,

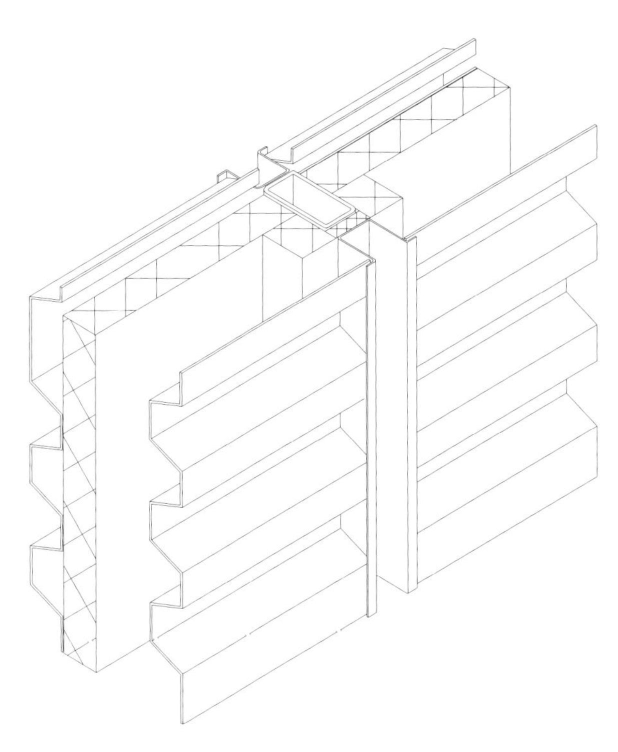

Profiles metal cladding with top hat sections: rising in use

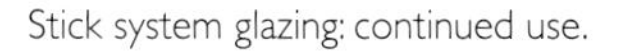

Stick system glazing: continued use.

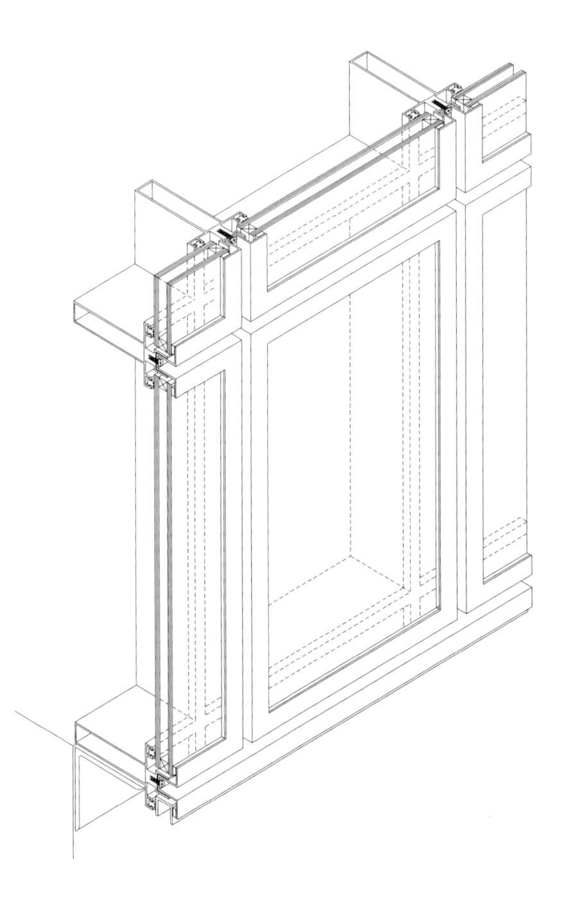

35A

Thin joint mortar system: This system is likely to become very popular, allowing brickwork to be curved vertically, using a cement-based mortar with adhesive additives, that increases the bond between brick and mortar.

Stainless steel mesh: increased use.

13A

62B

called a 'thermal break', that prevents heat or cold from being transmitted between inside and outside. This is used in glazed walling where the construction of a frame of mullions and transoms leads to a continuity of aluminium or steel from outside to inside. Plastic spacers of low thermal conductivity are positioned in a way that provides a thermal break but does not reduce the structural integrity of the construction.

Two generic types of thermal insulation have emerged for use in wall systems. These are 'rigid foam' and 'flexible quilt' types. The rigid foam type is formed in either polystyrene sheet or polyurethane foam. Since both are water resistant, they can be used in situations where the insulation can become wet without significantly reducing performance. Boards also form the structural core of metal-faced laminated panels. Flexible quilt is made from glass fibre quilt sandwiched between two layers of sheet material (often building paper) that hold it in place. The quilt can be cut to fit voids in panel frames more easily than board but is not water resistant, making it unsuitable for use externally. While heavier insulation helps to provide a more rigid, water resistant material, the lighter, less rigid types provide better thermal insulation. As a result, the choice of core is a balance between the needs of rigidity and thermal insulation. Both insulation types are used with both 'quick response' and 'slow response' construction.

## Responses in wall design to these issues over the past 15 years

The responses to these energy-based issues in the design of cladding have been in the development and widespread use of rainscreens, drained and ventilated cladding systems and a huge range of solar shading and daylight controls.

## Rainscreens

Rainscreen cladding is a development of the rainscreen for pressure-equalised walls researched during the 1960s. It was found that water commonly penetrates joints in walling because of the outside air pressure being greater than that inside the joint. Water arriving by a variety of means (gravity, wind, capillary action) was able to penetrate the fabric. Rainscreen panels solve this problem by creating a screen that takes away most of the water and allows the air pressure in the void behind the panel to be the same as that outside. Rain-

22E

22F

screen systems have three essential functions:

- To protect joints in the cladding assembly from the worst effects of windblown rain.
- To provide a decorative screen for a waterproofing system whose appearance is not suitable for an external wall. The same principle applies to roof cladding.
- To provide an outer protective layer to thermal insulation fixed on the external face of a high thermal mass construction such as concrete.

## Drained and ventilated systems

There has been a move away from 'hermetically sealed' systems, which rely entirely on a single weatherproof outer skin, towards 'drained' systems. This is common to virtually all cladding types. Drained systems accept that the air pressure differences between the outside of the wall and the inside will allow small amounts of rainwater to penetrate the outer seal. This effect can be countered through two means; the water can be drained away within the metal framing that supports the cladding panels, and the pressure differences can be equalised by ventilating the system by providing slots at drainage points, usually set at the base of the wall. This ensures that water is not trapped within the panel framing, nor is the water discharged at vulnerable points in the construction where staining and damage can occur on the face of the panel.

The use of ventilation within an assembly, on the external side of a thermal insulation layer, allows cladding systems to dry out once water has penetrated. Where water or water vapour penetrates a material in a wall or roof it is stopped from travelling further by the vapour barrier or waterproof membrane. At this point, it can be difficult for the water to dry out. A void is sometimes formed at the vapour barrier to allow the water to evaporate and prevent the adjacent material from being damaged. This is particularly important in the case of timber, where eventual rot can occur. The development of drained systems is a major step forward in the acceptance by cladding designers that all externally applied seals will inevitably leak small amounts at some stage of their lives and that it is better to design for that eventuality. Even systems with a single outer seal can be designed to drain away water within their construction.

Visible gasket seals on composite panels: continued use

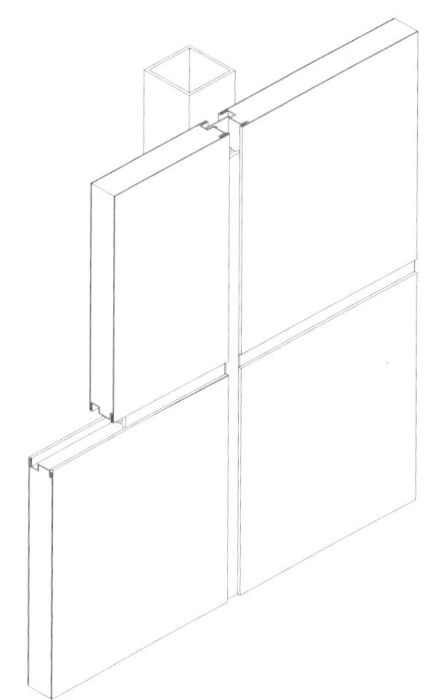

Structural gasket-fixed composite panels: falling from use

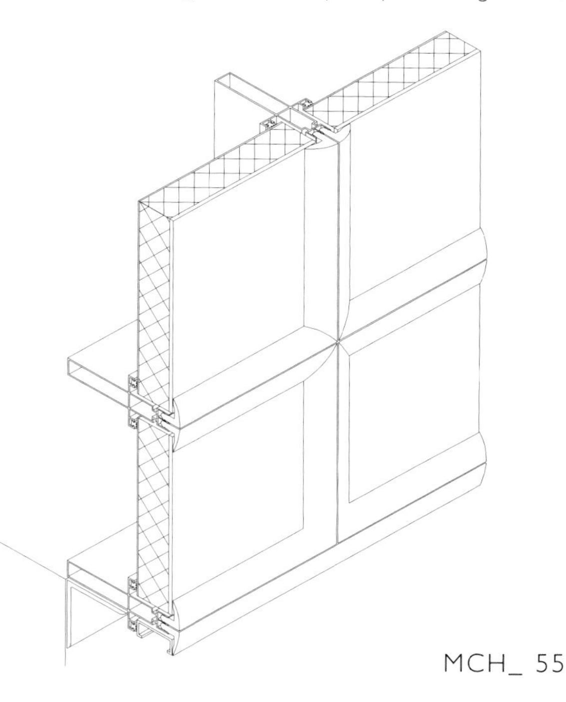

# Generic wall types in thin and layered facades

This section discusses changes over the past 15 years in the 17 generic cladding types set out in this chapter and further identifies a smaller set of 6 generic types from this list. Load- bearing wall types have changed little in recent years and are outside the topic of discussion in this book. The renewed interest in hydraulic limes to create large-scale masonry structures, with fewer movement joints, is still in its early stages. The application of this set of cladding types in 'thin' and 'layered' facades is discussed and how these facades are slowly developing from the use of technology from other industries.

The generic types are as follows:
1) Metal 1: fully supported sheet metal
2) Metal 2: profiled cladding
3) Metal 3: composite panels
4) Metal 4: rainscreens
5) Metal 5: mesh screens
6) Glass 1: stick systems
7) Glass 2: panel systems
8) Glass 3: frameless (patch plates and bolt fixed)
9) Glass 4: glass blocks
10) Glass 5: steel glazing
11) Pre-cast concrete panels
12) Masonry 1: facings to cavity walls (brick, block, stone)
13) Masonry 2: cladding (i.e. sealed joints)
14) Masonry 3: rainscreens (i.e. open joints)
15) Polycarbonate insulated systems
16) Timber 1: cladding to platform frame (i.e. sealed joints)
17) Timber 2: rainscreens (i.e. open joints)

Three types of material used in cladding systems have fallen into disuse over the past 15 years. These are glass reinforced polyester (GRP), glass reinforced concrete (GRC) and asbestos cement sheeting. All have been abandoned due to problems with their inherent properties. Pigments used in GRP fade in sunlight, GRC is difficult to apply in a way that ensures that fibres are correctly aligned in order to achieve the full panel strength, and asbestos sheet presents a public health hazard. In addition, patent glazing is no longer used to clad buildings as it is regarded as being no longer technically adequate. All other systems still in use have undergone technical development to improve performance, largely concerned with energy conservation.

Reports over the past 15 years have suggested that no clear pattern of use has emerged in the use of cladding systems. The choice of cladding system is very much affected by building type and the number of a particular type that are built in a given year. The full range of cladding systems in this book is in full use. The preference for one system over another is very much driven by the building type and particular performance criteria and budgets of individual projects. A clear pattern that is emerging is that the choice of cladding systems available has not affected the move towards the emergence of a new set of generic types, which are common to all the 17 cladding types.

From these generic types it is clear that the different types are supported by either a monolithic structure (or a backing wall to a frame) or an open framed structure. Each generic type is rarely supported by both structural types. Since framed construction is more recent, the more recently developed techniques of glazing are used in conjunction with frames. Traditional monolithic (load-bearing) structures continue to be used in conjunction with traditionally-based techniques. Structurally, the traditionally based cladding is characterised by techniques that use materials in short spans.

From the list, these short span generic types for monolithic structures (and backing walls to frames) are as follows:

- Metal 1: fully supported sheet metal
- Metal 4: rainscreens
- Masonry 1: facings to cavity walls (brick, block, stone)
- Masonry 2: cladding (i.e. sealed joints)
- Masonry 3: rainscreens (i.e. open joints)
- Timber 1: cladding to platform frame (i.e. sealed joints)
- Timber 2: rainscreens (i.e. open joints)

From the list, generic types for open framed structures are as follows:
- Metal 2: profiled cladding
- Metal 3: composite panels
- Metal 5: mesh screens
- Glass 1: stick systems
- Glass 2: panel systems
- Glass 3: frameless (patch plates and bolt fixed)
- Glass 4: glass blocks
- Glass 5: steel glazing
- Precast concrete panels
- Polycarbonate insulated systems

## New generic forms

Six new generic cladding types are identified below which are common to both non-glazed and glazed cladding systems. The first three types are used for small span applications with a backing wall. The second three types are used for large span applications without a backing wall.

Three types for small span applications with a backing wall:

1) Fully supported sheet with sealed joints: metal
2) Facings with sealed joints: facings to masonry cavity walls, glass blocks, masonry cladding, timber boarding
3) Rainscreens with open joints: masonry, timber boarding, metal, mesh screens

Three types for large span applications without a backing wall:

4) Self-supporting profiled sheet: metal
5) Stick systems: metal, glass, patch plate glazing, bolt fixed glazing, insulated polycarbonate
6) Panel systems: Pre-cast concrete, timber, metal composite, glass

## The application of these generic types in 'thin' and 'layered' facades

Modern single-sealed glass framing systems used in building construction, which rely on silicone bonding and rubber-based pressure seals, together with toughened, body-tinted glasses, are based on techniques developed by glass manufacturers for the car industry. Car manufacturers are relatively few in number, but they place very large orders for a restricted range of specially designed glass products such as windscreens. The building industry, in contrast, generates relatively small orders from a huge number of customers all of whom have quite different requirements. The current custom built nature of building construction does not encourage, in itself, glass manufacturers to invest significantly in new systems. As a result, most recent developments in glazing for buildings have followed in the wake of those developed for use in cars. Performance specifications for car glazing systems apply a complex set of criteria to a single very narrow 'wall' thickness. The use of glass with a structural capability, coupled with controlled light and heat transmission, is a strong influence in the design of facades. It reduces thickness of external walls, allowing floor areas to be maximised. It allows a single, competitively priced product, such as body tinted toughened glass, to fulfil a range of functions. In these 'thin' or 'compressed' facades, functions of weatherproofing, thermal insulation, air handling and glare control are compressed into a very thin wall.

Over the past 15 years, this principle has been developed to incorporate other components within a narrow depth. Louvered blinds and mechanical ventilation can now be incorporated. Heat from solar radiation can be partially absorbed by blinds set within the depth of a 100mm to 300mm (4" to 12") deep double glazed unit. Air passes through the unit, drawing away heat from the blinds. A recent example is an office development at Duisberg, Germany. The design team provided a thin external wall whose performance is a balance between different, and sometimes conflicting, criteria. In these thin facades, the integrated blind will absorb heat during periods of high solar gain when blinds must be orientated to act as a radiator of heat. This can conflict with occupants' requirements for daylight. Each facade will respond differently to the changing weather conditions, by either manual or electrical means.

An alternative approach for external walls is the separation and layering of the functions of weatherproofing, ventilation, thermal insulation, and daylight/glare control. With this 'layered' approach, specific layers in the envelope system are created to deal with specific tasks of excluding rainwater and controlling heat loss and heat gain, glare, and ventilation. To achieve this, a layered system superimposes generic types. A current disadvantage of this system is that a layered facade requires considerably more depth of 'wall', from around one metre (3 ft.) where external solar shading is provided, to about three metres (10 ft.), where the zone between the inner and outer wall becomes a usable space in a building in the manner of a conservatory or winter garden.

44H

44F

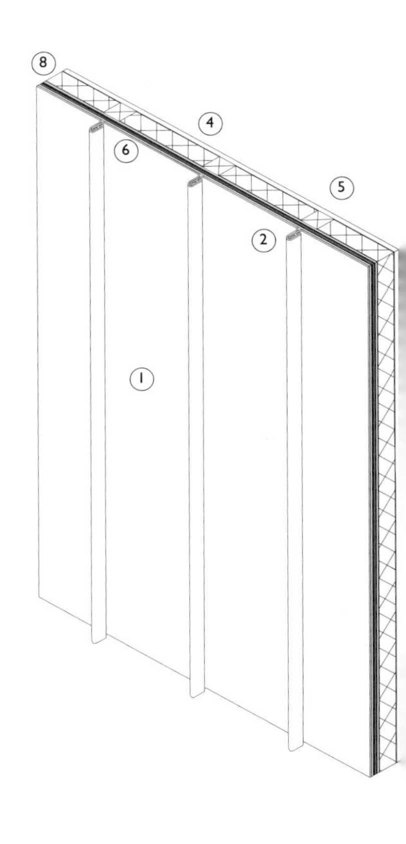


Sheet metal in walls uses a technology more typically used for roofing, where virtually all the same principles apply. Sheet metal such as copper, lead, and zinc, which is not rigid enough to span between structural members is fixed directly to a panel, providing full support across its surface. Plywood sheets or timber boards are typically used as support, although profiled steel decking is increasingly being used. This type of construction has the advantage of being adaptable to form complex shapes. It is essentially a handcrafted technique.

Once the shape of the wall has been created with the substrate, or 'decking' material, the surface is waterproofed with the metal sheet. Lengths of sheet are joined with a variety of seamed joints. The length of sheet used is between two metres and four metres depending on the type of metal, its thickness and its application. A proprietary separating layer is often used between sheet and substrate to allow thermal movement to occur freely. In the case of zinc, the back of the material must be well ventilated to avoid corrosion of the material from the underside. Open jointed timber boarding or proprietary matting are used for this purpose. The size of the sheets and their jointing are essentially the same as for sheet metal roofing except in one major respect: when used on vertical surfaces some metals such as lead have a tendency to 'creep' or sag more than on a sloping surface. This is countered by the introduction of additional fixings to support the sheet. Metals such as zinc and copper are less prone to creep and do not require additional fixings.

Sheet metal cladding is a highly effective weather barrier. However, it is essential where timber is used, to ventilate behind the metal surface to prevent moisture that might cause damage when trapped within the construction. In addition, the ventilation allows the metal sheet to be cooled, thereby reducing thermal expansion to a minimum.

## Developments

The essential trend in fully supported sheet metal in recent years has been a move towards fully prepared or 'pre-cut' systems that are site assembled. This is a move away from a site-based craft approach to a 'pre-engineered' approach where components are prepared in factory conditions. This has allowed this cladding type to be better designed to allow for thermal expansion and ventilation of the sheet metal, and has increased spans from around 400mm (16") to between 600mm (24") and 800mm (32"). Off-site preparation techniques allow joints to be designed with greater precision, providing greater confidence in the application of sheet metal cladding in large-scale projects. Because fully supported sheet metal is still assembled from lengths of metal sheet, it can still be site cut to accommodate complex geometries and unusual site conditions. This makes it especially useful as a cladding or outer covering to externally fixed insulation in buildings with a slow thermal response.

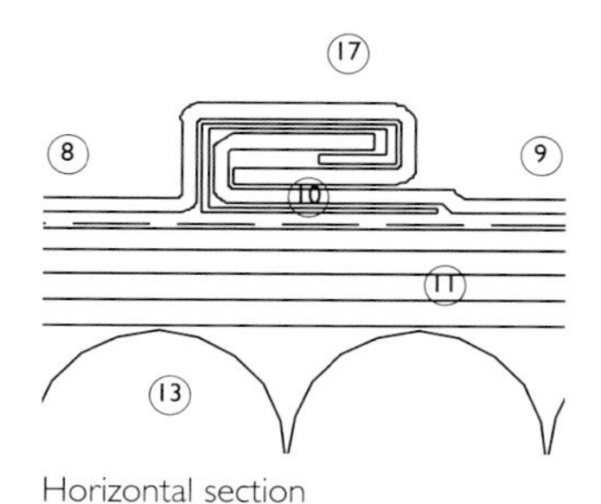


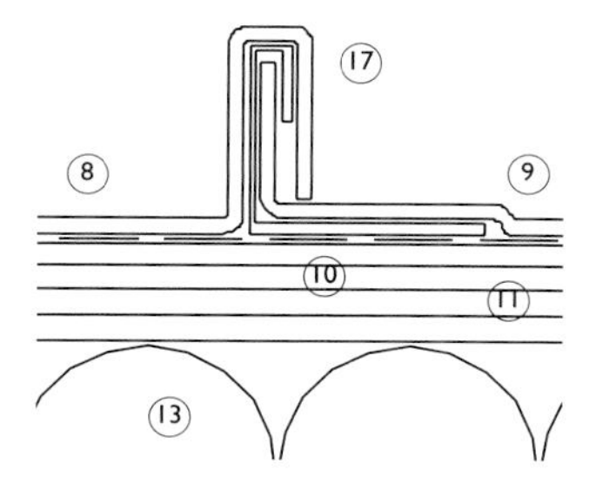


Horizontal section

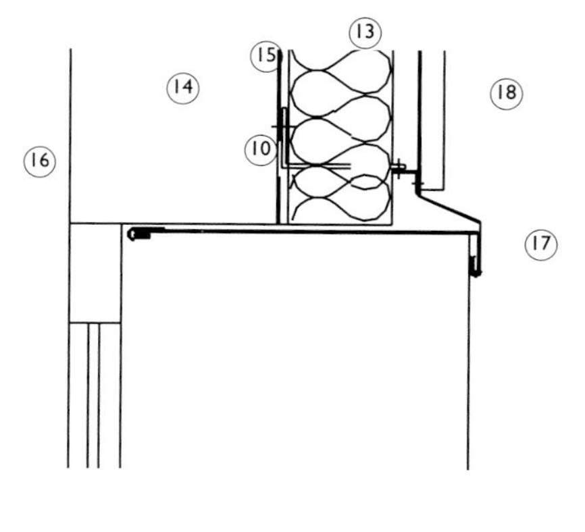


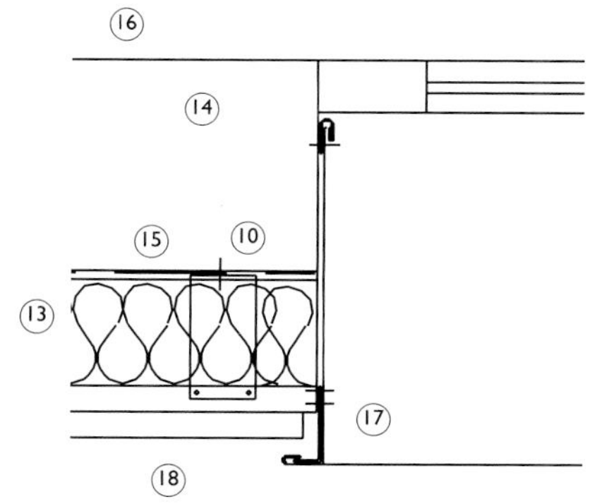


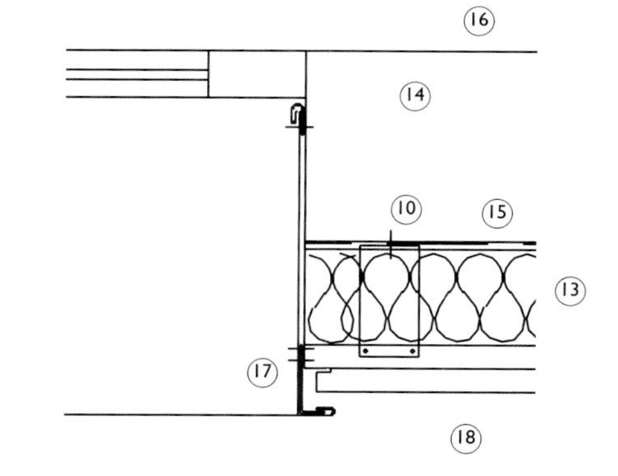


Horizontal section

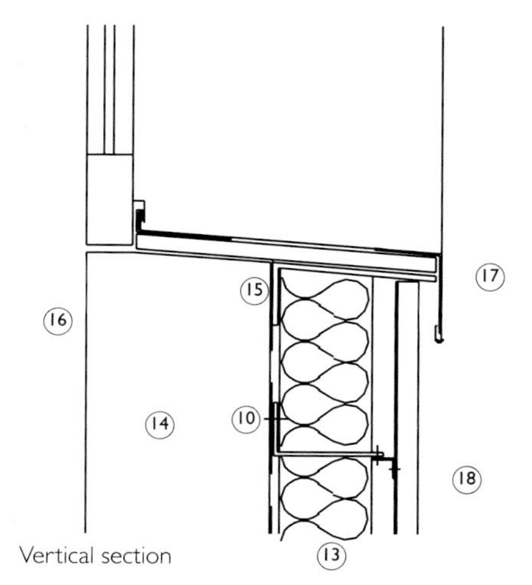


Vertical section

7C

A recent example of the renewed use of fully supported sheet metal is in the Jewish Museum in Berlin. The design team, led by Daniel Libeskind, used a cast-in-place concrete wall structure (a loadbearing concrete box) clad with insulation on the outside to retain heat but allow the structure to absorb and release heat slowly. The extra insulation has a thin outer skin of sheet metal, which provides a durable material for an external wall without the need for an additional structural external wall. The sheet metal can be fixed back directly through the insulation to the loadbearing concrete wall.

The ability of sheet metal to take up complex geometries has been exploited in the Guggenheim Museum in Bilbao, Spain. The project team, led by Frank Gehry, used a titanium alloy that would have a longer life than other metals and provided a slightly rippled, textured surface that complemented the large-scale smooth forms of the building.

Isometric view of assembly

1. Metal sheet
2. Standing seam joint
3. Waterproof membrane
4. Thermal insulation
5. Backing wall, typically timber/metal frame with plywood facing, concrete block
6. Vapour barrier
7. Drywall/dry lining
8. Plywood substrate

Details

8. Metal sheet 1
9. Metal sheet 2
10. Clips at centres
11. Plywood substrate
12. Breather membrane
13. Thermal insulation
14. Backing wall, typically timber/metal frame with plywood facing, concrete block
15. Vapour barrier
16. Drywall/dry lining
17. Standing seam joint
18. Metal sheet

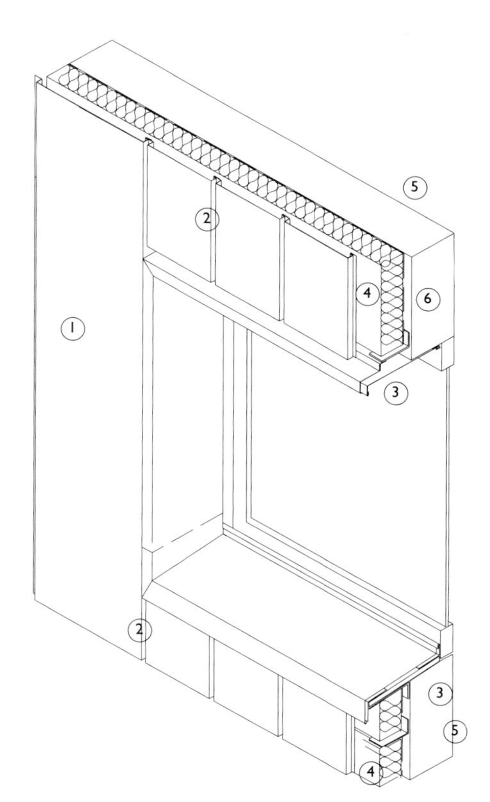

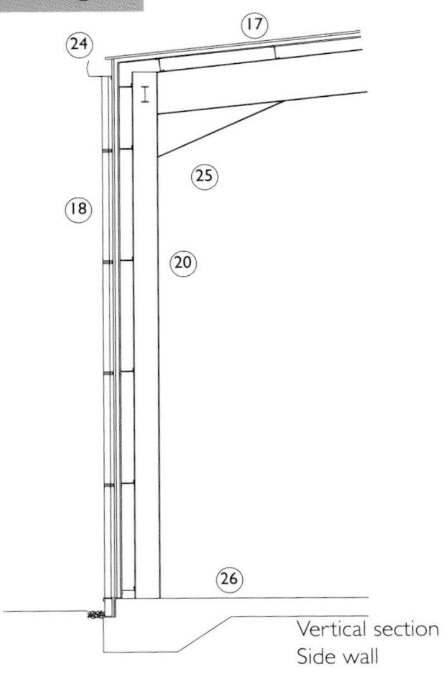


Vertical section
Side wall

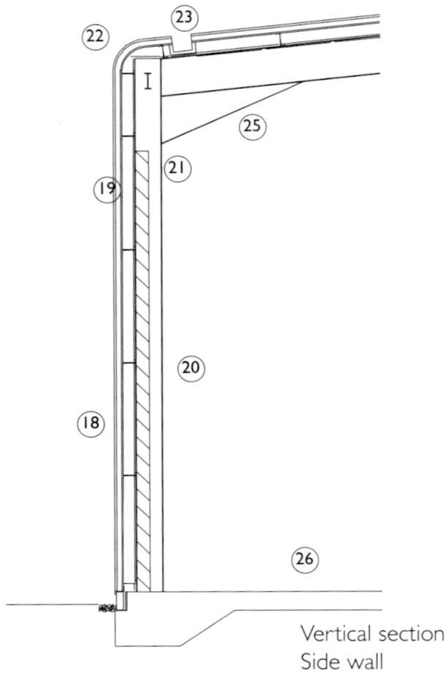


Vertical section
Side wall

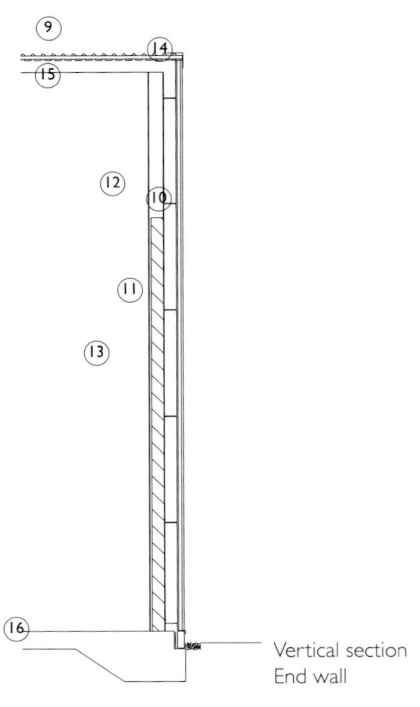


Vertical section
End wall

14D

Steel and aluminium profiled cladding provides a wall enclosure that spans between structural supports without the need for a backing wall. It usually has a proprietary coated finish to provide a weather- tight layer. It can span from around 3.5 to 6.0 metres (11ft 8in to 20ft) between supports and is one of the most economical types of metal wall because of its self-finish (the metal coil is pre-coated during manufacture) and structural rigidity (due to the folded shape).

Profiled sheeting is most commonly fixed back to sheeting rails, which are positioned vertically or horizontally to suit the orientation of the cladding. The rails span between primary structural members supporting the cladding. Liner trays can be used as an alternative to rails. These are metal trays used to form the inner lining of the metal wall. They clip together to form a continuous inner lining. Mineral fibre insulation is located in the void between the two skins to improve thermal performance. The cladding can be cut either on site or in a workshop depending on the level of complexity required.

Joints between sheets are formed by overlapping the edges and then screwing the sheets together. The amount of lap varies between systems, ranging from around 100 to 200mm (4in to 8in). Sheets are fixed back to a structural support with self-tapping screws, allowing large numbers of fixings to be completed quickly. The internal sheets, typically another profiled metal panel, are similarly attached to the other side of the sheeting rail, providing a void for the thermal and acoustic insulation.

An alternative configuration is to set the thermal insulation on top of the profiled deck. A waterproofing layer, usually either a single layer membrane or another metal sheet, is laid on top to provide the weatherproofing layer.

Window and door openings require pressed metal flashings around the edges of the exposed profile and are supplied by the manufacturer. The material is most economic when clad and glazed bays are separated. The life of profiled cladding is up to 25 years, depending on the finish and maintenance cycle.

## Developments

Spanning capabilities have increased over the past fifteen years due to the steady increase in profile depths. The material is frequently used with comparatively rigid structures such as portal frames that allow its use to be kept simple without the necessity for complex junctions or structural movement joints that increase expense. Increasingly, joints in long span sheets are sealed with an automated tool set on wheels that travels across the joints and 'zips' the material together along the continuous length of the joint.

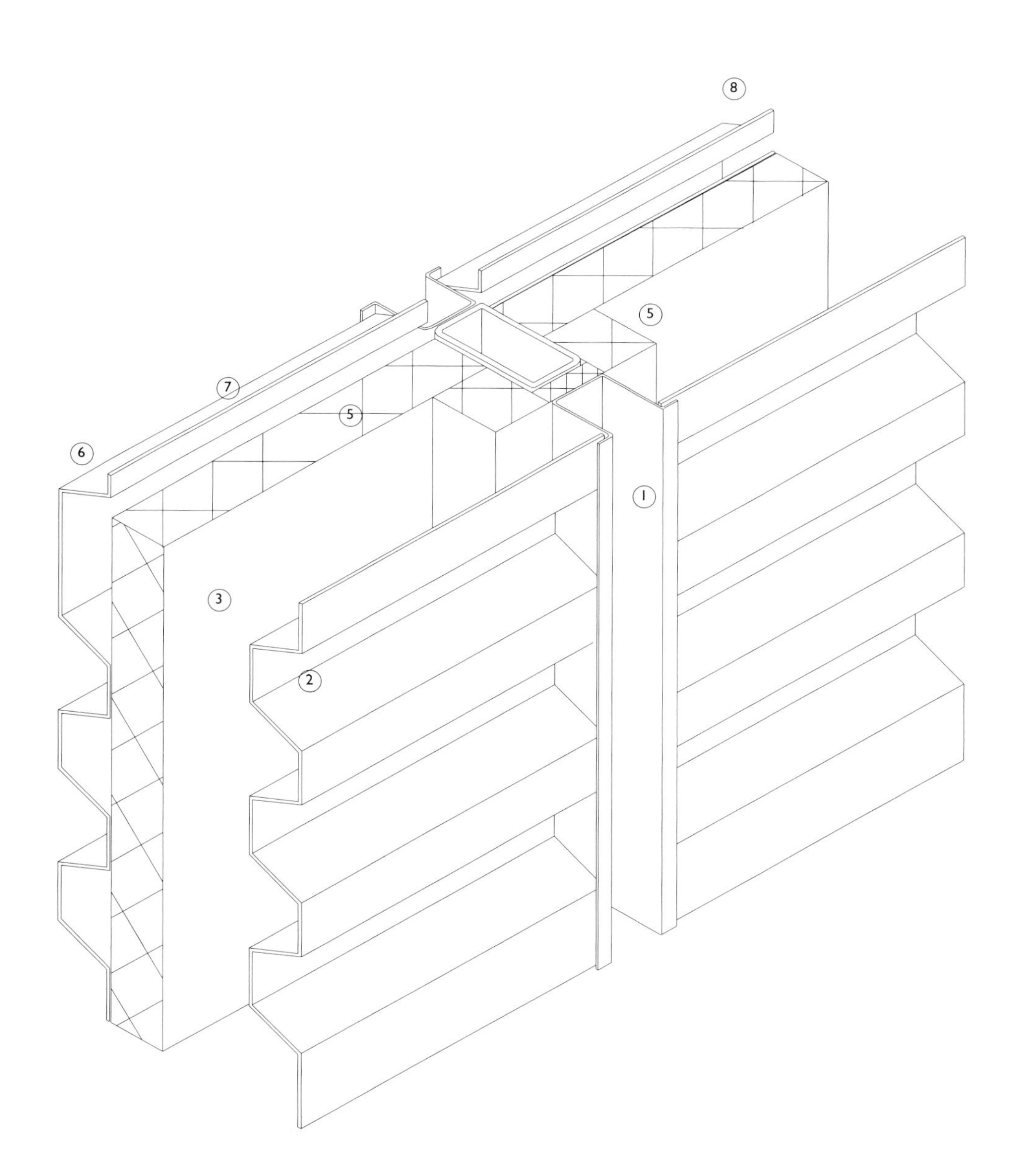


## Isometric view of assembly

1. Cover strip profile
2. Horizontally fixed profiled sheet (can also be vertically fixed)
3. Air gap
4. Breather membrane
5. Thermal insulation
6. Backing wall, typically timber/metal frame with plywood facing and waterproof membrane, or concrete block
7. Vapour barrier
8. Drywall/dry lining

## Vertical section through end wall: vertically fixed sheet

9. Roof, typically profiled metal sheet
10. Profiled metal sheet
11. Thermal insulation
12. Z section steel fixing rails
13. Backing wall
14. Edge cover strip
15. Structural frame
16. Ground slab

## Vertical section through side wall: vertically fixed sheet (two versions drawn)

17. Roof, typically profiled metal sheet
18. Profiled metal sheet
19. Thermal insulation
20. Z section steel fixing rails
21. Backing wall
22. Curved eaves profile
23. Concealed gutter
24. Exposed gutter
25. Structural frame
26. Ground slab

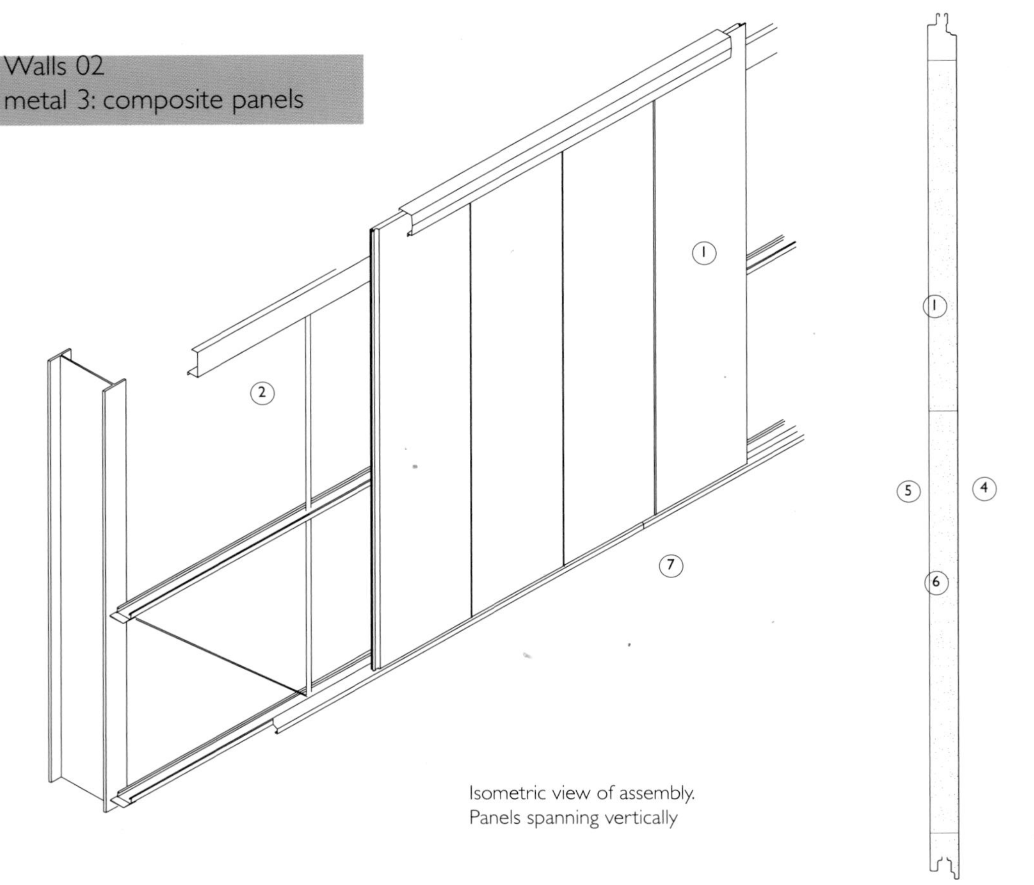


Isometric view of assembly.
Panels spanning vertically

Details

1. Composite panel
2. Panels fixed to primary or secondary structural steelwork
3. Polysulphide or silicone-based seal
4. Outer metal facing
5. Inner metal facing
6. Inner insulation core
7. Metal capping
8. Concealed fixing

5A

5B

Horizontal section.
Panels spanning horizontally.

2 5 3 1 6 7 4

Horizontal section.
Panels spanning vertically.

2 5 1 6 4

Composite panels are used as non-loadbearing, self-supporting panels that are capable of larger spans between structural supports than an equivalent profiled sheet. This allows them to span vertically from floor-to-floor, or be stacked horizontally between vertical supports. Vertical panel systems are fixed back to the building floor structure at top and bottom. Horizontally stacked systems are fixed back at their short sides to vertical supports such as columns or posts.

Composite panels are of laminated construction, consisting of an outer layer of flat or micro-profiled metal integrally bonded either side of a rigid core of thermal insulation. The panel provides a complete wall construction with a pre-finished outer and inner face. Until a few years ago, some panel types were stiffened with metal ribs but these cause both thermal bridging (from outside to inside) and 'oil canning' (dips in the metal surface). Until recently, it has been difficult to achieve a smooth flat surface with composite panels. A micro- profiled texture is often used as a way of breaking up reflections from the surface of the metal to avoid seeing oil canning across the surface. Stiffer metal sheet materials, such as Alucobond (two sheets of thin aluminium bonded to rigid plastic interlayer) have been developed, but they remain expensive. Improved lamination techniques have resulted in smooth flat panels employing thin sheets around 0.7mm thick.

The rigid core of insulation is commonly formed in either polystyrene sheet or polyurethane foam. While heavier insulation helps to provide a more rigid panel, the lighter, less rigid types provide better thermal insulation. As a result, the choice of core is a balance between the needs of rigidity and thermal insulation. Because they are pre-finished in the factory, composite panels can provide a very high quality surface finish. Panels are fabricated entirely off site and are fixed in place as complete units on site. Panel systems require far fewer fixings than other methods, such as profiles or metal sheets, and

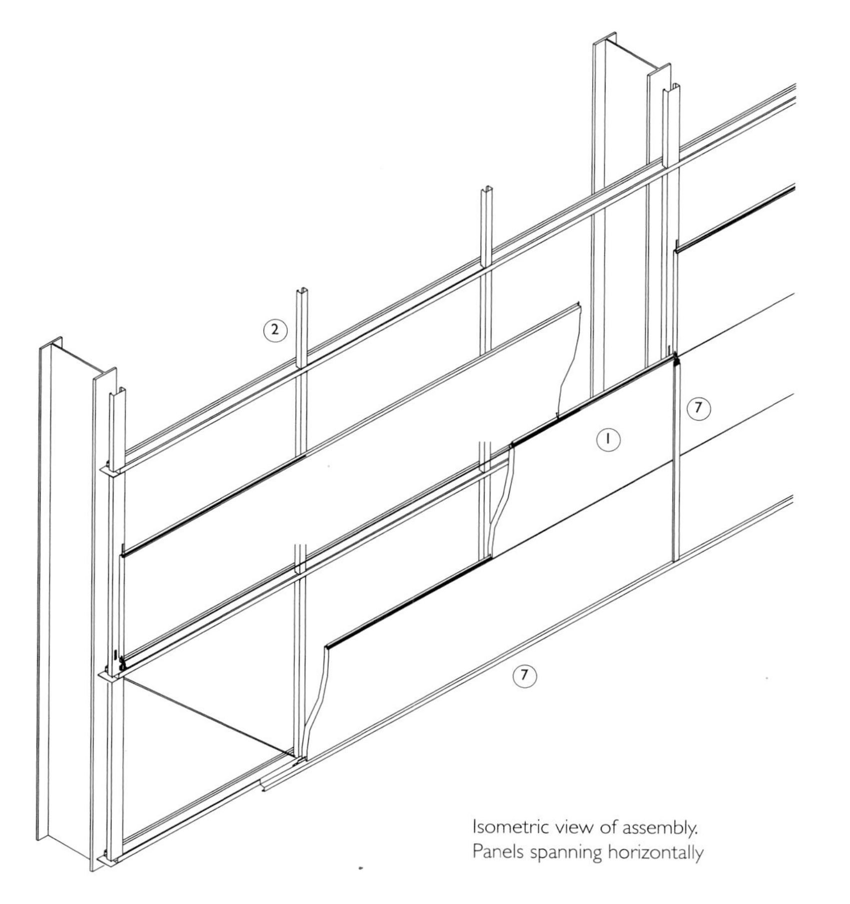

Isometric view of assembly.
Panels spanning horizontally

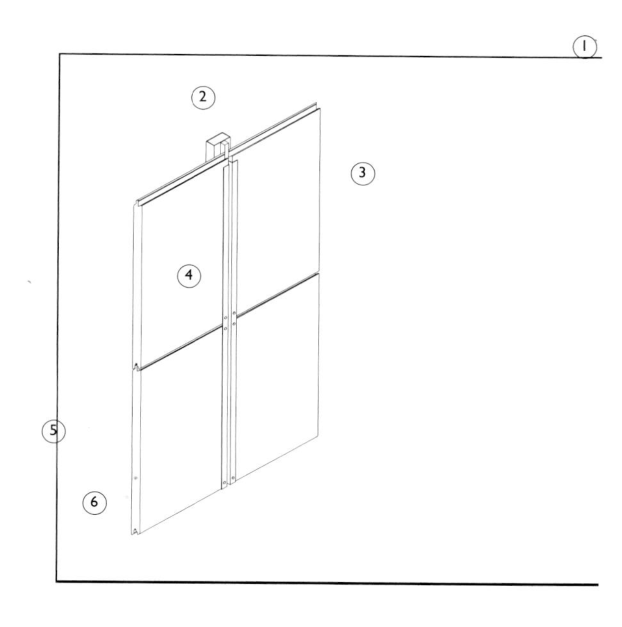

are quickly erected on site. The maximum horizontal span between vertical supports is currently around four metres (13 ft.). Panel widths are determined by the width of sheet metal available. Standard maximum widths of coiled steel sheet are available at either 1200mm or 1500mm (4ft or 5ft).

This technique allows the integration of a window either within a wall panel or as a separate glazed panel. Replacing panels is difficult because joints have 'tongue and groove' profiles in order to provide an economic method of interlocking the panels. More sophisticated, fully interchangeable panels allow glazed and opaque panels to be exchanged and fixed with a minimum of effort. Junctions between panels allow panels to accommodate thermal and structural movement, preventing the metal facing from separating from the core of insulation through the process of delamination through solar gain. Panels can be accurately curved to a range of shapes and at higher cost. All are marketed as proprietary systems but can be designed as custom elements.

## Developments

A difficulty encountered in these panels 15 years ago was in the delamination of the bond between the insulation core and the outer metal skin. This would often happen because of different rates of thermal expansion between the outer skin and the inner core. Bonding techniques, which include adhesives, have now solved this problem. This has allowed composite panels to become longer with ever-greater spans between structural supports. Panels spanning up to 15 metres (48 feet) horizontally are now in use. Although there remains some difficulty in producing curved panels economically, increasingly, these panels slot together with concealed fixings in the manner of tongued and grooved timber boards, fitting together either vertically or horizontally.

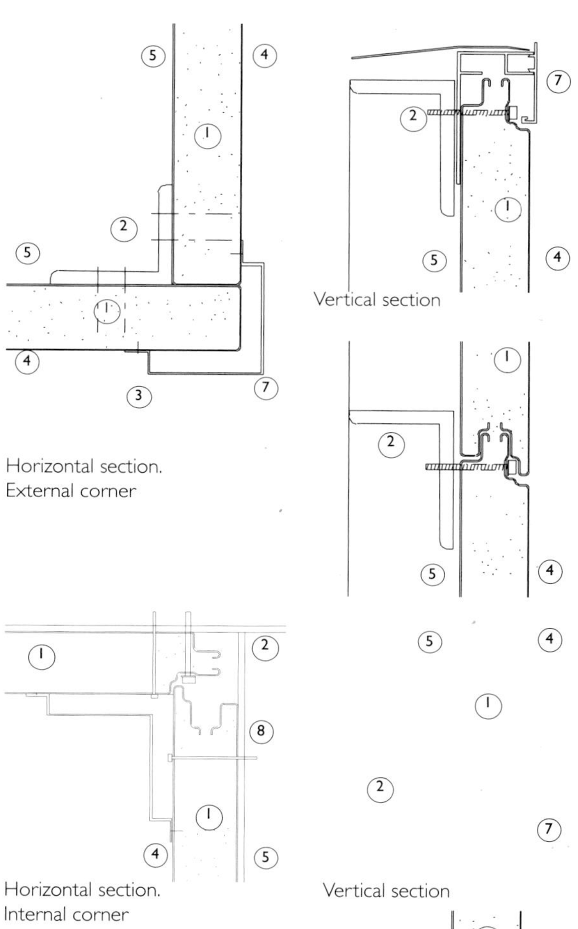

Horizontal section.
External corner

Vertical section

Horizontal section.
Internal corner

Vertical section

62A

Details

9. Backing wall or structural wall supporting rainscreen
10. Support frame
11. Support bracket
12. Metal rainscreen panel
13. Open joint
14. Closed cell thermal insulation
15. Finish to inner wall
16. Roof finish
17. Window frame inserted into opening in backing wall or structural wall
18. Window cill
19. Pressed metal coping
20. Continuity of waterproofing layers of wall and roof
21. Vapour barrier
22. Roof deck

Metal rainscreen cladding consists of a steel or aluminium panel mounted in front of, and supported by, an insulated backing wall. This outer screen drains away most of the rainwater and shields the wall behind from the worst effects of the exterior climate. The small amount of rain that does pass through or behind the screen is repelled by a waterproof layer fixed to a backing wall. The void between the two layers is ventilated to the outside air to help evacuate any moisture accumulating there. The backing wall provides the necessary structure and thermal and acoustic insulation.

The outer layer can also be used as a means of screening an economic waterproofing material behind, which otherwise would not be suitable for use as wall cladding. Rainscreen panels can be used to clad anything from a blockwork wall to a PVC membrane on a supporting surface. Gaps can be left between the panels since no seal between outer panels is required.

Metal rainscreens provide a metal wall system that uses the material in sheet form. Consequently, the sheet must be rigid enough to stay flat and avoid 'oil canning'. Panels can be removed easily in order to inspect the waterproof layer and carry out any maintenance work. The panels also protect the insulation or waterproof layer to the backing wall. The vertical supports sometimes incorporate plastic components to reduce any noise created by the thermal expansion and contraction of the metal panels.

There has been much emphasis on the flatness of rainscreen panels (particularly in metal) as the main reason for their use, but this is only one of the benefits. Rainscreens also have the advantage of being a very economical way of cladding a low-cost backing wall, such as concrete block, with a material that provides both weather protection and a more durable finish than a backing wall. Metal panels are easily cleaned and maintained. Complex, non-rectilinear geometries can much more easily be clad in metal panels, where the weatherproofing layer behind can be assured in the most economic way without regard to appearance, since this layer is covered by the panels. Rainscreen panels are also increasingly used with single membrane weatherproof layers that are fragile and need protection from solar radiation (UV light) as well as from accidental puncturing.

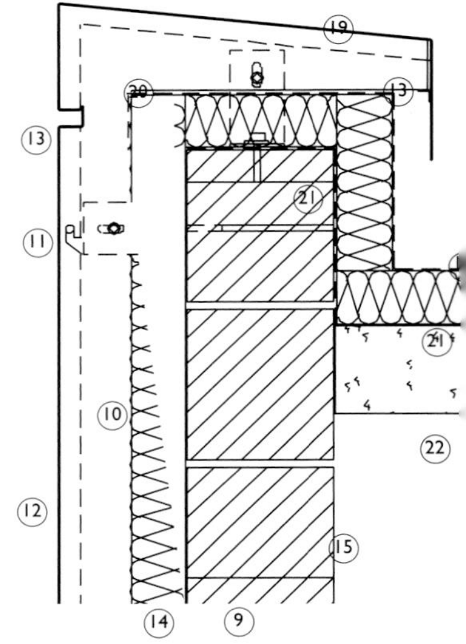


Vertical section

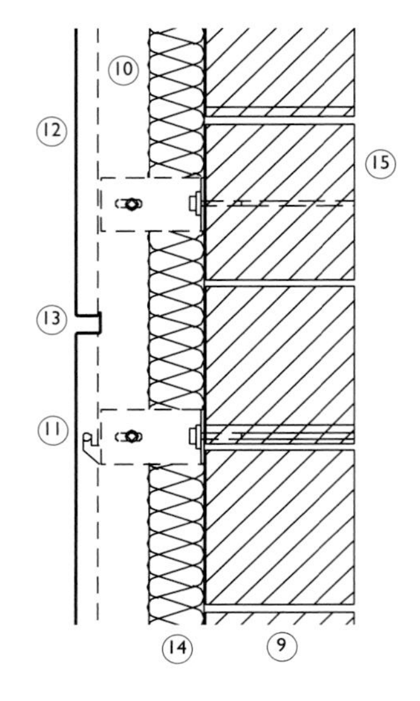

## Isometric view of assembly

1. Backing wall or structural wall supporting rainscreen
2. Support frame
3. Support bracket
4. Metal rainscreen panel
5. Open joint
6. Closed cell thermal insulation
7. Waterproof membrane
8. Finish to inner wall

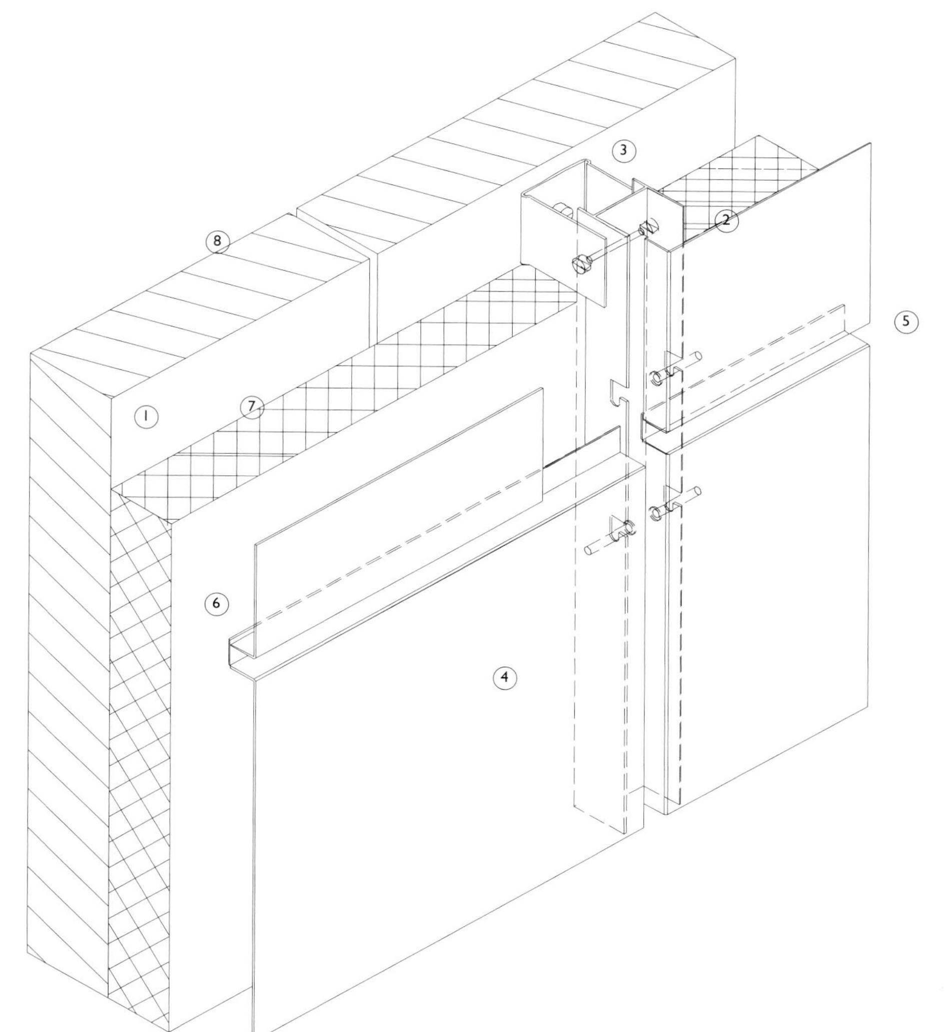

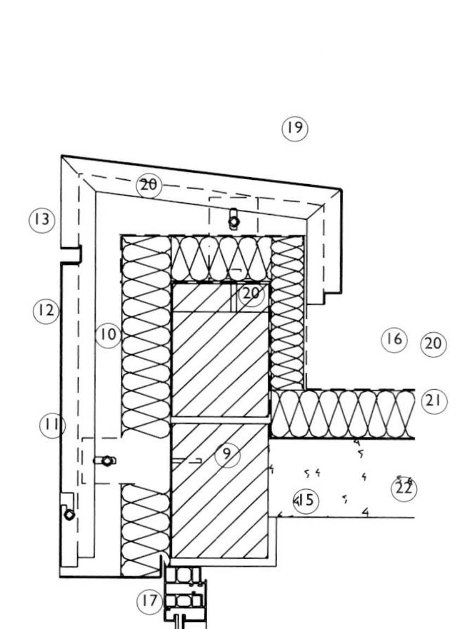

Vertical section

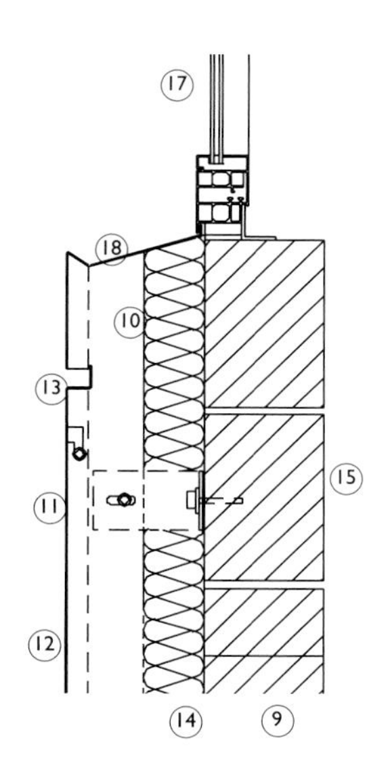

15E

Horizontal section

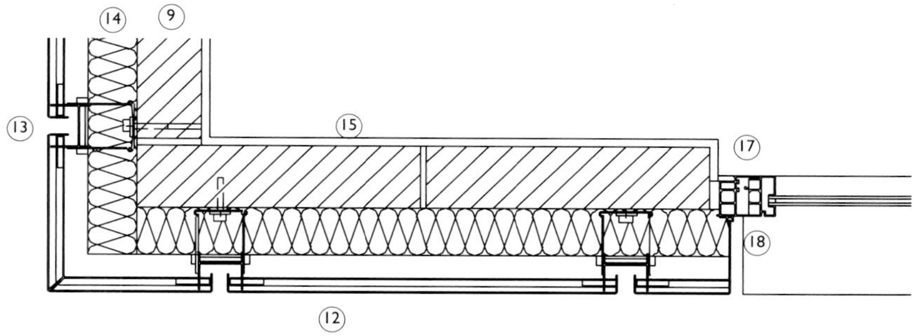

17

Vertical section:
Santa Monica Place, Santa Monica, California, USA.

*Frank O Gehry Associates.*

Chain link fencing is used to create a continuous translucent texture across an entire facade. The mesh has been sign-painted to give a silk-screened effect.

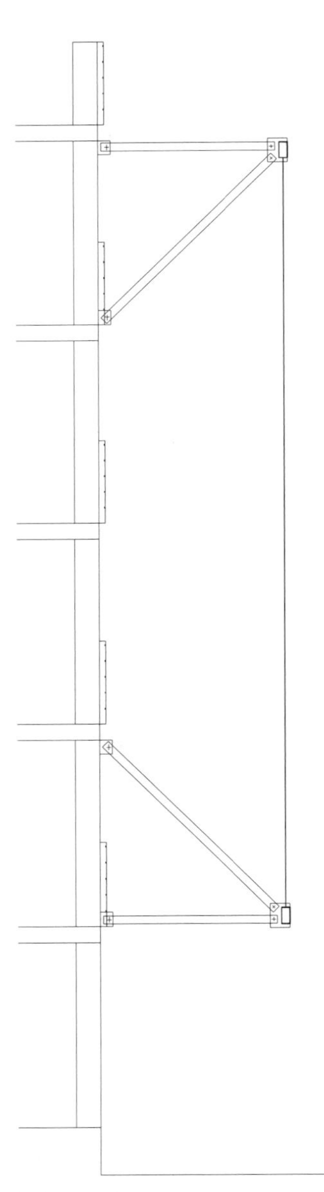

The use of metal mesh screens is a recent development in building construction. A range of wire or sheet metal based materials is used to provide a translucent screen that provides a continuous texture across parts of a building facade or roof. They are not expected to provide any significant resistance to rainwater penetration. The wire-based types are often produced in stainless steel for architectural use, providing high resistance against corrosion. Sheet-based materials are usually.supplied in a pre-coated finish, such as galvanising or paint, to resist corrosion. The most commonly used materials are as follows :

·Woven wire-cloth. This has a fine texture. It is used mainly for filtering in industry but its fine texture makes it suitable for small screen panels of low transparency, primarily for internal use.

·Crimped wire screens. These have wires crimped down the length of the material and straight wires across its width.

·Chain linked metal, usually used for fencing, is produced in either a galvanized finish or in stainless steel. Its flexibility both along the length of the material and across it make it ideal for forming curved shapes in the manner of a fabric roof.

·Welded wire mesh. These are available in a range of wire thicknesses and weaves. Different weaves can be mixed in a single facade application to give a range of textures.

·Conveyor belts. These are destined primarily for use in factory machinery. They comprise rods in one

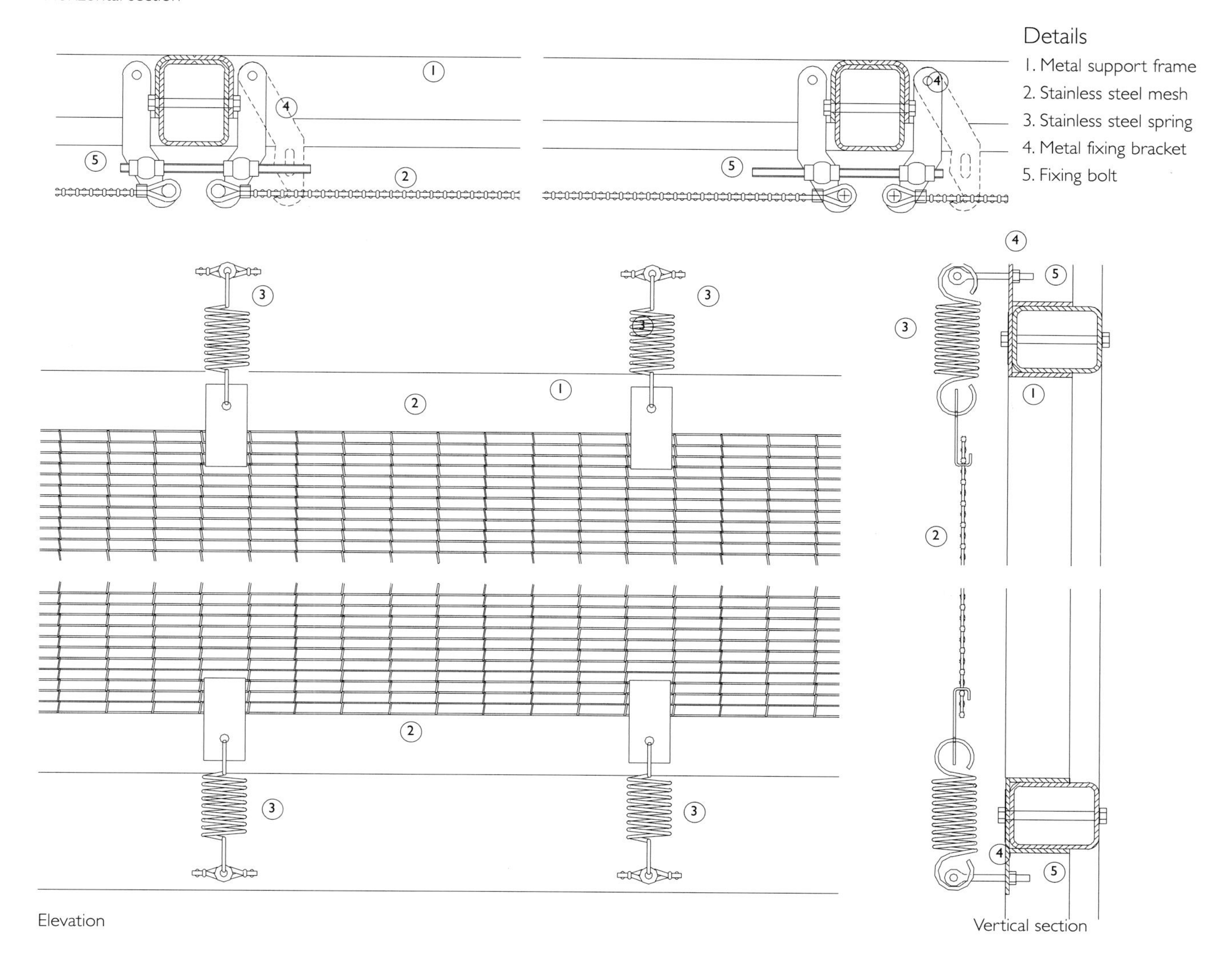


direction woven together with wire in the opposite direction. This allows them to bend along the length of the material but remain rigid across its width. This one directional stiffness has been exploited in curved screens, such as that used in the Bibliotheque Nationale, Paris, France.

·Raised expanded metal sheet. It has higher strength and rigidity than wire products, allowing for the use of greater widths than most wire-based products.

·Perforated metal sheet. This is often used to form screens that are more rigid. A large range of size, shape and frequency of the punched holes is available in sheets to give different visual effects.

Meshes are manufactured in large quantities for many different industrial applications and are therefore relatively inexpensive. The available types vary enormously in texture and gauge of material and there is sufficient variety to enable choices to be made that are appropriate to different construction methods and budgets. The wire-based materials are either stretched or held in tension by cables, springs or brackets. The method used is very much dependent upon the visual effect desired. The sheet metal based products are fixed with metal brackets back to supports. The examples illustrated here give an indication of the sparse and lightweight nature of the associated detailing.

20A

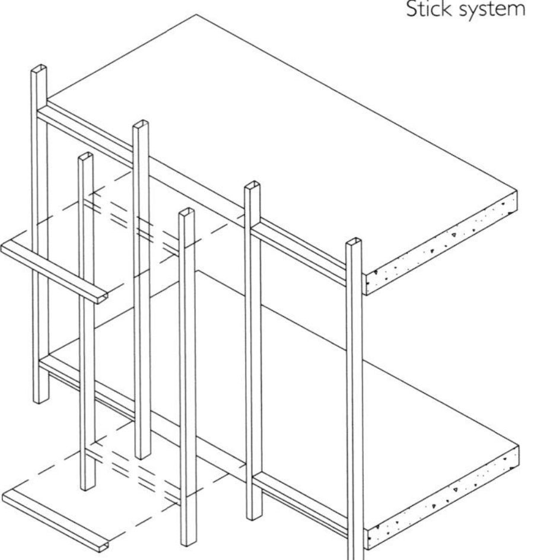


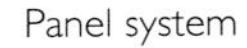


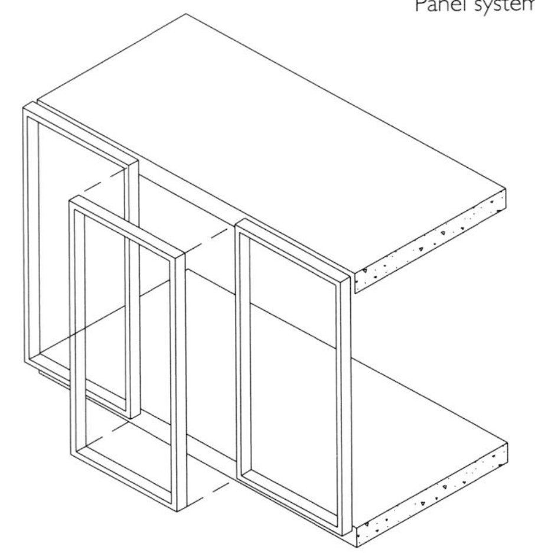

8J

8D

The building design team commonly considers the term 'cladding' to include all non-loadbearing walls except those in glass. Glazed walls are commonly referred to as 'curtain walling'. However, this distinction is based on an assumption that glazed walls are built in a way that is fundamentally different from other materials. Whilst this was the case up until the 1960's, it is certainly not the case today. This distinction between glass and other materials hinders a better understanding of the nature of cladding. An example of this difference is in the use of the term 'structural glazing'. It is used to describe both bolt-fixed glazing and patch plate glazing. In fact, neither system uses the glass to support the building structure and few applications allow the glass to be supported independently. Genuine structural glazing, still in an early stage of development, is not referred to as such since manufacturers have adopted the term for something different.

The principles of glazed cladding systems are shared with those of metal and reinforced concrete. Glazed walls are designed as either 'stick' systems, which are assembled primarily on site, or panel systems or 'unitised' systems, which are manufactured in a factory setting and then assembled on site. Aluminium is the most common material used for framing, since it can be extruded to make the complex profiles required in glazed walling. Equivalent steel sections, which are rolled, are unable to match the precision of aluminium extrusions. Steel systems are often limited to applications such as large windows and fire resistant glazing, but their thinner sight lines give them an enduring appeal.

Double glazed units are used to increase thermal and acoustic insulation. A comparable performance in the support framing is achieved by providing a thermal break between the outside and inside faces of the frame. This stops any thermal bridging that might produce condensation on the inside face of the framing.

In 'stick' systems, the glass is fixed back to the carrier framing with a continuous metal strip, called a pressure plate. The pressure plates are fixed to the outside face of the framing and hold the glass in place. Joints on the outside face of the supporting frame are sealed, preventing water from penetrating the joint. An alternative method, called 'structural gasket' glazing, uses a rubber strip to combine the functions of seal and pressure plate, but this method has largely fallen into disuse due to the rapid fading of the gasket colour but also a general shift in the aesthetics of modern design. Recent developments in rubber-based products may see a modest reintroduction of this method.

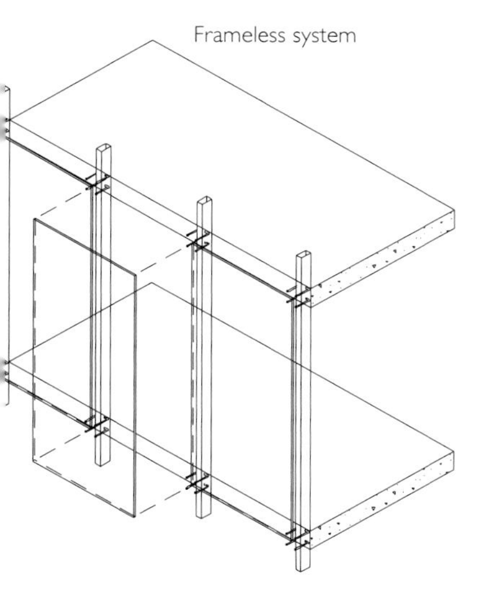

71D

In 'unitised' systems, the glass is fixed back to an independent frame. Panels are lifted by cranes, which can avoid the need for full scaffolding in tall buildings. This method also allows a damaged panel to be removed without affecting adjacent panels, since the entire frame is removed and replaced. Similarly, glass can be replaced from the inside of the building much more easily in unitised systems. Joints between panels allow seals to be made behind the face of the panel where they are less vulnerable to attack from ultraviolet light, which causes the deterioration of sealants. Like stick systems, most panel systems use two seals, separated by a cavity which serves to drain away any rainwater that may penetrate the outer seal. This flexibility allows the possibility of interchanging panels on low-rise buildings but this approach to the use of unitised systems is still in its early stages.

In practice, the difference between stick and unitised systems is generic only. Some stick systems have an element of prefabrication, and some unitised systems use a partial site-based approach. In the following section, the assembly of glass-based generic types is discussed.

98E

98A

98D

98C

Details
1. Transom
2. Mullion
3. Extruded aluminium section
4. Single glazed or double glazed unit to suit application
5. Pressure plate
6. Rubber-based seal
7. Fixing screws at 300mm centres
8. Mild steel structural member
9. Metal-faced or opaque glass-faced insulated panel
10. Floor slab
11. Floor finish
12. Ceiling finish

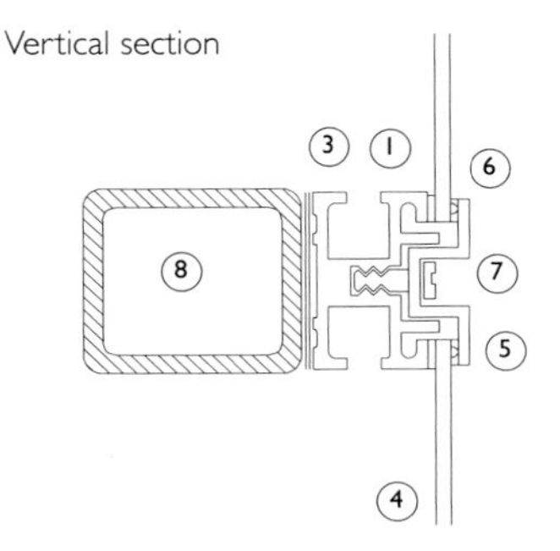


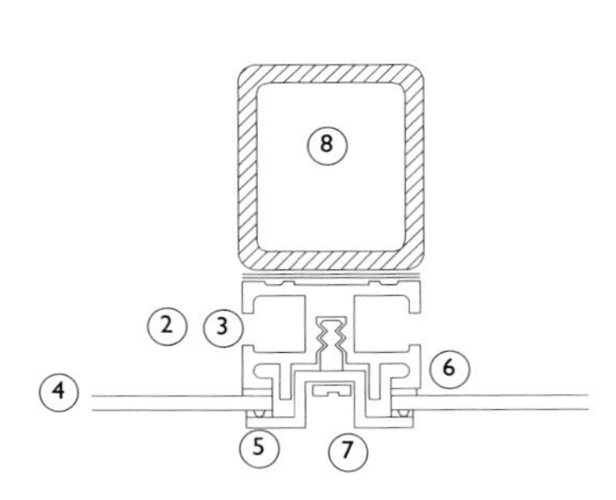


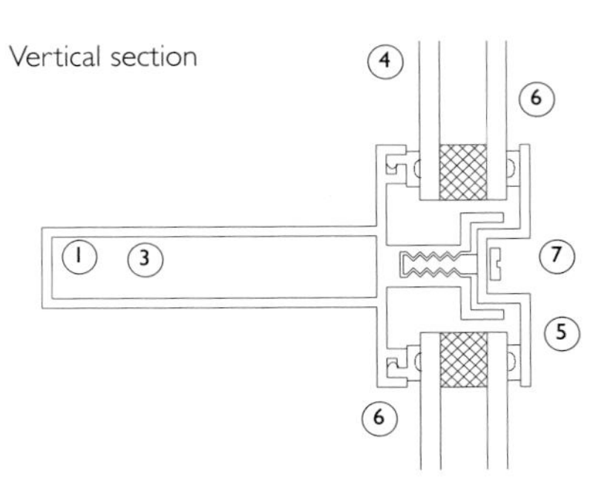


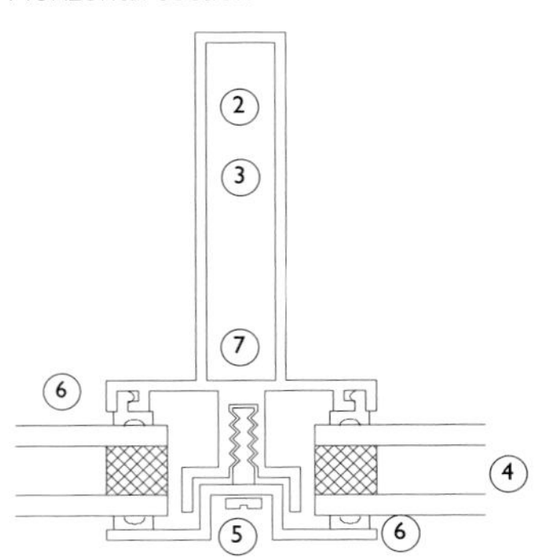


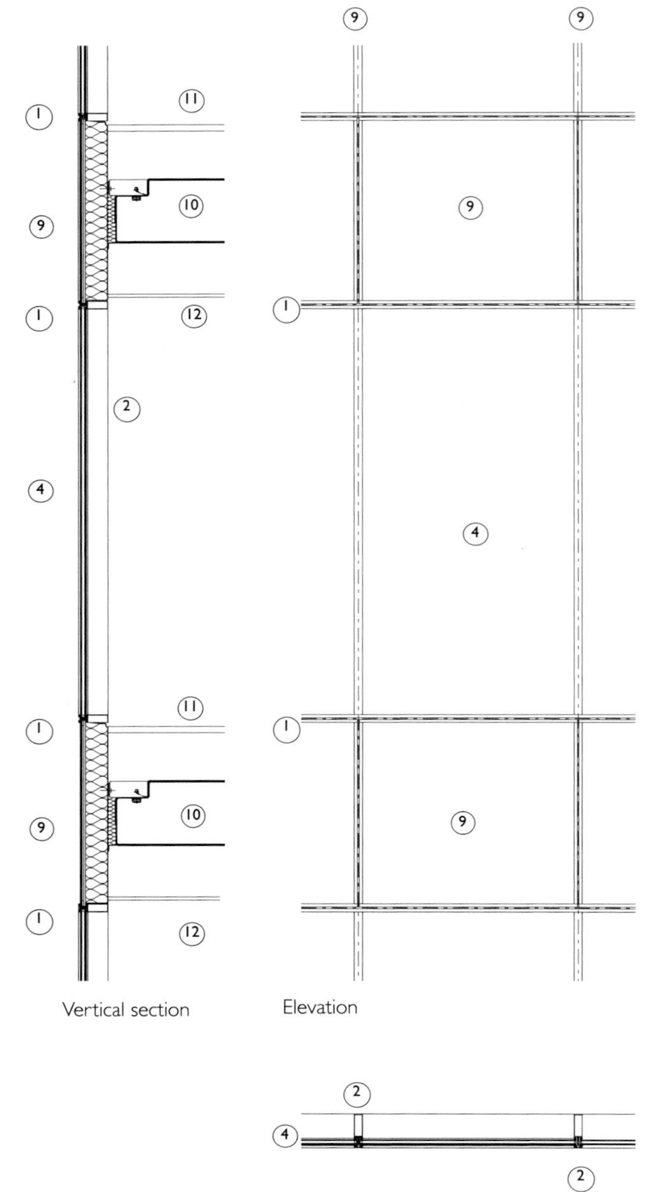


Vertical section

Elevation

Horizontal sections

Stick systems consist of mullions (vertical framing members) arranged at regular centres, which are often fixed vertically to the edges of the floor so that they stand forward of the slab. Transoms (horizontal members) are set between the mullions. The support framing is usually made from aluminium extrusions, preferred for its lightness, workability and precision. Glass sheets, double glazed units or metal panels are fixed to this supporting framework. These are held in place by a continuous aluminium plate fixed to the outside of the glass and connected back to the mullion with threaded bolts or screws, usually at around 300mm (12in) centres.

Components are cut to size in a workshop and assembled together on site. The aluminium pressure plates are pre-fitted with a neoprene or EPDM sealing gasket that allows the plate to apply pressure to the glass without breaking it. The rubber strip also seals the joint between the metal framing and the glazed panels, making the junction watertight and airtight but also allowing for thermal movement between the supporting building structure, the glazing and other components.

Unfortunately, airtight seals can result in a pressure build-up but the introduction of a pressure-equalisation system can prevent this condition from occurring. Slots in the transoms and mullions allow air to enter the drainage channels to prevent capillary action drawing any water behind the glass. These slots also drain away any moisture that does penetrate the seal back out to the exterior of the building.

## Metal panels within glazing systems

Stick systems can incorporate metal panels where glass is not an appropriate material. They usually have an integral layer of thermal insulation. Panels are fabricated in a workshop and are fixed to the framing being assembled on site. Metal 'cassettes' are made either as 'sheet metal' or 'box panel' type. Sheet metal panels have a layer of insulation fixed to the inside face. These are used in combination with an inner lining of plasterboard or a backing wall of concrete block. Box panels are made by welding two metal trays together to form a metal box. Thermal insulation is set between the two layers. The two faces of the box provide both internal and external finish-

35B

35C

es. A thermal break is typically located where the internal and external trays meet. This method allows the integration of a window as a separate glazed panel within the system. Maximum panel sizes for both types of panel are restricted by the manufactured size of steel coil from which panels are made, currently between 1200mm and 1500mm (4ft and 5 ft).

## Structural gasket glazing

This system uses the rubber gasket both to support and to seal the glazing. The structural gasket uses the shape of the rubber extrusion to support the glass. Structural gasket glazing is also called the 'zip up' system, from the way the gasket is fixed in a continuous metal strip to hold the glass sheets in place.

This technique was inspired by a method used in the car industry to hold car windscreens in place. One of the first applications in building construction was with the General Motors Technical Centre in Warren, Michigan, completed in 1956. Examples that are more recent include a factory in Tadworth, England, from 1972, and the Patera Building System, developed in the early 1980's.

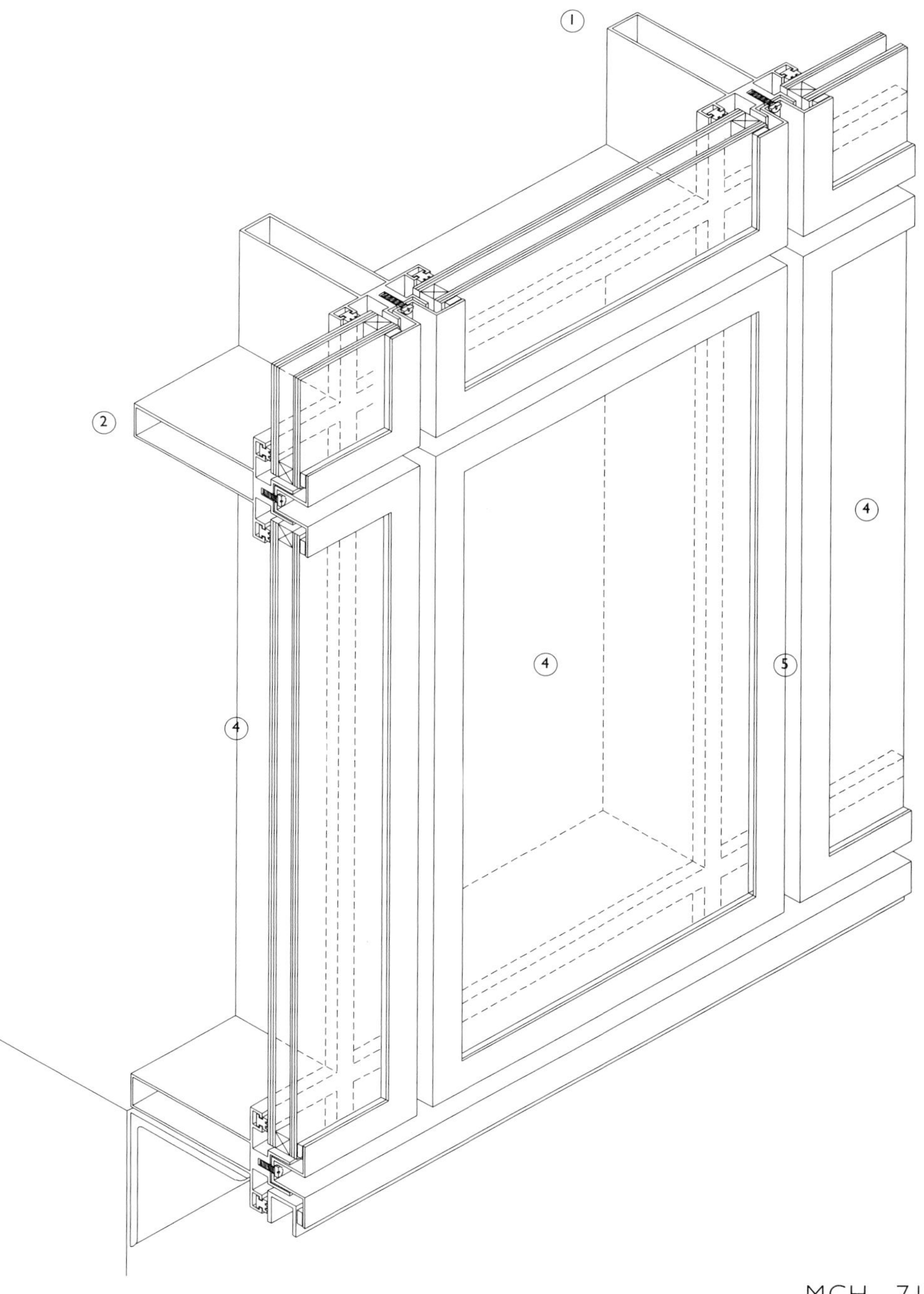

9 I A

Details

1. Transom
2. Mullion
3. Extruded aluminium section
4. Single glazed or double glazed unit to suit application
5. Silicone bond
6. Optional securing clips to suit application
7. Rubber-based baffles
8. Rubber-based seal
9. Fixing screws at 300mm centres
10. Floor slab
11. Metal-faced or opaque glass-faced insulated panel
12. Floor finish
13. Ceiling finish

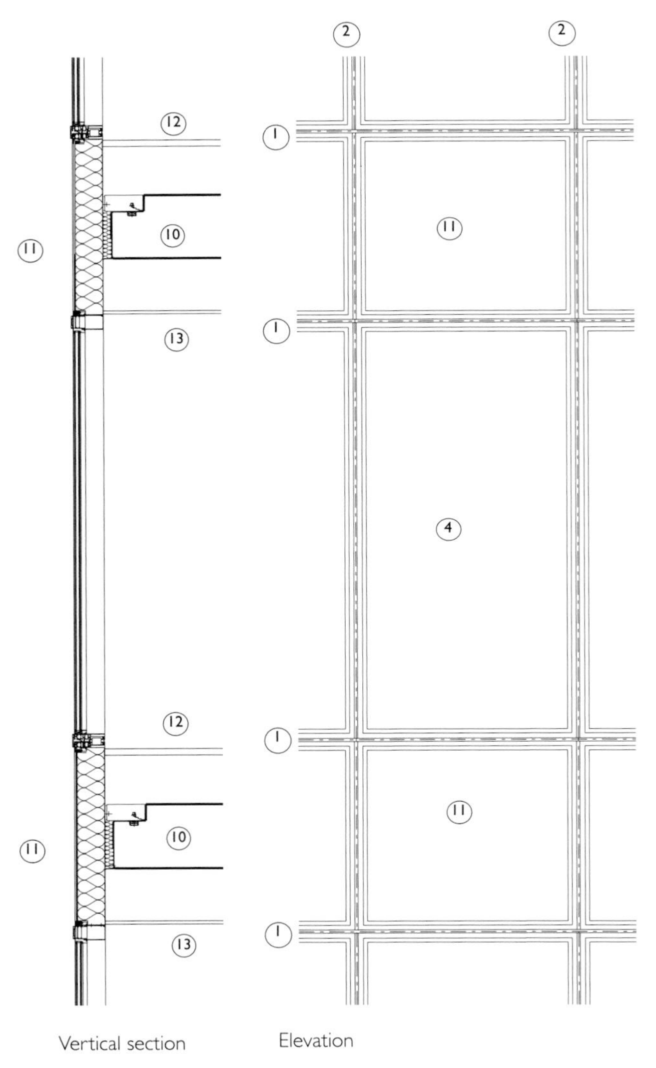


Vertical section

Elevation

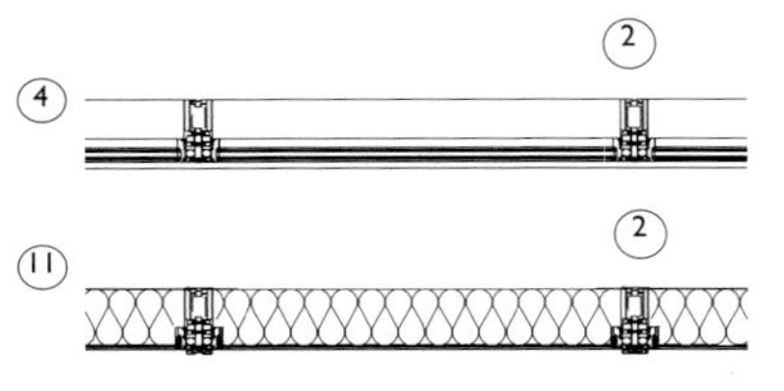


Horizontal sections

Panel systems consist of glass panels fixed to a supporting metal frame. The glass is fixed either by a silicone bond or by a pressure plate mechanical fixing. Self-supporting glazed panels do not require a secondary framing to provide support. However, each panel must be sufficiently rigid for it to be installed with the glass in place, making the metal frame sometimes appear more bulky than the stick system. Some panel systems interlock when put into place to improve stiffness and reduce the sightlines and some systems can even match the thinner sightlines of stick glazing. The frame can be partially concealed by bonding the glass panel directly to its supporting aluminium frame, rather than using a pressure plate. The glass is bonded using a silicone bond and sealant rather than a pressure plate. This method is known as 'silicone bonded' glazing. The use of this technique gives large areas of almost uninterrupted glass, with small recesses between panels. Since the silicone bonding needs up to one week for the silicone to cure, this technique is very suited to panel systems rather than stick systems. However, some stick systems are beginning to use silicone-bonding techniques by adhering glass panels to an aluminium subframe and mechanically fixing these with a concealed pressure plate on site.

In addition to the pressure plate fixing, there are two other generic types of panel system: two-edge glazing and four-edge glazing. In the two-edge system, the glass is restrained top and bottom by a pressure plate system. The two vertical sides are silicone bonded. In the four-edge system, all four sides are silicone bonded. Panels are made in factory conditions and transported to site where they are lifted into place by crane and fixed to the building frame.

Joints between panels are either of the 'closed' or 'open' type. Closed joints in both a vertical and horizontal direction use neoprene or EPDM gaskets to seal the joint at the surface of the glass. Open joints, which are more commonly used, have a gap between panels with a baffle, set into a slot recessed within the depth of the panel, acting as a weather seal. Wind-blown rain entering the slot between two panels is checked by lapped baffles. Rainwater that penetrates the slot runs down the face of the baffle and is directed out at the junction with the horizontal joint at the base of the panel. The baffle forms the first barrier against wind-blown rain. Any water that passes around the edges of the baffle is drained away in the rear half of the vertical slot. Sealant is applied to the joint on the inside face of the panel where it is protected from the effects

9 I B

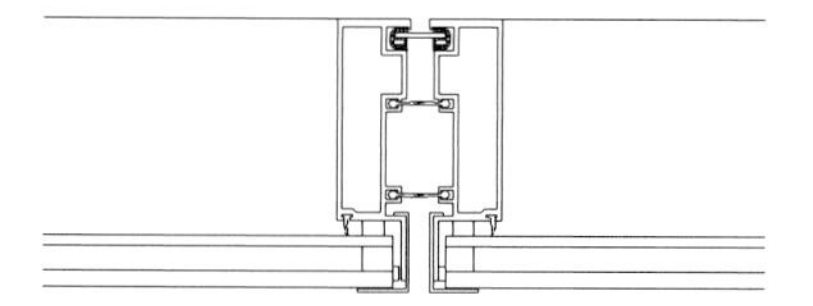
Horizontal section

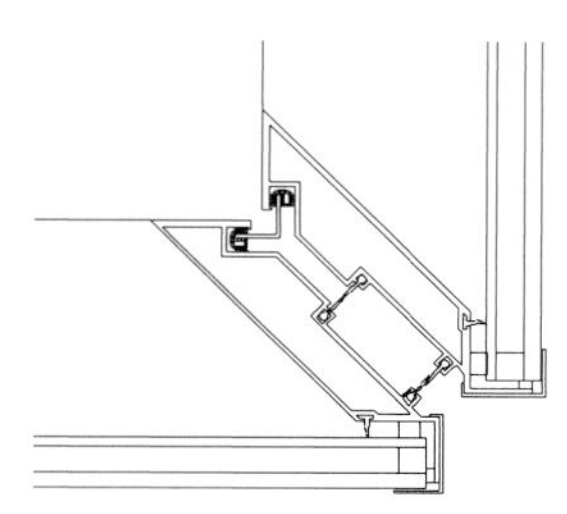
Horizontal section

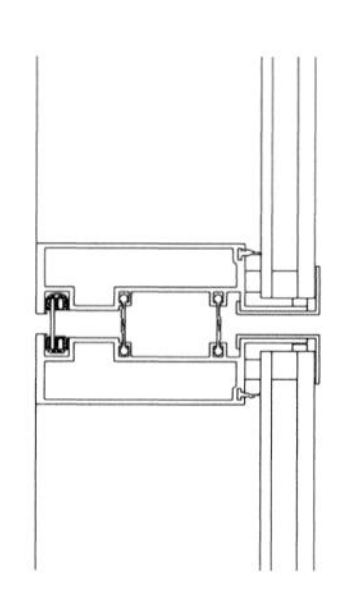
Vertical section

of ultraviolet light. Horizontal open joints are stepped to allow the rainwater to drain out from the baffle strip and the slot behind whilst preventing rain from penetrating the joint from outside. A continuous flashing is used at the junction of the vertical and horizontal joints. A rubber-based material is used for the flashing in order to accommodate movement between adjacent panels.

Where each panel must be able to be removed without damaging the waterproofing of adjacent panels, an open joint system is often preferred. Panels are usually one-storey height, spanning from floor to floor. Each panel is supported by a floor slab or edge beam, at either the top or the bottom. Panels are either top hung or bottom supported. Slotted fixings are used in order to allow for any vertical movement between the panel and the supporting structure. The cleats are sometimes fixed to a slotted channel set into the floor slab in order to accommodate horizontal movement. Where silicone bonded glazing is used, the bond between frame and glass is visible through the glass, and panels are often screen printed around the edge on the interior face of the glass to conceal this effect.

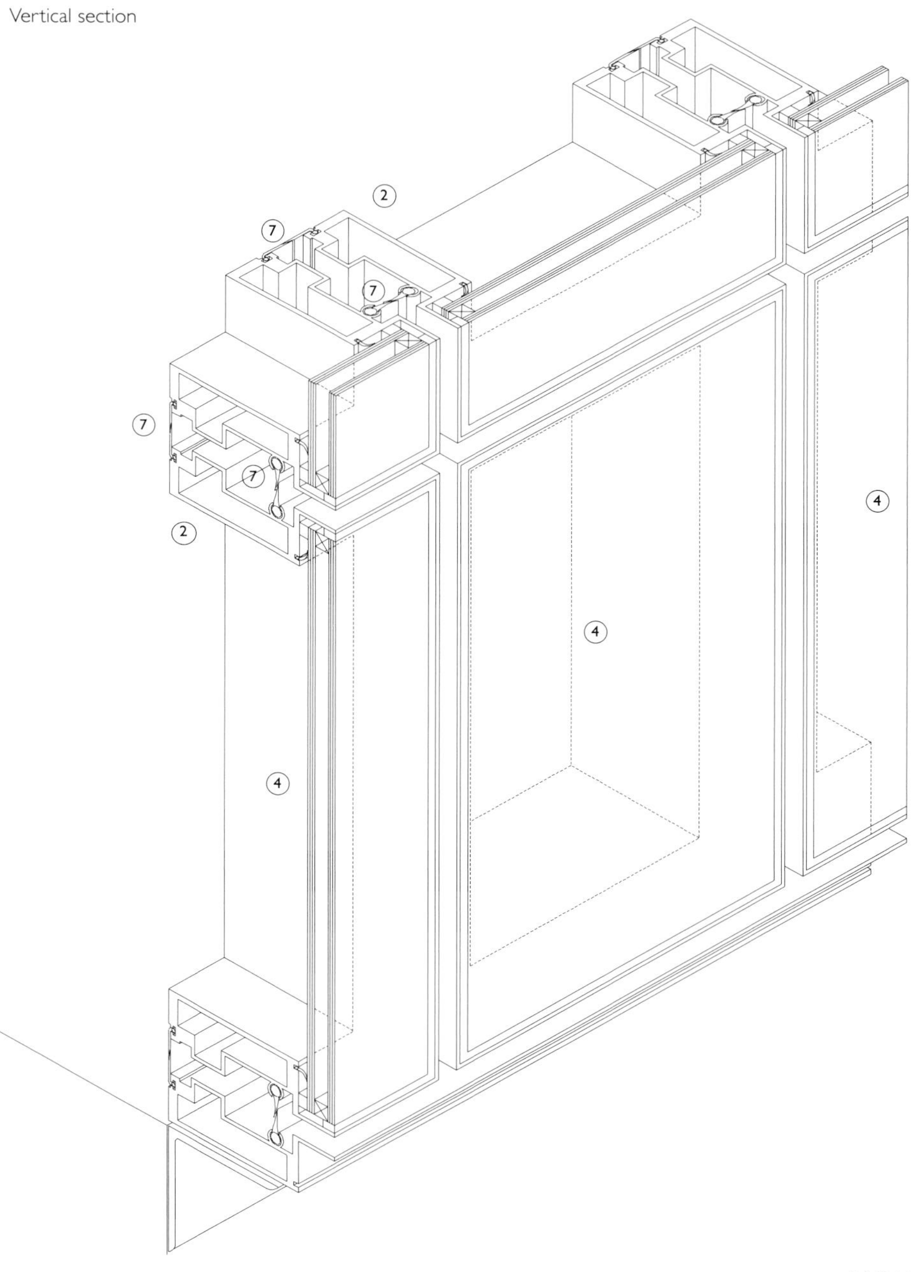

Details
1. Stainless steel patch plates
2. Single glazed or double glazed unit to suit application
3. Silicone seal
4. Glass fin
5. Support bracket
6. Bolt

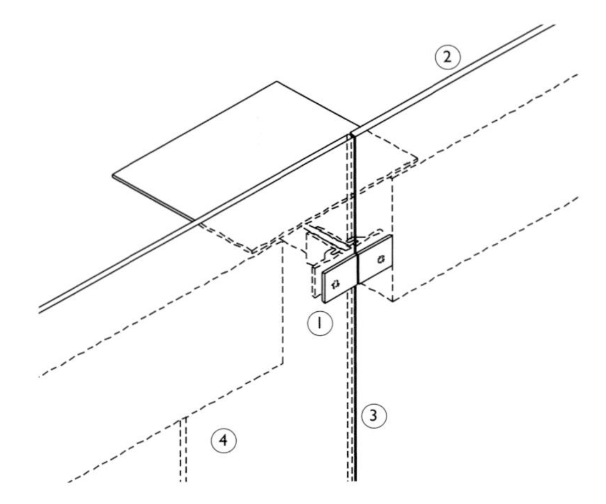


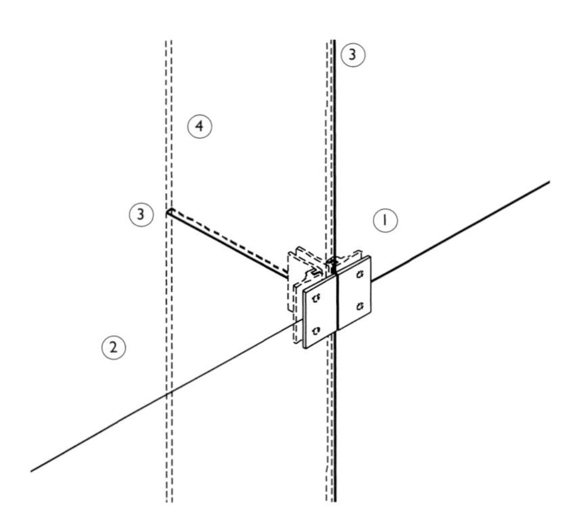


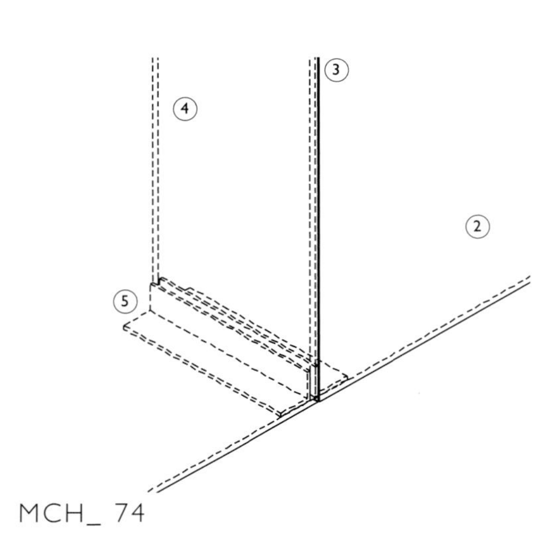


Frameless glazing has the advantage of giving a greater sense of transparency than other glazing systems. It developed during the 1960's as 'patch plate glazing' and has been developed recently as a variation called 'clamped' glazing. Both are essentially intended for single glazed uses.

## Patch plate glazing

Patch plate glazing uses a steel plate fitting (usually stainless steel) set at each intersection of four glass sheets. The patch fitting is a corner bracket assembly secured with bolts that pass through the glass. The bolts are tightened to provide sufficient friction between the patch fittings, and sheets of glass. Wind loading is resisted mainly by glass fins, used instead of mullions, which are fixed directly to the patch plate fixings. The glass fins provide both support and bracing. The whole assembly is either suspended from a hanger at the top, or supported at its base. No horizontal support is supplied from transoms between sheets of glass. The gap between each glass sheet is sealed with silicone sealant that allows movement to occur between sheets.

Toughened glass is generally used, where the glass sheets are drilled prior to the glass being toughened. In recent years, laminated glasses can be successfully drilled through the interlayer with a clean cut. Some applications require fins to be made from toughened glass laminated together, making the fabrication process ever more complex. Panel sizes are restricted to around 4200 x 2100 mm (14 ft x 7ft), which is maximum size for readily available toughened glass. However, in practice these sizes are rarely achieved due to the thickness of glass needed to withstand wind loads (around 20mm thick depending on conditions), making the assembly both heavy and expensive.

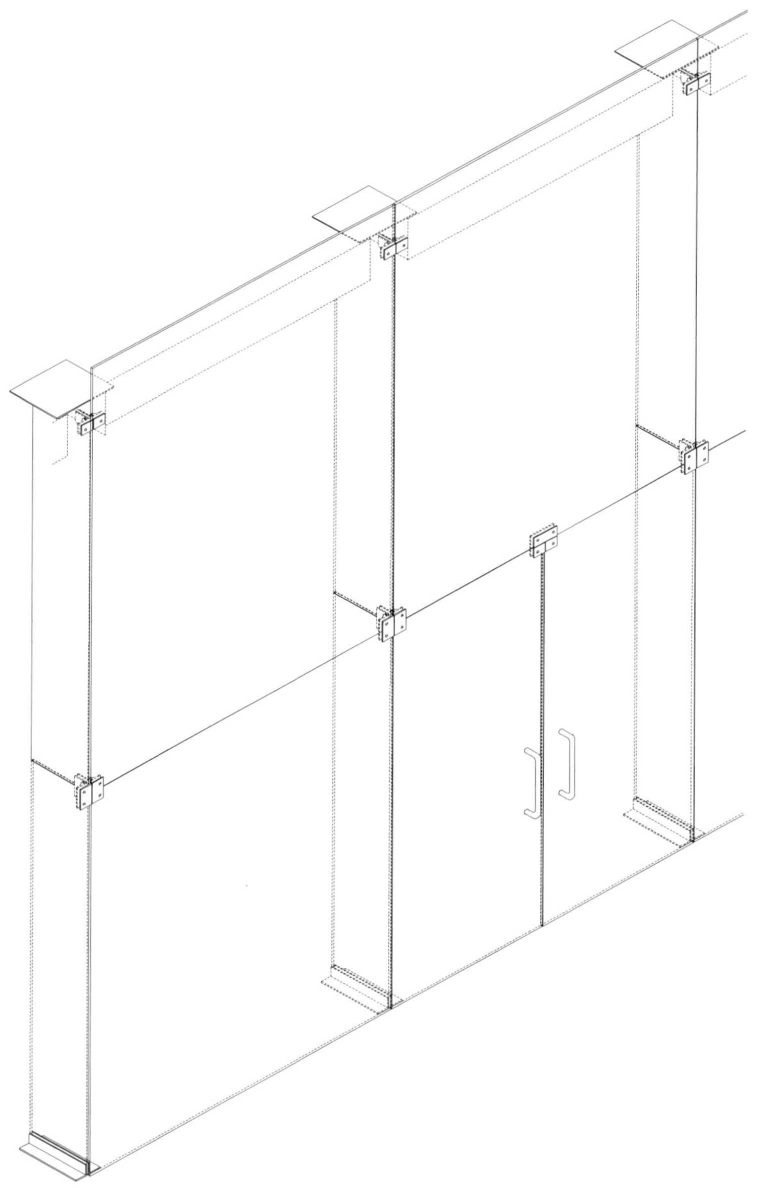

## Clamped glazing

In order to avoid drilling the glass, with its associated costs, laminated or toughened glass sheets can be fixed together with stainless steel brackets either side of the glass to form a clamp or 'shoe'. Glass and clamp are separated by a rubber-based gasket. The glass is usually supported at points along its bottom edge and restrained on either the top or side edges. It is difficult to achieve enough friction between clamp and glass to support the glass only on its side edges whilst allowing the glass to move with thermal expansion and structural movement.

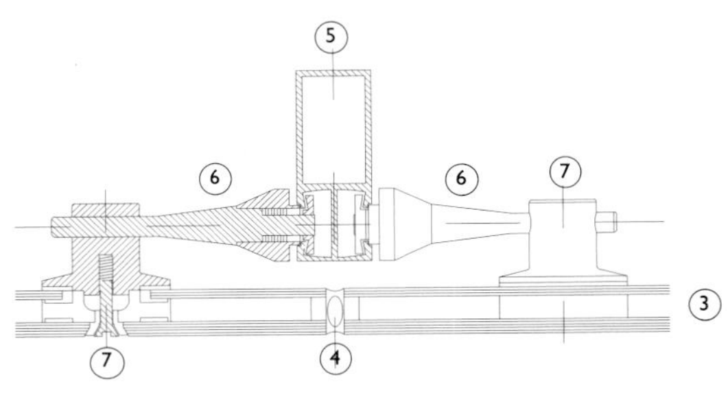


Horizontal section

Details

1. Cast steel connector
2. Mild steel or stainless steel angle bracket
3. Single glazed or double glazed unit to suit application
4. Silicone seal
5. Mullion
6. Support bracket
7. Bolt fixing
8. Steel connector rod secured to primary structure

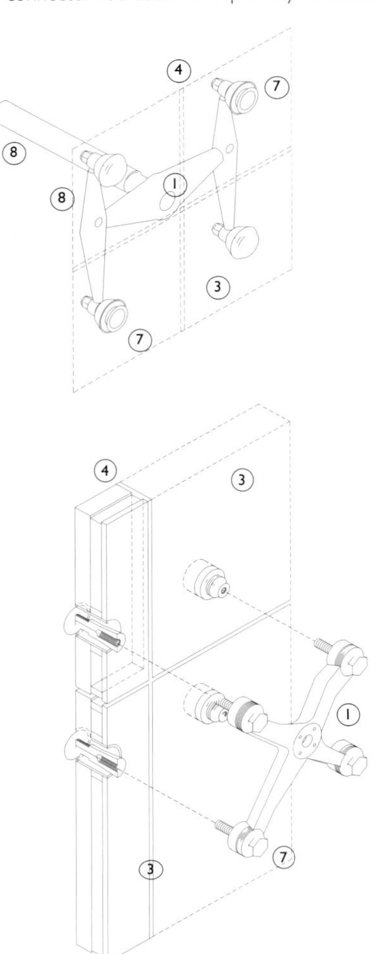


Bolt-fixed glazing is a development of the patch plate system. Whereas the patch plate system is restricted to single glazing, bolt fixed glazing can accommodate double glazed units. This system uses single sheets of glass or double-glazed units that are fixed back to supports with a bolt in each corner. This method allows the supporting mullions or transoms to be visually less obtrusive than in other glazing systems, providing a higher degree of transparency.

The glass is fixed directly to the bolts, which in turn are fixed to supports that must provide resistance to wind loading as well as supporting the glass panels. Cantilevered brackets supporting each corner are fixed back directly to a mullion. The joint between the bolt and the glass is sealed with a compression gasket. The gap between each panel is sealed with a silicone sealant. Bolt-fixed glazing can be used horizontally to form a glazed roof. Double glazed units can be used by introducing a metal or nylon spacer between the glass sheets to transfer loads through the glass. Bolt fixed glazing techniques vary in the amount of tolerance and adjustment in the glazing to align with structural supports. Fixings vary from a simple short length of steel angle in each corner, each with a single bolt, to a fully cast node connector with arms to connect four bolts back to a single structural support. The bolt connectors can be either set flush with the outer face of the glass by countersinking the fixing into the glass, or can be set proud of the glass with a steel disc (larger than the size of the hole), to avoid the cost of countersinking.

## Developments

Bolt-fixed glazing has developed through the advances made in the design of the fixing bolts themselves. Fixing bolts and attachments to a main structural support comprise a mixture of metal castings, machined bolts and flat plates to suit a range of fixing conditions. Bolt fixings are generally smaller than they were ten years ago in order to be less visible in elevation. Bolts can also incorporate variable geometry fixing to cope with a range of joint angles within a single wall assembly. The introduction of double glazed units has provided a bolt fixed assembly that has much better thermal insulation but which

71B

has a thick black edge to the unit. This results in a much heavier 'sightline' around panel edges making it appear significantly less transparent.

Bolt fixed glazing provides a high level of transparency to glazed walls. Because fewer components are used, each part must be engineered to perform several functions at once. Fixings must support the glass, resist wind loads and provide a weatherproof seal. In addition, each piece is highly visible thus requiring close attention.

Flexible fixings have hinged connections to accommodate movement in a flexible supporting structure such as steel tension cables. Rigid fixings are used where a small degree of movement in the glazed wall is expected. Bolt fixed glazing can be supported by rectangular mullions for use in divisible office space. The space between the mullion and the glass can be sealed.

80

8M

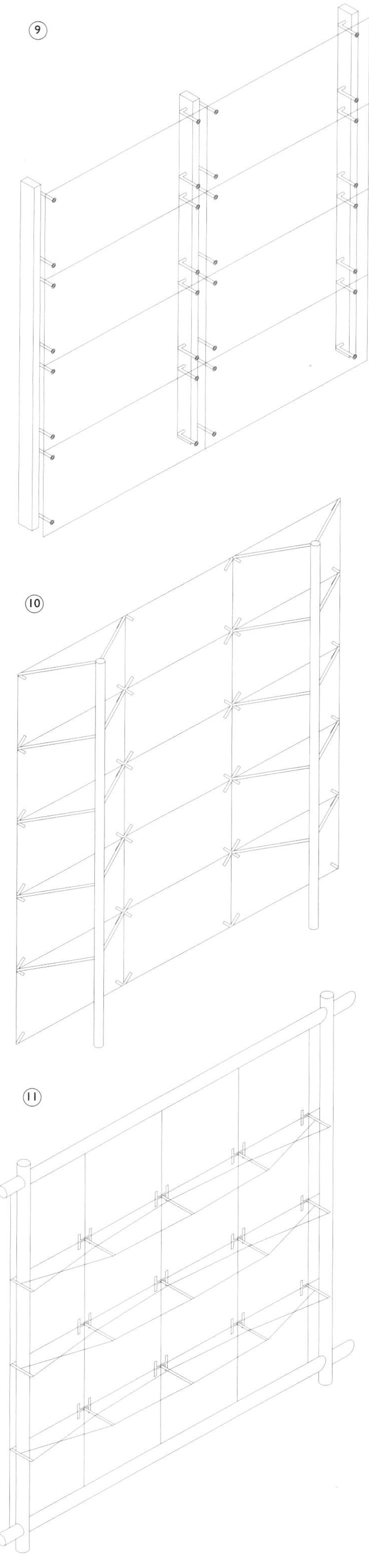


Typical supporting structures

9. Vertical support behind each row of bolt fixings

10. Cantilevered brackets to reduce frequency of vertical supports

11. Horizontal cable truss to further reduce frequency of vertical supports

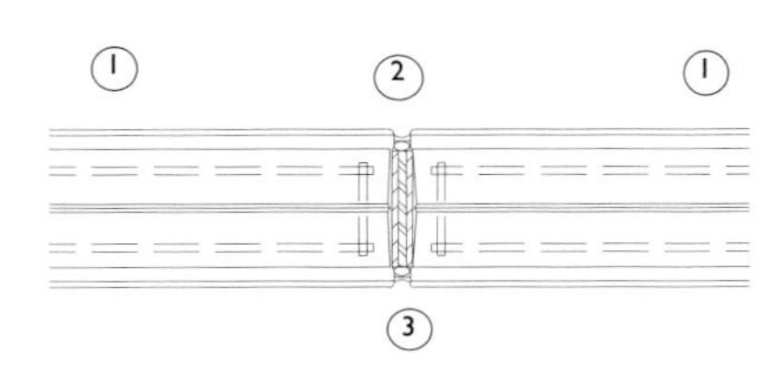


Horizontal section

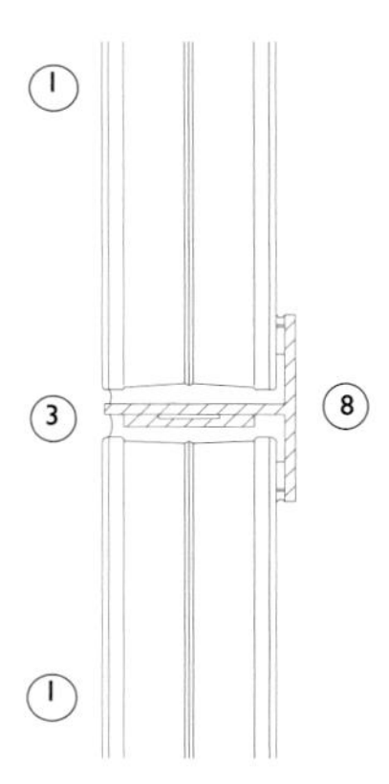

Vertical section

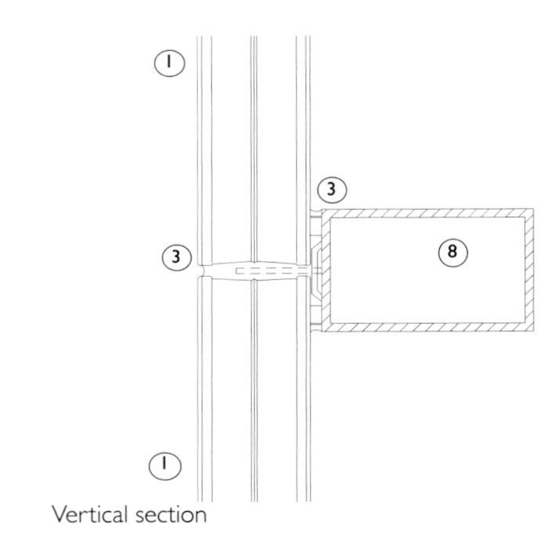


Vertical section

## Details

1. Glass block
2. Bedding reinforcement
3. Bedding compound, mortar or silicone-based bond.
4. External cladding
5. Metal flashing
6. Thermal insulation
7. Enclosing wall or adjacent wall
8. Steel or reinforced concrete support frame

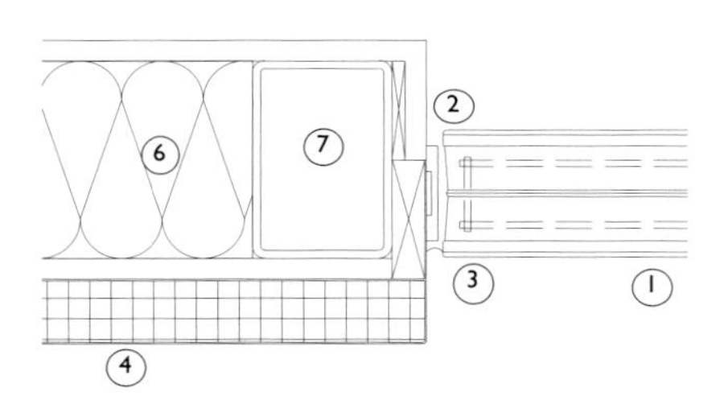


Horizontal section

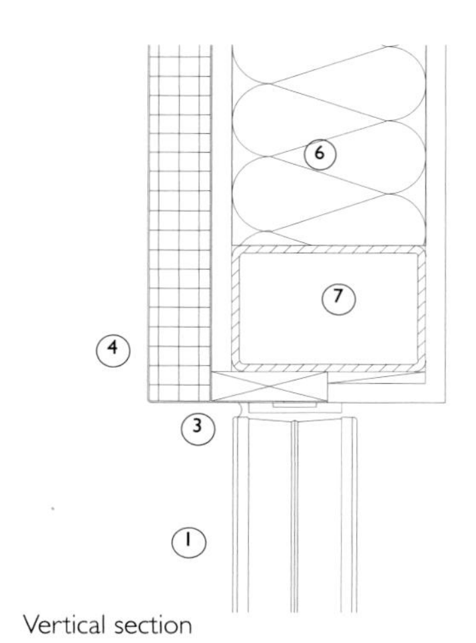


Vertical section

Glass blocks were initially completely site assembled to form walls. Nowadays they are made into panels, approximately 2 x 2 metres (6ft 6in x 6ft 6in) in a stack-bond configuration. Panels are supported by a frame made of any suitable material such as steel or concrete. The joints between blocks are reinforced with steel rods, because the glass blocks lack inherent structural stability. Blocks are bedded in either a cement mortar or a silicone bond. The air cavity in each block provides some thermal insulation. Expansion joints must be provided between panels.

Although glass blocks have inherently poor thermal insulation, thermal bridging in walls of this material is inevitable where a single skin of blocks is used, as is commonly the case. A single skin of glass blocks provides a translucent wall with a diffused light that still has an element of transparency. However, improvements in glass technology have assured that sandblasted, acid etched and screen printed glasses are steadily taking the place of glass blocks in building because they can offer the same light transmission and light diffusion qualities but in a more controlled and robust material.

## Developments

A recent development has been the introduction of steel angles and T-sections that form larger frames without revealing the smaller panel size. This allows large panels of blocks to be fully prefabricated rather than being built on site, allowing construction to proceed at a faster rate and with better quality control than would be expected of this traditionally site-based technique.

The textured, cast quality of glass blocks has recently been exploited in cast glass often formed into channel-shaped profiles around two metres in length (6ft 6in). This is a development of a method pioneered in the 1960's. Channels are set together vertically and fixed top and bottom. The vertical joint between channels is then sealed with silicone.

75A

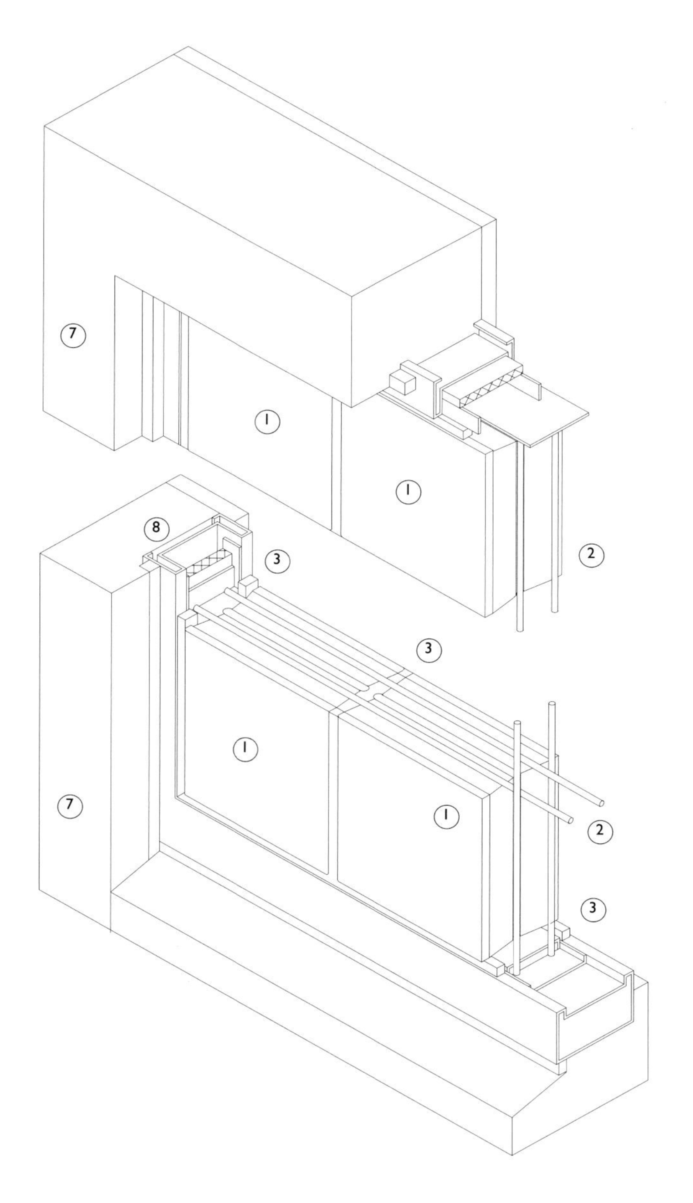


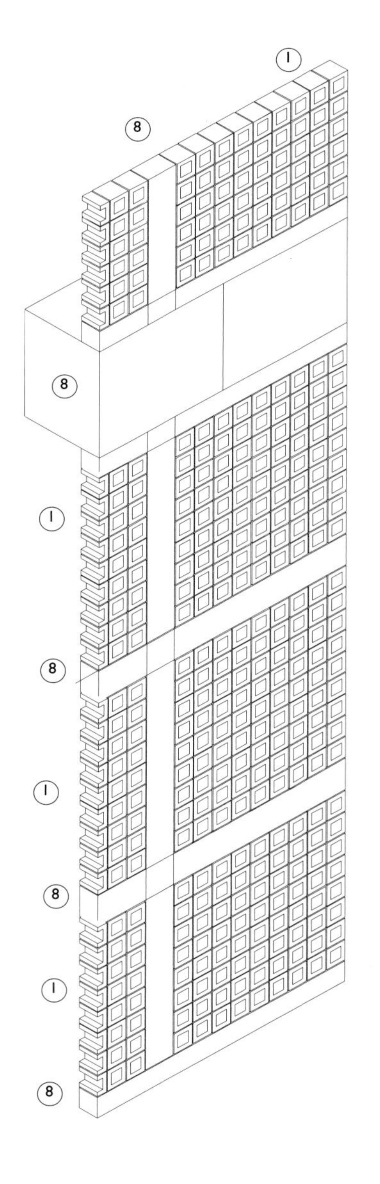

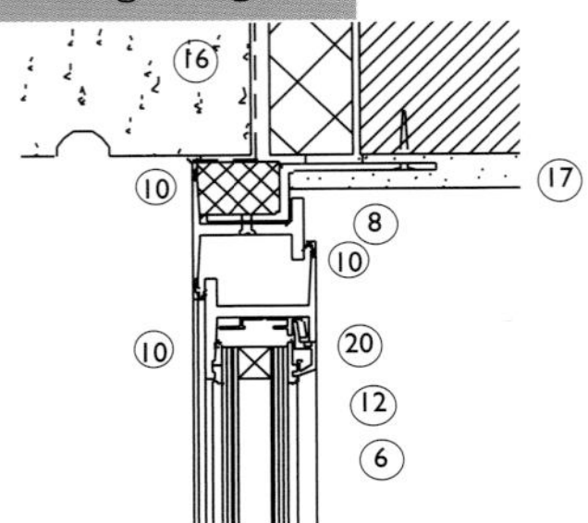

Vertical section with double glazed unit

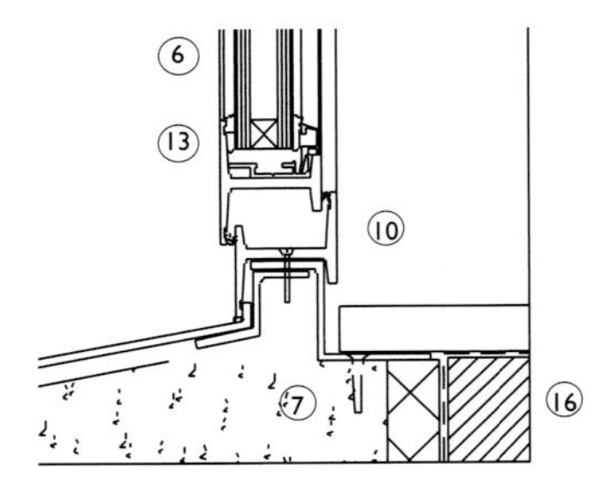

## Details

1. Outside
2. Inside
3. Rolled steel glazing section
4. Transom
5. Mullion
6. Single glazed or double glazed unit to suit application
7. Fixing bead
8. Fixing lug
9. Projecting transom
10. Rubber-based seal
11. Fixed light
12. Inward opening light
13. Outward opening light
14. Window cill
15. Condensation tray
16. Damp proof course (DPC)
17. Internal finish
18. Drip
19. Packing
20. Aluminium clip to secure double glazed unit
21. Steel cill

Horizontal section

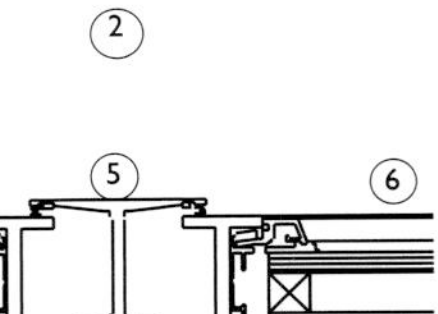

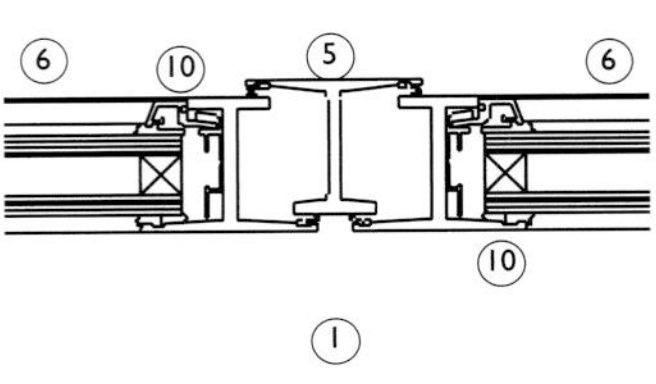

Horizontal section

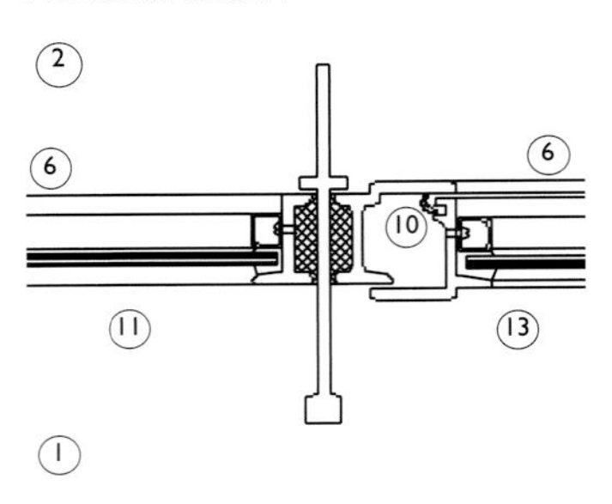

12A

12B

12C

Frames for steel glazing are made from cold rolled sections, rolled sufficiently accurately to allow the forming of a reliable metal-to-metal contact at junctions. Most sections are based on a standard width (in elevation) of around 25mm (1in). Sections with both equal and unequal flanges are used. The depth of sections varies to suit the size of window without changing the thickness seen in elevation (the sightline).

Corners are usually welded and glazing bars are tenoned and riveted in position to make the frames rigid whilst keeping the sections slender. Frames are hot dip galvanized and delivered to site unpainted. Peripheral components such as cills and condensation channels are made from galvanized pressed steel sheet. Steel glazing is usually supplied either in a galvanized finish or with a proprietary coating such as polyester powder coating.

Glazing can be single-glazed or double-glazed, though the system was developed primarily for single glazing. Single glazing is fixed with sash putty traditionally, but this is being replaced by pre-formed aluminium glazing beads. Putty glazing is usually restricted to one- or two storey applications where this site intensive method requires no special equipment. Glazing beads may be clip-on or screw-on type, of angle or of channel. Narrow section double glazed units can be accommodated in the standard window sections and be fixed with specially profiled glazing beads. With standard 12 x 12mm beads, the available glazing platform is usually around 11mm deep. Thicker units require an insert frame.

Windows are fixed either by projecting lugs secured to the fixing surface, or directly through the frame if a concealed fixing is required. Bedding and pointing is done with a mastic material. At the window head the lintel is throated to avoid water running back to the joint between the window and the head of the opening. The jamb is better protected by providing a rebate in the face of the wall, but this can be impractical where working tolerances of the wall material and steel windows are not compatible. An extruded neoprene or EPDM gasket can be inserted into a groove provided in the window frame, sealing it against the window opening to give better weather protection.

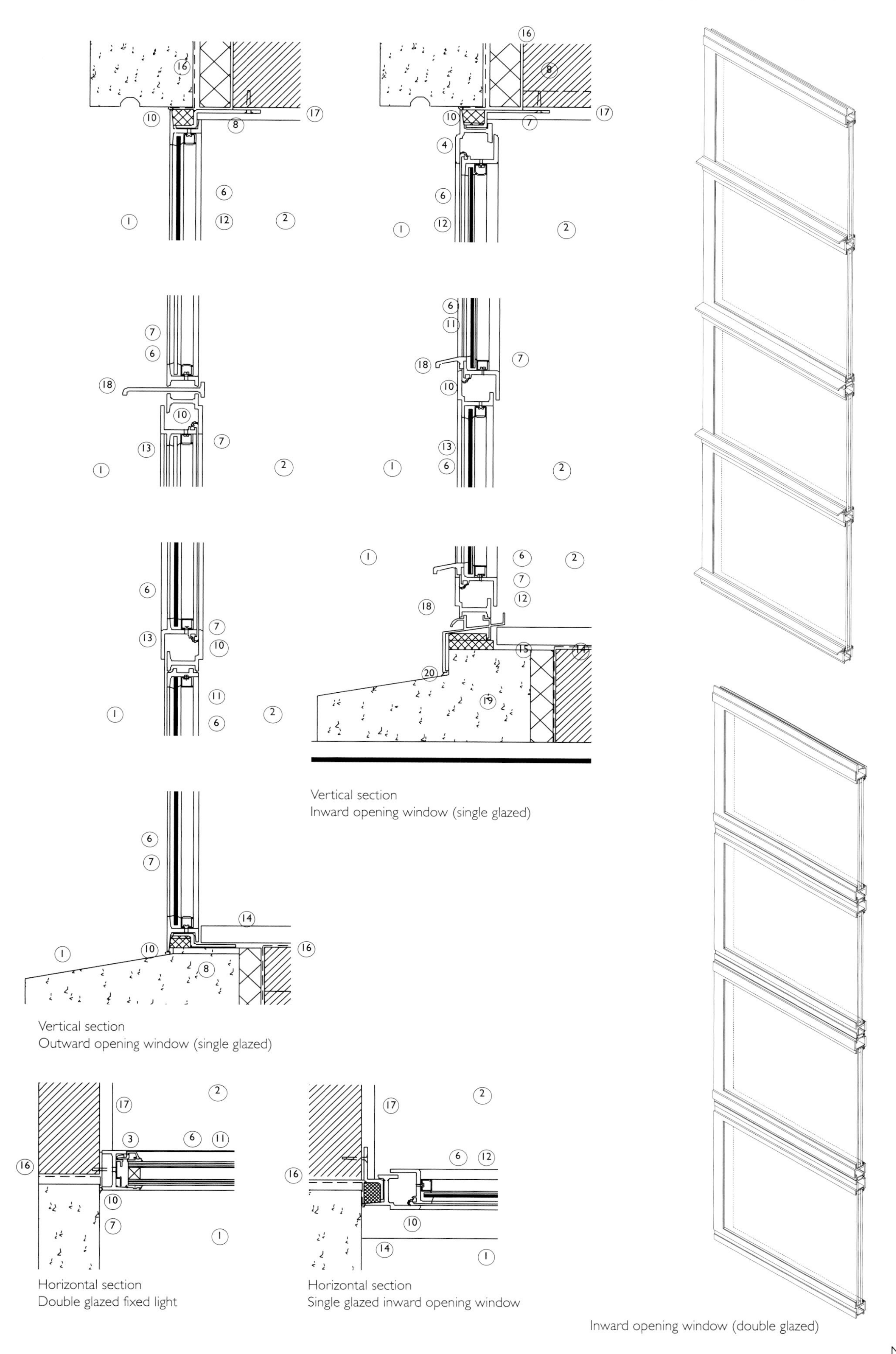

Outward opening window (double glazed)

Vertical section
Inward opening window (single glazed)

Vertical section
Outward opening window (single glazed)

Horizontal section
Double glazed fixed light

Horizontal section
Single glazed inward opening window

Inward opening window (double glazed)

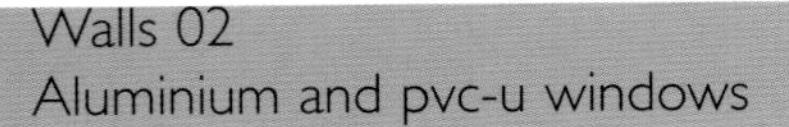


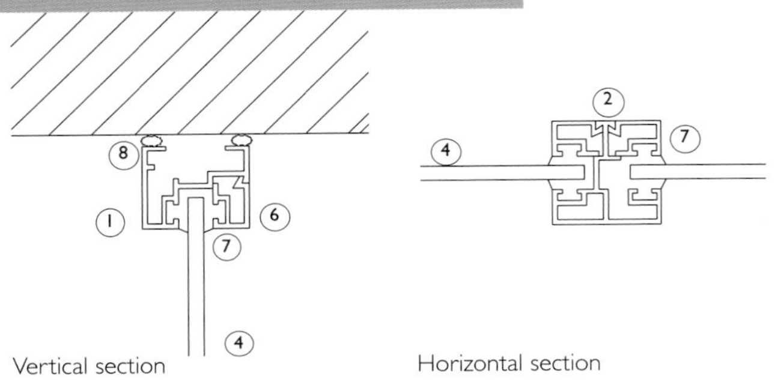


Vertical section

Horizontal section

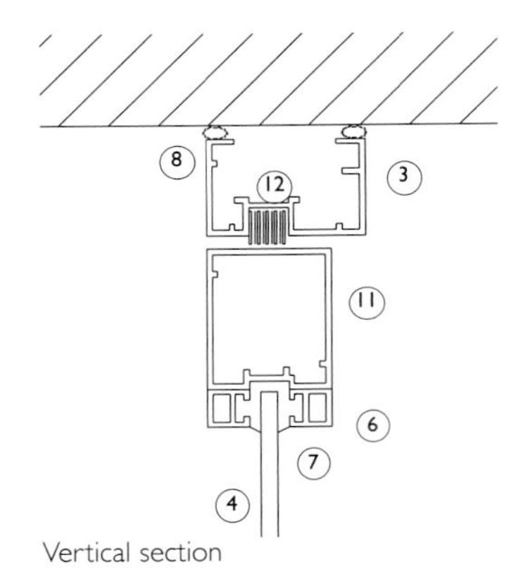


Vertical section

Details

1. Transom
2. Mullion
3. Extruded aluminium section
4. Single glazed or double glazed unit to suit application
5. Rubber-based structural gasket
6. Glazing clip
7. Rubber-based seal
8. Silicone-based seal
9. Fixing screws at 300mm centres
10. Mild steel structural member
11. Door section
12. Nylon brush

42B

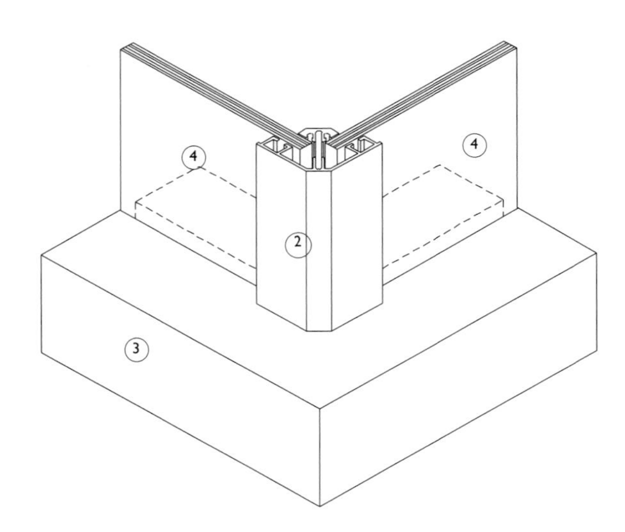


Aluminium and PVC-U windows are less rigid than steel and as a result neither is used as rolled sections. Instead, extruded profiles are preferred which gives them greater stiffness but a bulkier appearance than steel sections. Windows are typically delivered to site as pre-glazed, prefinished items.

## Aluminium windows

Aluminium windows are available in the following formats: sliding, side hung (casement or french), bottom hung (hopper), top hung (awning), centre pivoted and sash (double-hung). In addition, the tilt-and-turn window combines side hung and bottom hung window into a single mechanism to allow the window to be opened in two ways. When side hung the window can be cleaned from the inside, avoiding the need for external cleaning gantries or cradles. Aluminium windows, with a thermal break, have a similar U-value (thermal transmission) to either PVC-U or timber.

## PVC-U windows

An essential characteristic of PVC-U windows is that they require very low maintenance making them very popular for use in housing. PVC-U windows are made as side hung and tilt-and-turn. Their use is more restricted than aluminium since the material is not as rigid and requires thicker sections with more ribbing. Some manufacturers reinforce some frames with pressed galvanized steel sections to reduce their size. Unlike metal windows, PVC-U does not oxidise, and therefore no additional finish is necessary. PVC-U windows have a U-value comparable with that of a solid timber frame. Unfortunately, PVC-U tends to have a strident white colour (although now other colours are available they tend to fade) and the range of sections available is limited.

## Combination

Some manufacturers make windows that combine aluminium, PVC-U and timber in a single frame. Aluminium and PVC-U profiles can be used to clad a timber section that provides rigidity. This configuration combines the strength and warmth of wood on the interior with the lower maintenance of the exterior material. Other types have an outer frame of PVC-U on the external face and aluminium or timber on the inner face.

Vertical section

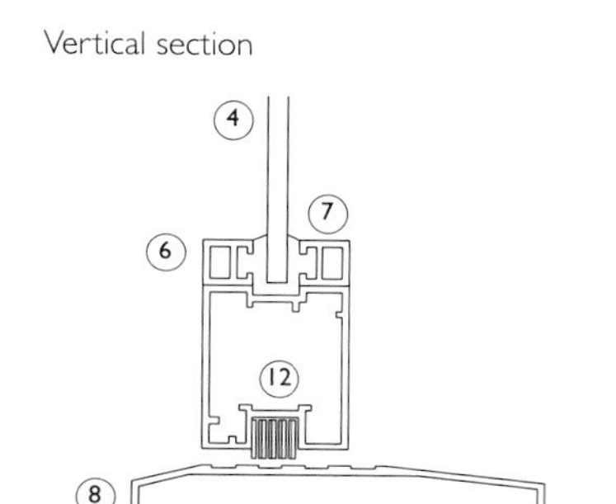


Horizontal section

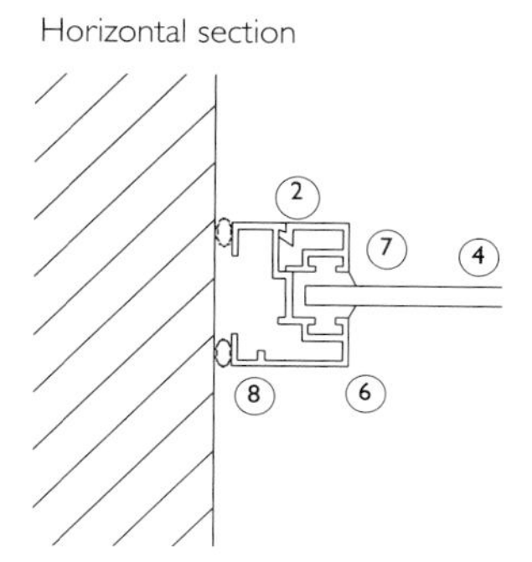


Vertical section

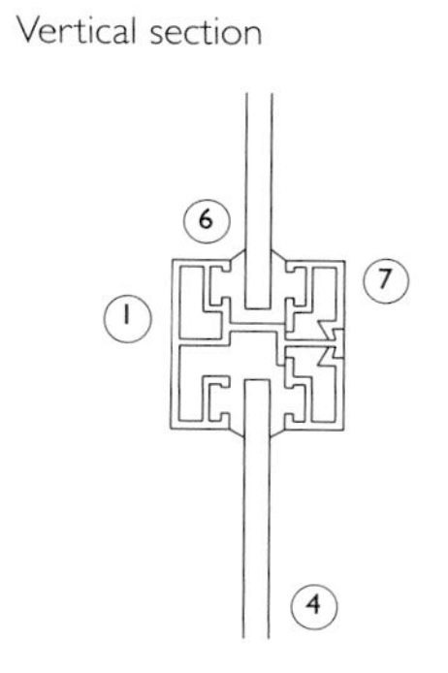


Vertical section

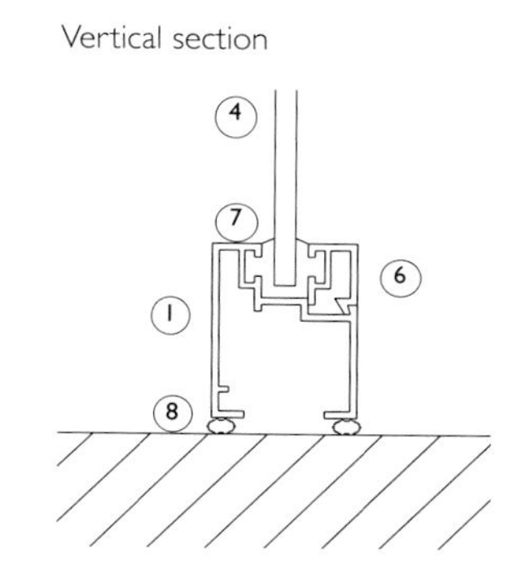


102C

102B

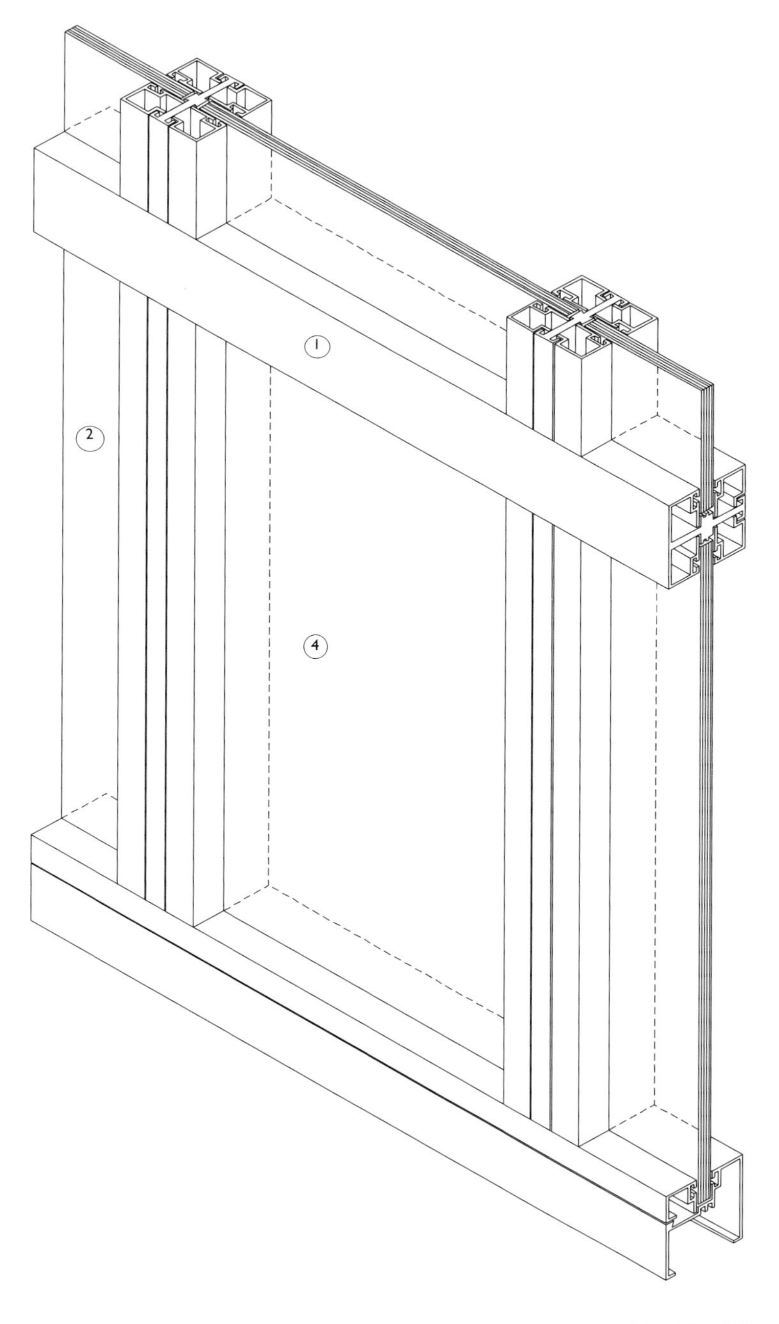

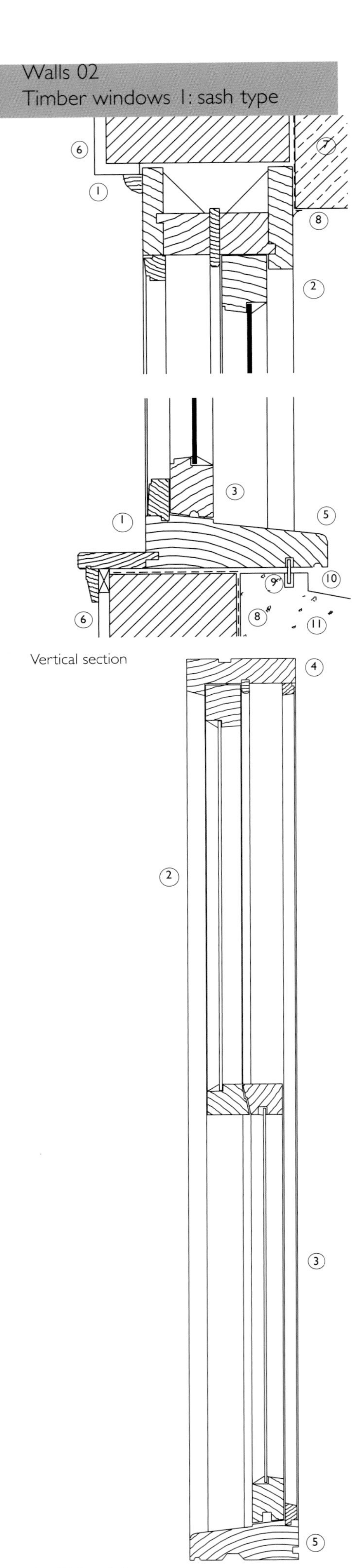

Vertical section

53

The design of the traditional timber sash window has been refined and developed over many years. The principle of sliding sash lights, as opposed to inward or outward opening hinged casements, has been used since at least the 17th century. The advantage of the sash window is that it does not impinge on the internal space when open and retains its position to provide a variable amount of ventilation. The window can be cleaned simply by reversing the position of the sashes.

Window sashes are counterbalanced by weights, usually lead, concealed in the jambs. The sashes run in guides formed by the staff bead, the parting bead and the front lining, enabling the sashes to be easily removed for maintenance. The fact that the sashes are counterbalanced allows them to move freely between the guides without jamming.

The sash box (containing the counterweights) is normally hidden behind a rebate in the wall construction. The sash window continues to be developed today with the introduction of PVC-U, which is used for both frame and sashes, and spring balances replacing counterweights. The disadvantage to the sash type is that it is more difficult to achieve a reliable seal between the sash and the frame. Consequently, their performance, like that of the sliders, is below the performance of hinged and pivoting types of windows.

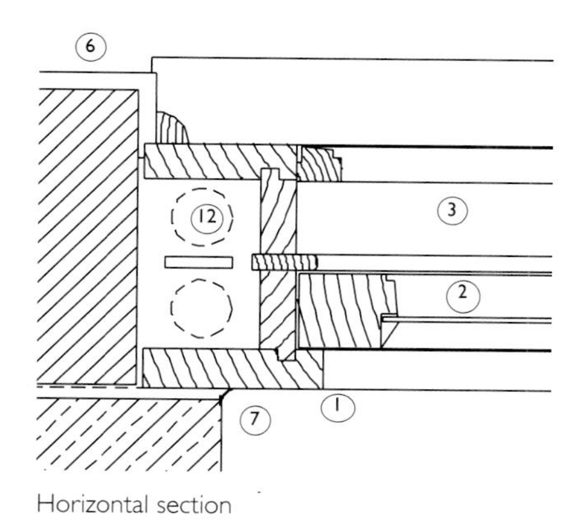

Horizontal section

Details
1. Window frame
2. Outer sash
3. Inner sash
4. Heads
5. Cill
6. Internal finish
7. Outer wall facing
8. Damp proof course (DPC)
9. Weather bar
10. Drip
11. Sub-cill
12. Counterweights

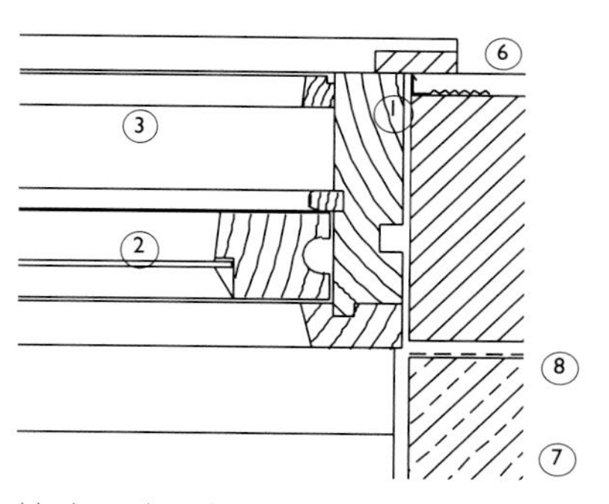

Horizontal section

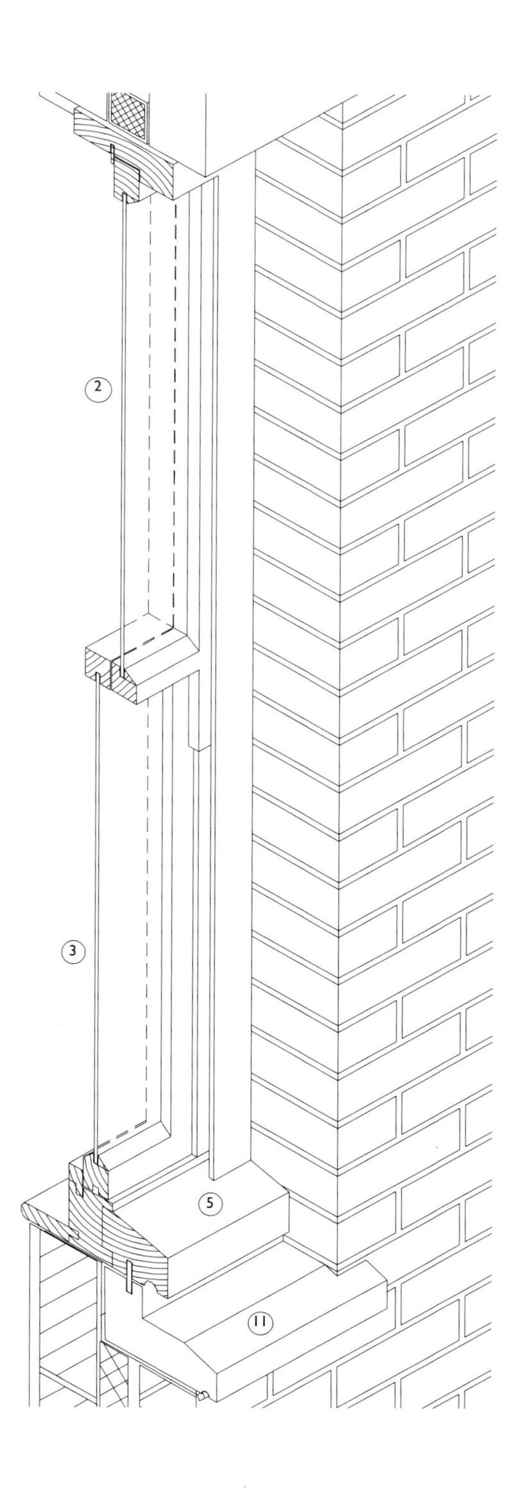


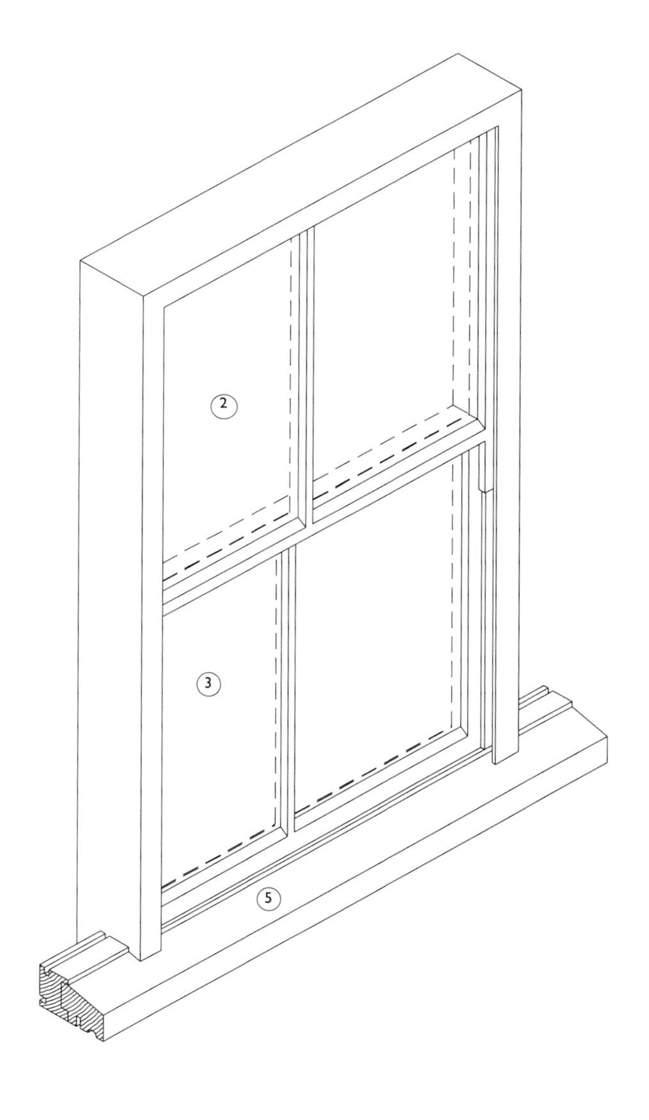

# Timber windows 2: hinged and pivoted units

52B

Details

1. Window frame
2. Outside
3. Inside
4. Head
5. Cill
6. Sub-cill
7. Routed timber glazing section
8. Line of transom
9. Line of mullion
10. Single glazed or double glazed unit to suit application
11. Fixing bead
12. Rubber-based seal
13. Fixed light
14. Outward opening light
15. Damp proof course (DPC)
16. Internal finish
17. Drip
18. Outer wall facing
19. Weather bar

52D

Vertical section
Opening light

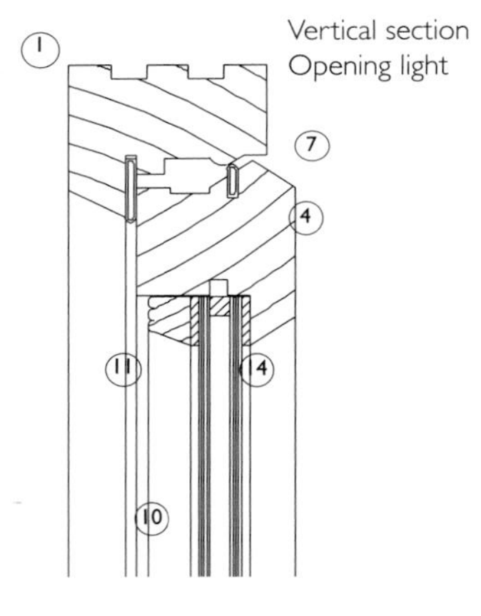


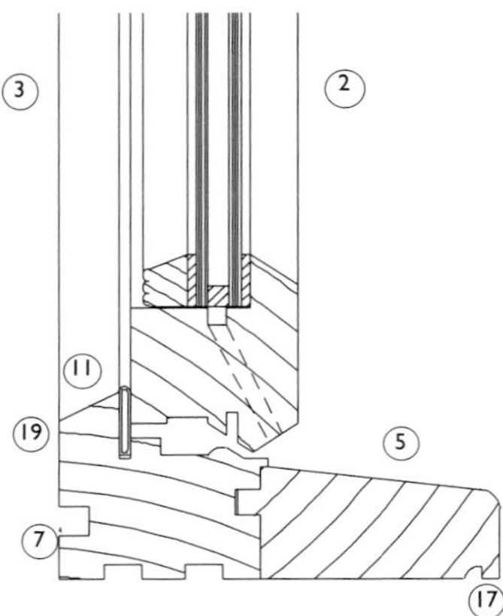


Hinged and pivoting windows tend to employ similar timber sections regardless of the specific orientation of the opening light. Sash and sliding windows, however, use very different sections. Hinged and pivoting units can be top hung, bottom hung or side hung and are available as either inward or outward opening.

Inward opening windows are inherently vulnerable to water penetration at the bottom of the frame, while outward opening frames are weaker at the head of the frame. For this reason, drainage channels are provided within the frames, together with weather seals and weather bars, in different combinations as required. Similar sections are used for windows and frames on all sides of a casement. Seals and bars are attached separately.

Although numerous standard size timber windows are produced, a huge number are made to order to replace, or to match, existing casements. Since these products are widely available, the designer has considerable freedom in the economic design of timber windows. Timber casements are constructed using joinery techniques although their design is informed by advances in other glazing systems. In recent years technology transfer has helped to improve the performance of traditional casements without altering their outward appearance.

Rainwater that penetrates the gap between opening light and frame is drained by the grooves in the light at the head and jambs, and in the frame at the cill, and is discharged at the bottom. Weather seals are added behind channels to ensure water does not pass any further as well as to reduce air leakage.

Single panes and double glazed units are used in timber casements although heavier glazed units require stiffening. This extra support is typically provided in the form of steel brackets rebated into the opening lights. Timber frames may be painted or stained.

Hinged and pivoted casements are usually fixed into wall openings at the jambs, while large frames, particularly those that span from floor to ceiling, are often attached on all four sides. Windows are secured with face-fixed lugs or screws if the casement is installed after the opening is formed. Where the window is built-in as part of a masonry wall, frames can be secured by flat cramps that are bedded into the mortar joints.

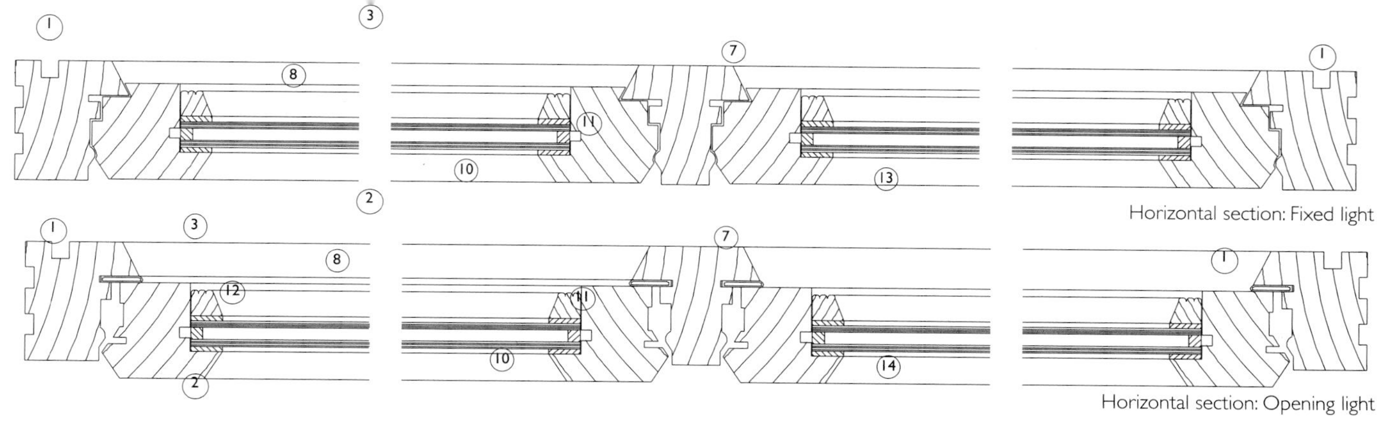


Horizontal section: Fixed light

Horizontal section: Opening light

51B

51A

Vertical section
Fixed light

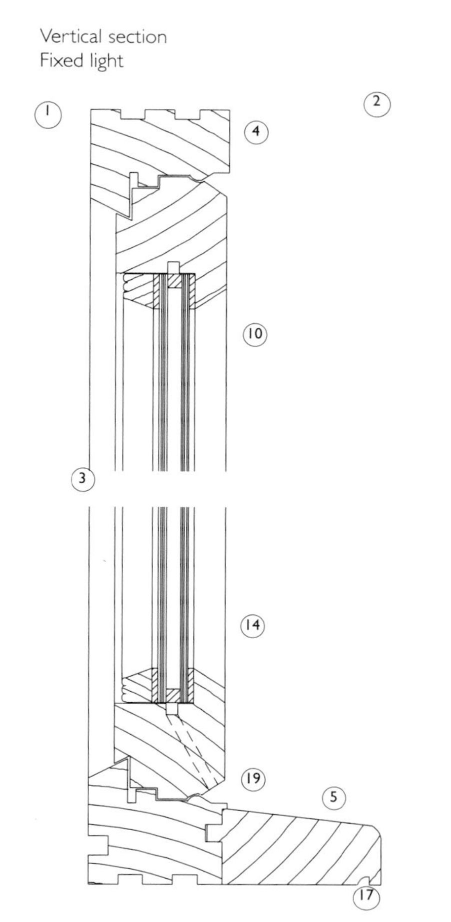


Elevation

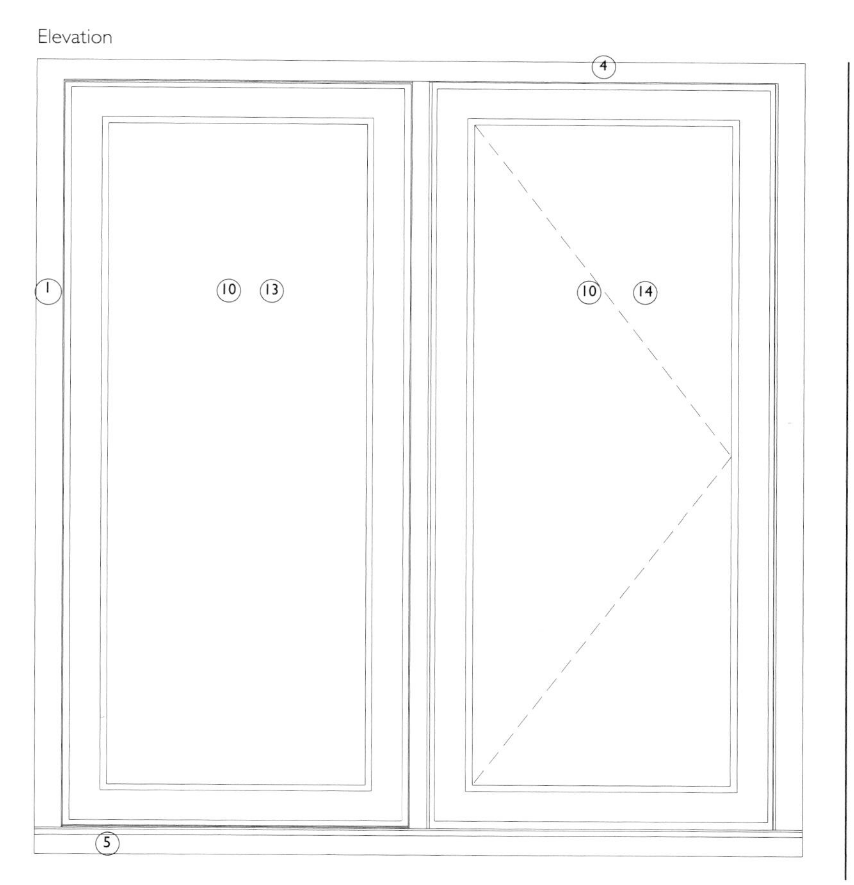

1 E

1 F

Cast-in-place concrete (in situ concrete) used for wall construction offers many advantages such as good stability, fire resistance, acoustic performance and even weather-tightness due to the density of the material. Insulation is often required to achieve an acceptable thermal performance.

However, the major problems with cast-in-place concrete walls are to do with achieving and maintaining a good quality of finish. Rigorous control of material and workmanship is required. This involves ensuring that concrete batches are consistent in colour and texture and that shuttering is properly cleaned and prepared before each use. Small changes in the water content, for example, can result in different pours having different appearances. Concrete that has been too heavily vibrated will have too many fine aggregates close to the surface giving a blotchy appearance. Concrete that has been too little vibrated may not fill up the mould leaving pits in the surface of the wall.

In spite of these technical difficulties, it is possible, with care, to achieve good fair-faced concrete. The imprint of the shuttering, whether it is plywood, steel or board-marked, as well as the positioning of boltholes, will have a major impact on the wall's appearance.

On external walls, the weathering of concrete is an important consideration. Concrete absorbs water at different rates depending on its location within the building and consequently weathers differently across its surface, causing blotches. Streaking is caused where rainwater runs off one surface onto another, depositing the dirt it has collected. As a result, areas of concrete in close proximity to windows and ledges require particular attention. Cills and overhangs are detailed to throw water as far as possible from the wall area beneath.

If fair-faced concrete cannot be achieved it is possible to paint, tile, render (exterior cement plaster) or clad the wall. Often this approach is taken from the outset, with the cladding acting as a rainscreen or sealed membrane to a rough concrete wall behind.

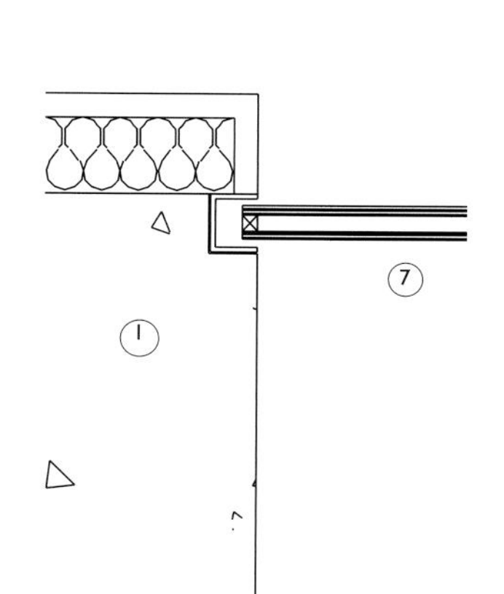


Details

1. Concrete
2. Roof
3. Metal lining to gutter
4. Throating
5. Slot formed as part of casting concrete.
6. Drip
7. Single glazed or double glazed unit to suit application
8. Metal cill, though precast concrete could be used
9. Internal floor level
10. External floor

Vertical section

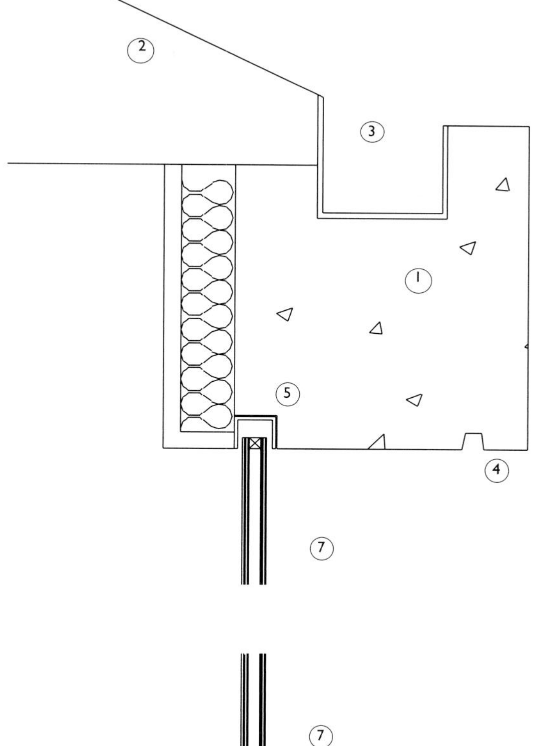


Vertical section

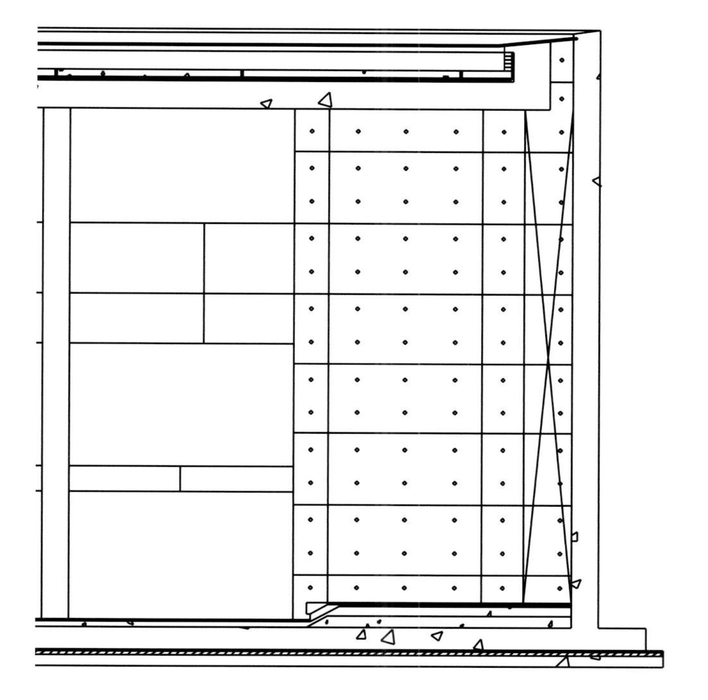

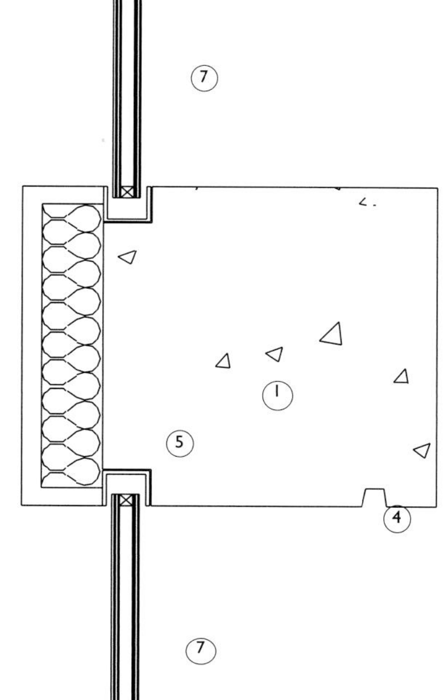


Vertical section

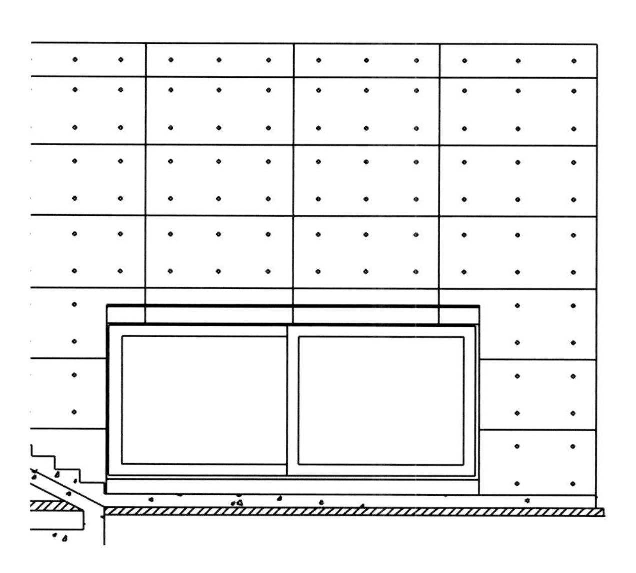

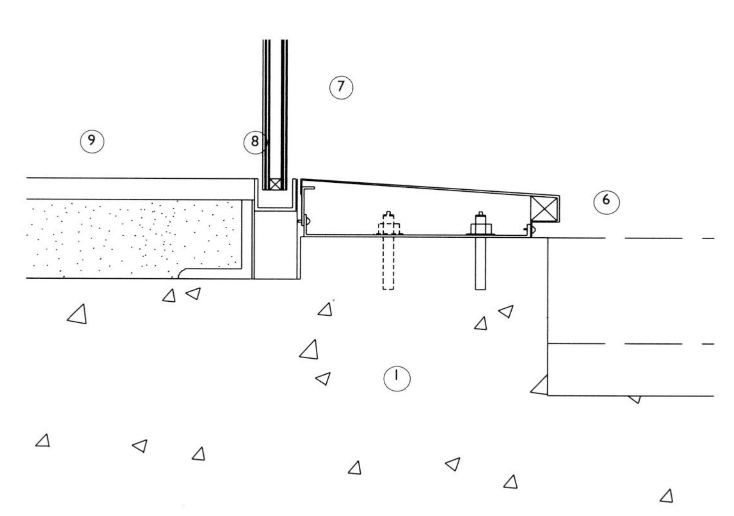

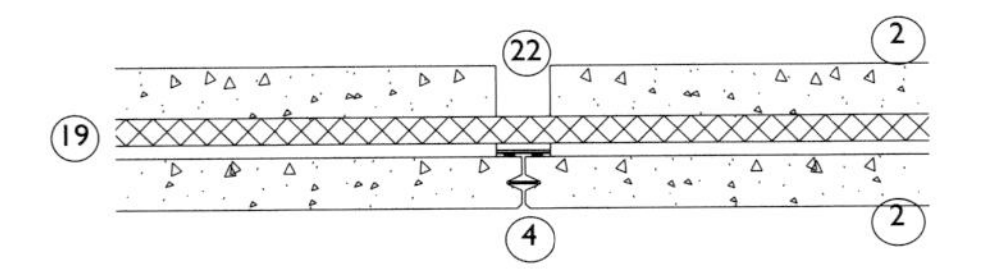


Sectional views of assembly
1. Concrete floor deck
2. Precast concrete panel
3. Optional dowel fixings
4. Vertical joint (typically a baffle) between panels
5. Horizontal joint (typically a lap) between panels
6. Drywall/dry lining
7. Precast concrete panel
8. Concrete floor deck
9. Steel dowel
10. Steel angle
11. Rubber-based baffle
12. Drywall/dry lining

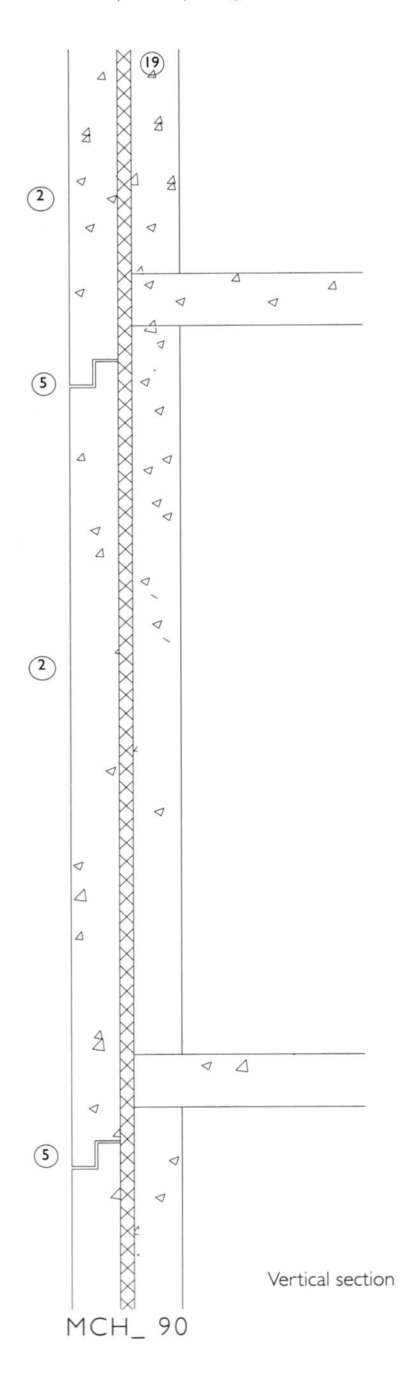


Vertical section

19

Precast concrete panels are usually made storey height, spanning from floor to floor. Each panel is supported by a floor slab or edge beam, usually at the top or bottom. Panels are top hung or bottom supported, held either by concrete nibs that are integral with the panel, or by stainless steel brackets. Where nibs are used, the panel is held in place by non-ferrous dowels or angle cleats. Fixings are usually set near the corners. Slotted fixings are used in order to allow for both thermal movement between the panel and in the supporting structure and for adjustment during construction. The cleats are sometimes fixed to a slotted channel set into the floor slab in order to accommodate horizontal movement.

Panels are lifted into place on site by crane, using hooks that are fixed to threaded tubes set into the concrete. The hooks are unscrewed when the panel is in place, and the holes are plugged. The hooks are usually fixed to the inside face of the panel. Because concrete cladding panels have many more joints than in situ cast walls, which are essentially monolithic, their design is critical.

Joints between panels are either of the 'closed' or 'open' type. The closed joint is used where panels are fixed back to columns or ends of walls that provide the role of 'backing wall' at this point for the outer seal. Open joints tend to be used at panel-to-panel junctions where there is no column.

Closed joints in both a vertical and horizontal direction are made waterproof with sealants injected into the gap between panels. The sealant bonds itself to the two edges of the concrete panels. The sealant is required to prevent wind-blown rain from entering the joint. A backing strip of closed cell polyethylene foam is placed to ensure the correct depth of sealant is achieved. Vertical open joints are sealed with neoprene or EPDM baffles, set into slots recessed within the depth of the panel. Rainwater that penetrates the slot runs down the face of the baffle and is directed out at the junction with the horizontal joint at the base of the panel. The baffle forms the first barrier against wind-blown rain. Any water that passes around the edges of the baffle is drained away in the rear half of the vertical slot. Sealant is applied to the joint on the inside face of the panel. The sealant is protected from the effects of ultraviolet light that causes its breakdown. Horizontal open joints use steps to allow the

Isometric and sectional views of assembly
1. Concrete floor deck
2. Precast concrete panel
3. Optional dowel fixings
4. Vertical joint (typically a baffle) between panels
5. Horizontal joint (typically a lap) between panels
6. Drywall/dry lining

Isometric detail
7. Precast concrete panel
8. Concrete floor deck
9. Steel dowel
10. Steel angle
11. Rubber-based baffle
12. Drywall/dry lining

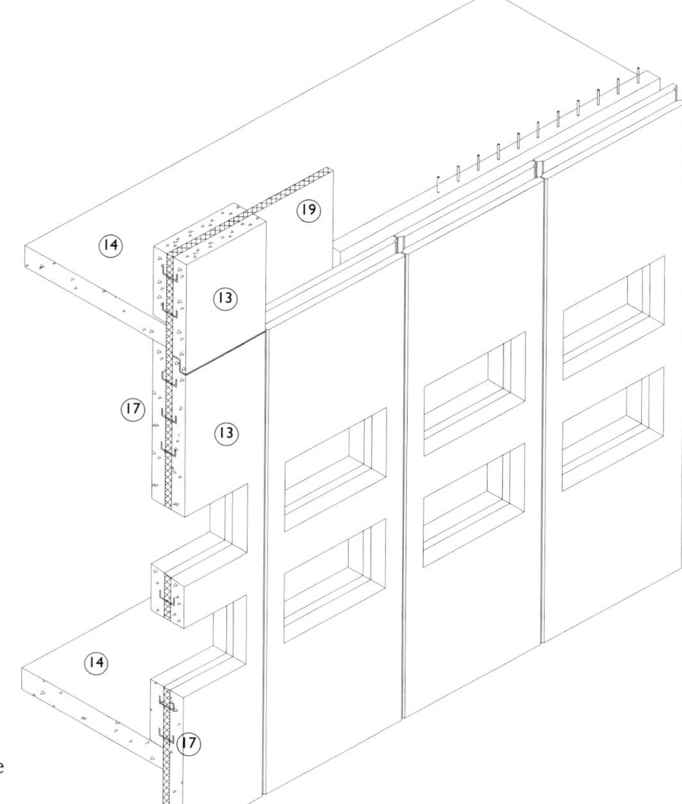


Details
13. Precast concrete panel
14. Concrete floor deck
15. Steel dowel
16. Steel angle
17. Rubber-based baffle
18. Drywall/dry lining
19. Rigid thermal insulation
20. Soffit insulation
21. Metal flashing
22. Thermal break
23. Rubber-based seal
24. Silicone-based seal
25. Stone or tiled facing
26. Metal faced panel
27. Window panel
28. Optional metal framing or concrete block construction as backing wall
29. Roof construction

rainwater to drain out from the baffle strip and the slot behind whilst preventing rain from penetrating the joint from outside.

## Developments

The use of precast panels remains very much as cladding rather than as components in loadbearing construction. Pre-cast panels are perhaps more suited to loadbearing construction than any other building material, since they can be prefabricated to almost any shape to provide a weather-tight, fire-resistant wall with high acoustic insulation and thermal mass. Until the building collapse at Ronan Point in London in the 1970's, pre-cast concrete panels were used as components in a loadbearing wall construction. Panels were 'stitched' together to form a monolithic construction. When the building industry returns to this potential, we will see huge strides made in this material. The lack of close tolerances in forming a panel and its characteristics of thermal movement make it more compatible with a concrete frame than with a steel frame.

### Isometric section.

Panels are often lapped over one another at the horizontal junctions and butted together at the vertical junctions. Cladding panels are rarely lapped on both the vertical and horizontal joints, as this would make it more difficult to construct. The double lapping is more associated with monolithic construction. In this example, the open vertical joint is sealed with an inner baffle and an outer silicone-based seal.

### Vertical section.

Panels are located using steel dowels and fixed into position at the head with short lengths of steel angle. A cast-in channel is often set into the ceiling soffit of the floor slab to avoid the need for drilling directly into the concrete.

### Vertical section.

Typical use of pre-cast panels as a background for a decorative finish such as ceramic tiling or stone. The facing material is bonded to the panel. Separate glazed and metal panels can be used in adjacent areas of external wall.

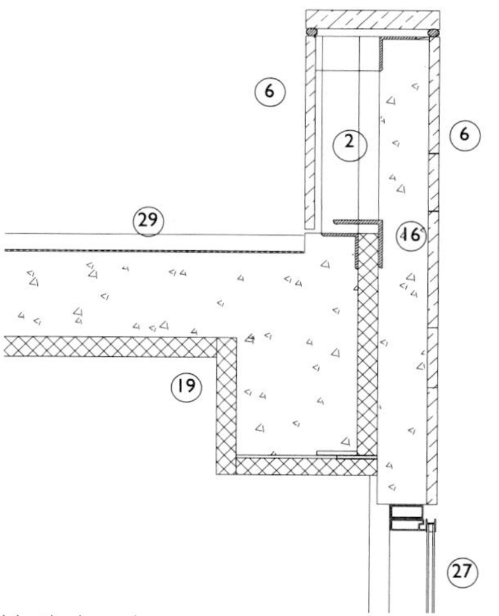


Vertical section

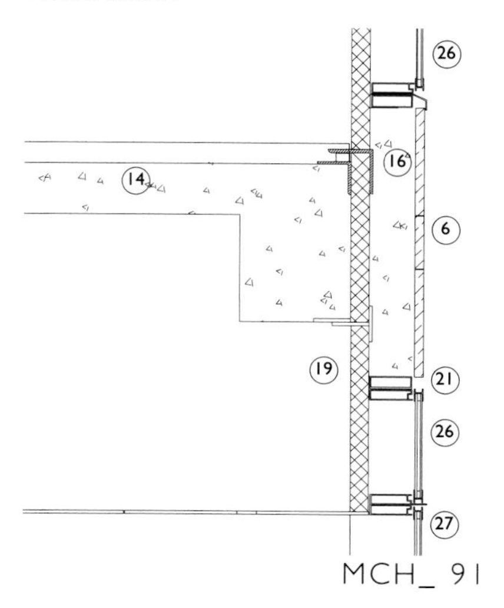

Details
1. Loadbearing brick wall
2. Timber framed window
3. Internal plaster finish or dry lining/drywall
4. Thermal insulation
5. Precast concrete lintel
6. Precast concrete cill
7. Brick arch
8. Damp proof course (DPC)
9. Weather bar
10. Seal

23B

23A

Horizontal section

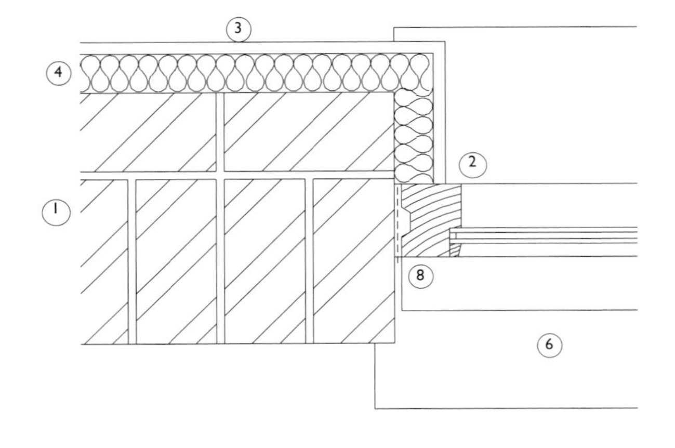


The issues in the construction of loadbearing walls are common to all the materials used: mainly brick, concrete blockwork (concrete masonry units) and stone. The most striking developments have been in brick construction, where a single, relatively porous, material can be used in a monolithic form in large-scale structures. This section focuses on the construction issues in terms of brick, and then shows the variations in concrete blockwork and in stone.

## Brick

In loadbearing brick construction, the strength and porosity of the brick and mortar must be evenly matched, to achieve an homogeneous structural action. The walls must be thick enough to form an adequate structure as well as preventing rainwater that is absorbed by the brickwork from reaching the inner face. Experience has shown that traditionally 330mm (1ft 1in) walls are thick enough to prevent rainwater penetration, although most are at least 215mm (8in) thick. Strong dense mortars should be avoided in order to ensure sufficient porosity in the joint, reducing shrinkage, so that cracking is eliminated and allowing the wall to accommodate expansion and contraction caused by cycles of wetting and drying.

Modern brickwork requires movement joints to cope with expansion and to compensate for the lack of flexibility in modern mortars. When used as cladding, movement joints also take up differential movement between panels of brickwork and the structural frame. By using traditional flexible lime putty mortar, it is possible to eliminate movement joints, although this requires careful research into the properties of the bricks and mortar and an analysis of overall structural behaviour. Lime putty mortar has a lower strength than modern mortars, but is not as brittle, and creeps slowly in response to applied loads. The exposed surface of the mortar sets through a process of carbonation, leaving the core in a putty form. Any movement will crack the joint slightly and expose fresh, unreacted lime putty to the air, precipitating the carbonation and setting of the newly exposed material.

Where impervious bricks, such as engineering bricks, are used, they are typically smooth faced and do not assist adhesion between materials. The dense mortars required to provide impermeability in the joints

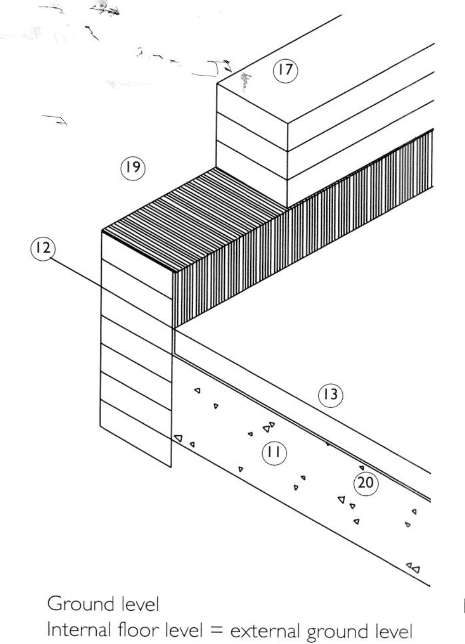


Ground level
Internal floor level = external ground level

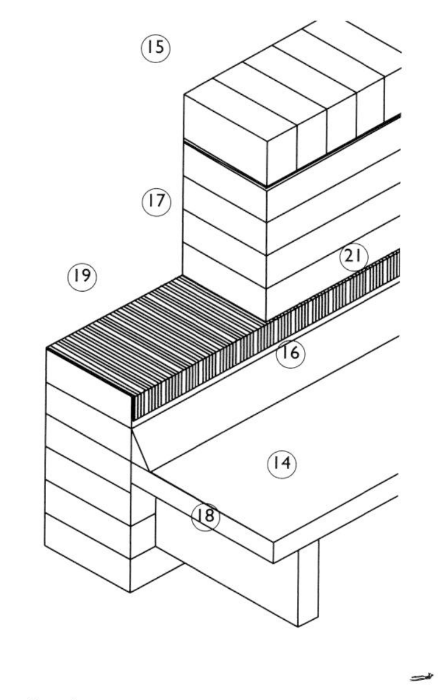


Parapet level

23C

undergo a large initial shrinkage and as a result, there is a tendency for cracks to appear in the joints allowing rain penetration. Since the bricks do not absorb the water running down the wall, it will rapidly enter such cracks. Water unable to escape from these cracks, due to the difficulty of its drying out in the impermeable brickwork, will freeze and expand in cold weather, causing serious damage. Hence, successful impervious brickwork requires a high quality of workmanship. To reduce shrinkage, the proportion of cement is reduced and the loss of workability is compensated by adding lime for flexibility.

As with other forms of masonry wall construction, the use of a damp proof course at ground level is essential in loadbearing masonry construction. Although some materials, particularly dense stones, have very low water absorption, care should be taken to ensure that structures for floors or roofs are properly protected against damp from within the wall construction. Isolating the structure, typically concrete or timber, with an impervious material such as slate or bituminous damp proof course, can achieve this.

Isometric details

11. Ground floor/slab
12. External ground level
13. Screed
14. Waterproofing layer to roof
15. Brick parapet
16. Angle fillet
17. External wall
18. Roof substrate
19. Damp proof course (DPC)
20. Damp proof membrane (DPM)
21. Overflashing required to DPC (not shown)

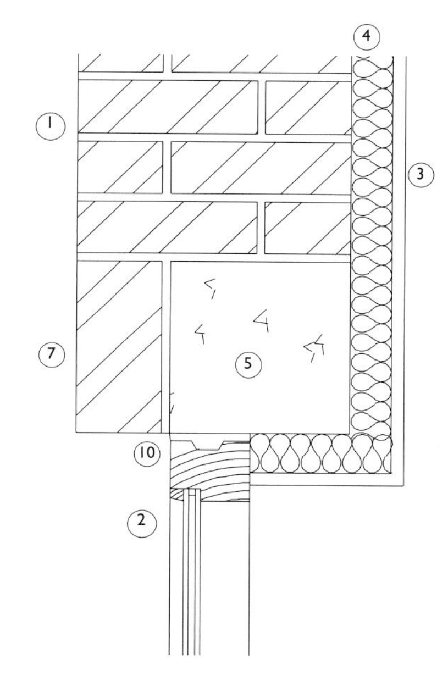


Vertical section

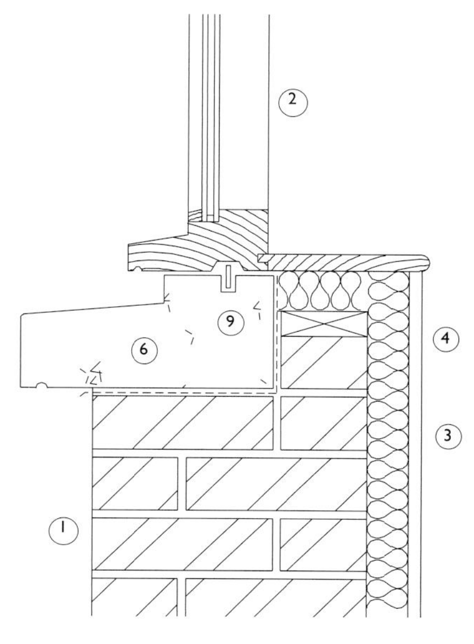

80B

Solid stone, in the form of ashlar blocks, is hardly ever used in modern construction in a loadbearing capacity since the thickness required makes it extremely expensive. In loadbearing construction, it is more commonly used as a structurally integral facing to brickwork or blockwork wall. The stone forms part of the loadbearing construction, but a less expensive masonry material is used in the depth of the wall where it is not visible. On these visible faces, the traditional craft of bonding and jointing stones is used to provide a very durable facing that requires a minimum of maintenance. Stones are most commonly bonded with a cement mortar with a flush joint on its visible face.

80A

Horizontal section

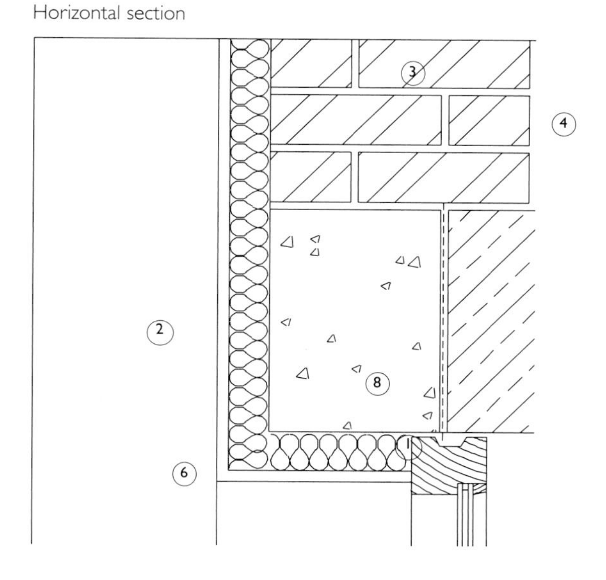


Vertical section

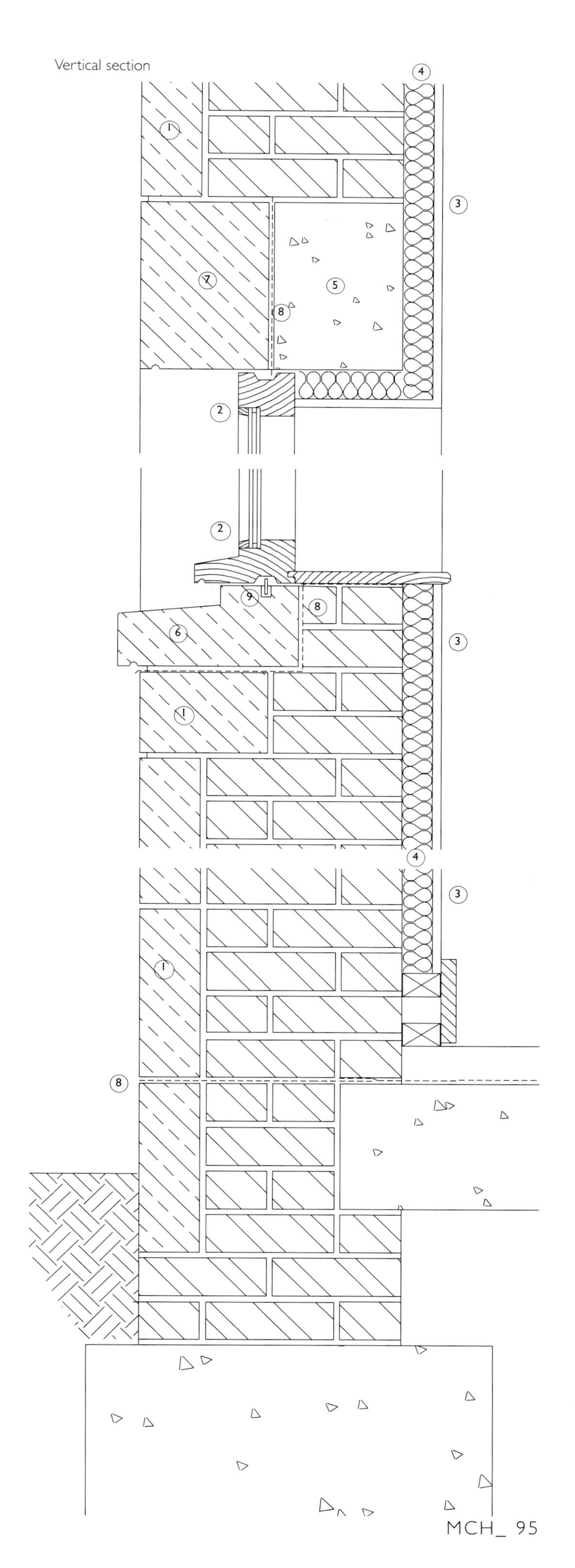


## Details

1. Loadbearing stone wall
2. Timber framed window
3. Internal plaster finish or dry lining/drywall
4. Thermal insulation
5. Stone lintel
6. Stone cill
7. Stone arch
8. Damp proof course (DPC)
9. Weather bar
10. Seal

## Isometric details

11. Loadbearing stone wall
12. DPC
13. Thermal insulation
14. External ground level
15. Damp proof course (DPC)
16. Damp proof membrane (DPM)
17. Ground floor/ground slab
18. Screed
19. Strip foundation shown. Type used depends on loads and ground conditions.

Ground level
Internal floor level higher than external ground level

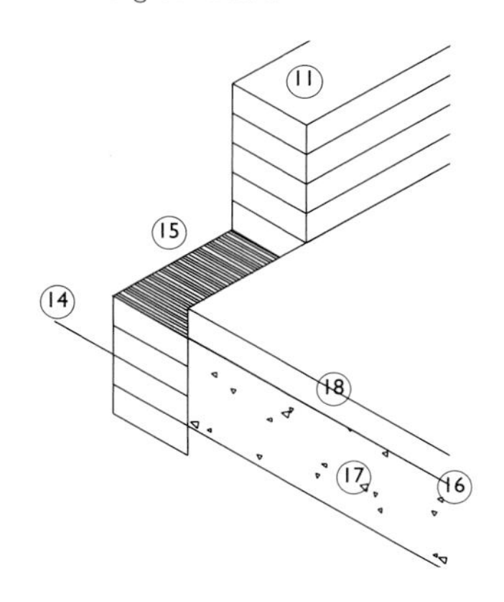


Ground level
Internal floor level equal to external ground level

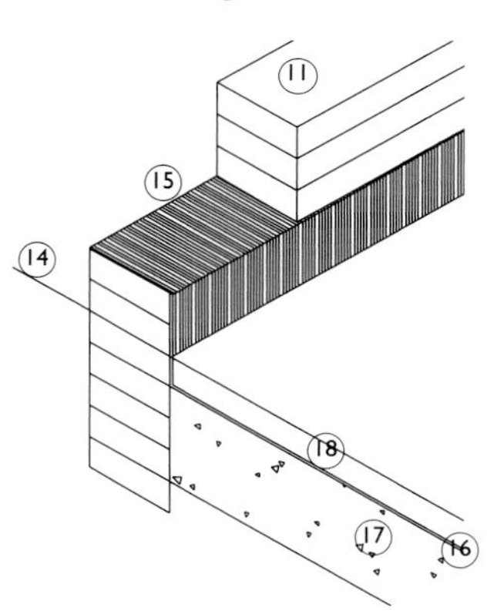

22A

33A

The principles of loadbearing blockwork are similar to those of load bearing brick construction. Walls are commonly at least 240mm (10in) thick. Strong dense mortars are avoided in order to ensure sufficient porosity in the joint and to reduce shrinkage so that cracking between blocks and mortar is eliminated. The wall can thus move in response to cycles of wetting and drying out. This is an economical form of construction that is often given a water-resistant render finish. Fair-faced finishes are usually restricted to cavity wall construction, where the penetration of rainwater through mortar joints is less important. Movement joints are provided to enable structural movement without cracking the masonry, since concrete and mortars are relatively inflexible.

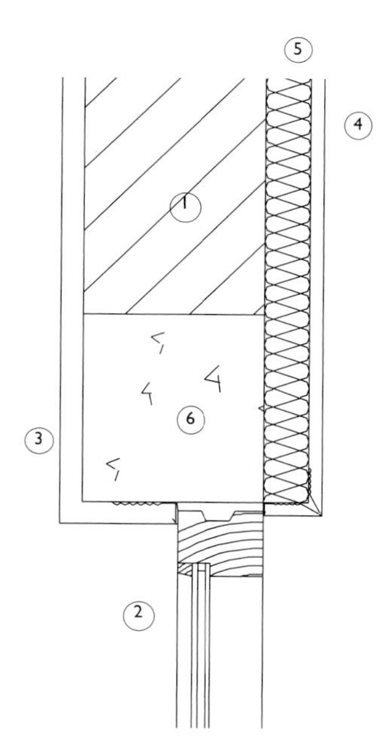


Vertical section

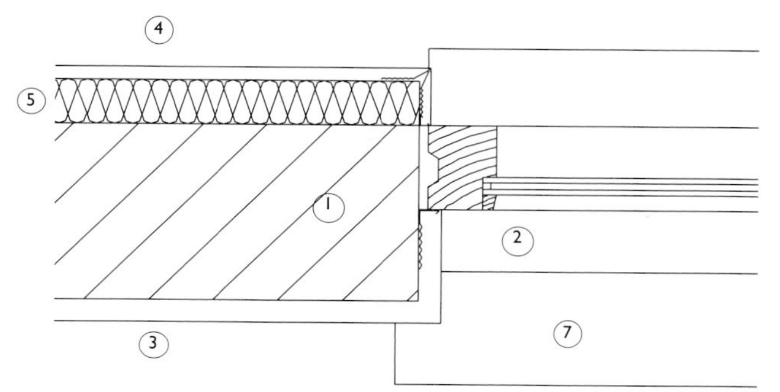


Horizontal section

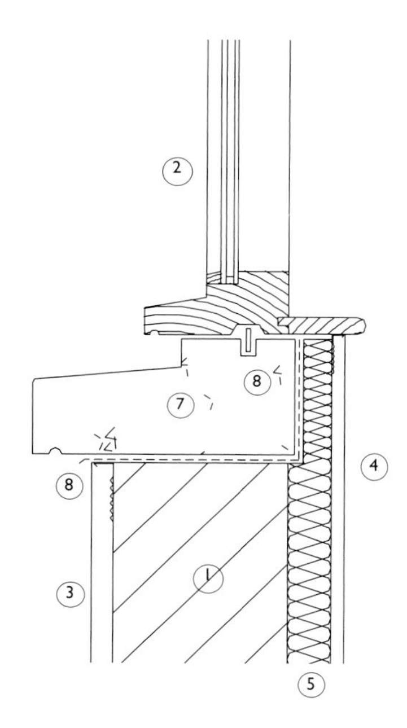


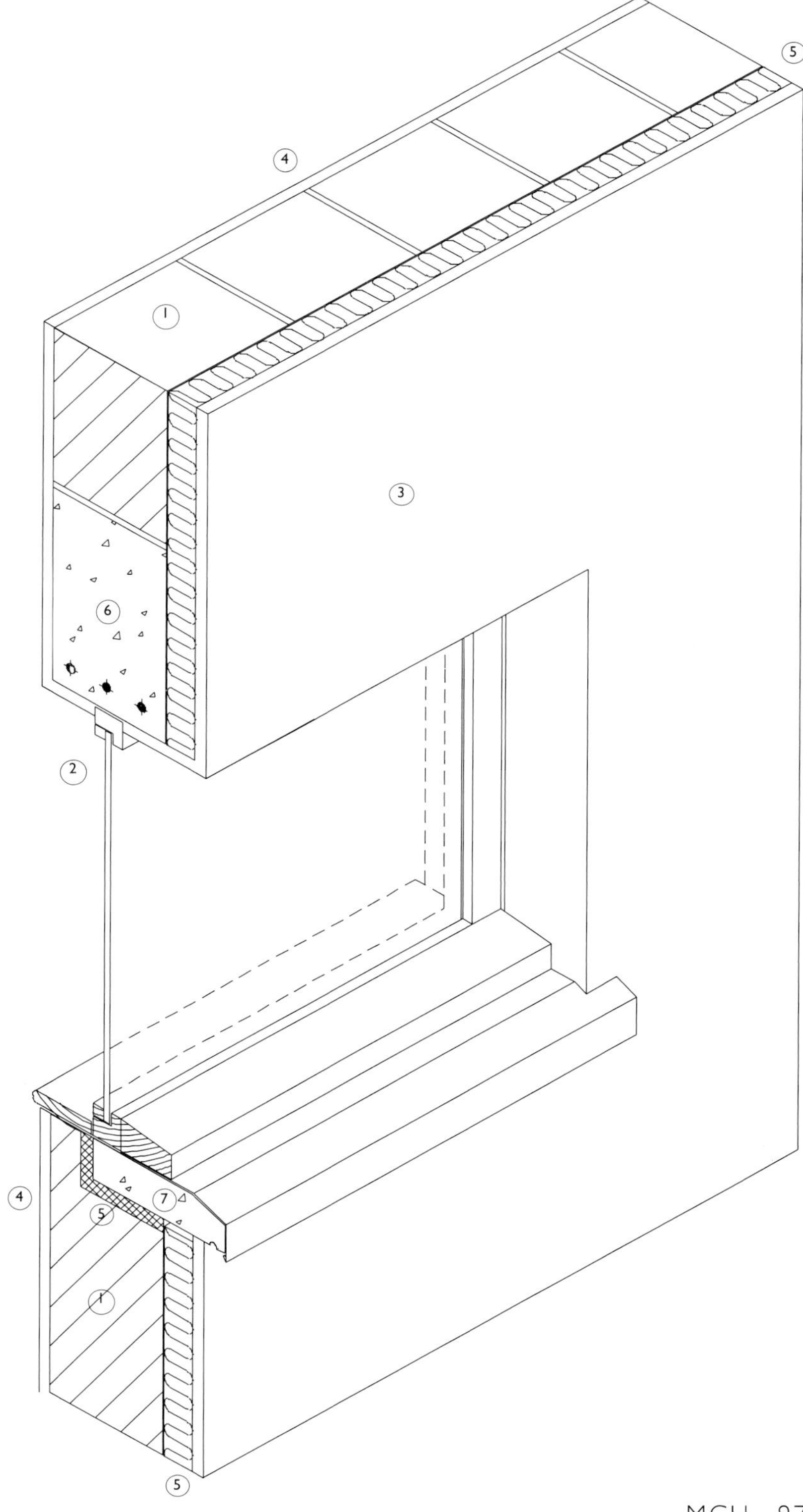


## Details

1. Loadbearing block wall
2. Timber framed window
3. External render finish
4. Internal plaster finish or dry lining/drywall
5. Thermal insulation
6. Precast concrete lintel
7. Precast concrete cill
8. Damp proof course (DPC)
9. Weather bar
10. Seal

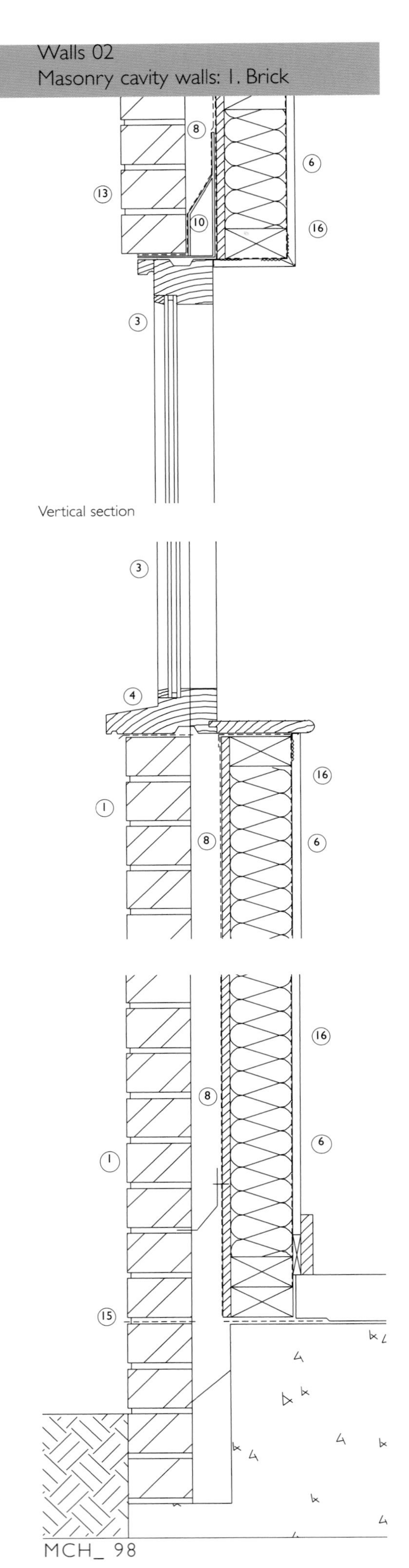

Loadbearing brick walls are seldom built today due to their high cost. Now most brickwork is used in the form of an outer facing, or 'skin', to a cavity wall. As such, it is as a cladding with no loadbearing function. In cavity wall construction, the outer masonry leaf provides some stability to the inner leaf. Cavity wall construction still allows the traditional craft of bonding and jointing masonry units. It provides a very durable facing requiring a minimum of maintenance yet providing an appearance of durability and stability. The most commonly used masonry materials for cavity walls, as with loadbearing masonry walls, are brick, stone and concrete block. A thickness of 100mm is typically used for all materials with a flush jointed cement mortar.

Masonry facings are seldom waterproof in a cavity wall, but overall they have greater resistance to rain penetration and better use of thermal insulation than an equivalent loadbearing wall. Thermal insulation is usually provided by fixing rigid insulation board to the cavity face of the inner skin so that the environment inside the building can benefit from its thermal mass. Sometimes the entire cavity is filled with foam insulation, but extreme care must be taken that water cannot reach the inner leaf. A concrete blockwork inner leaf is often used where structural or acoustic requirements dictate, but frequently the inner leaf is composed of wood or metal studs with sheathing on either side. The cavity allows any water that penetrates the outer stone leaf to drain away safely at the bottom of the wall through weep holes. The outer facing is tied back to the inner leaf with metal (galvanised or stainless steel) wall ties bonded into cement joints. During construction, care should be taken that mortar droppings do not block the weep holes at the base of the wall or cover the ties, thus directing water from the outer to the inner leaf.

The void between leaves is ventilated to allow the cavity to dry out where rainwater penetrates the external envelope. Weep holes are provided in the external facing at the

## Details

1. Outer brick wall (or 'leaf')
2. Inner blockwork wall (or 'leaf')
3. Timber framed window
4. Timber cill
5. Cavity closer
6. Internal plaster finish or dry lining/drywall
7. Thermal insulation in cavity
8. Air cavity (sometimes omitted where insulation fills cavity)
9. Inner precast concrete lintel
10. Pressed steel lintel
11. Steel angle
12. Outer precast concrete lintel or brick flat arch
13. Precast concrete cill
14. Damp proof course (DPC)
15. Seal
16. Timber framed inner wall with flexible quilt insulation

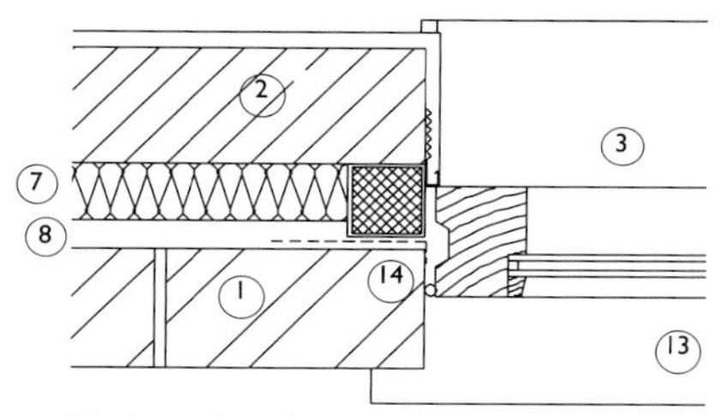

Horizontal section

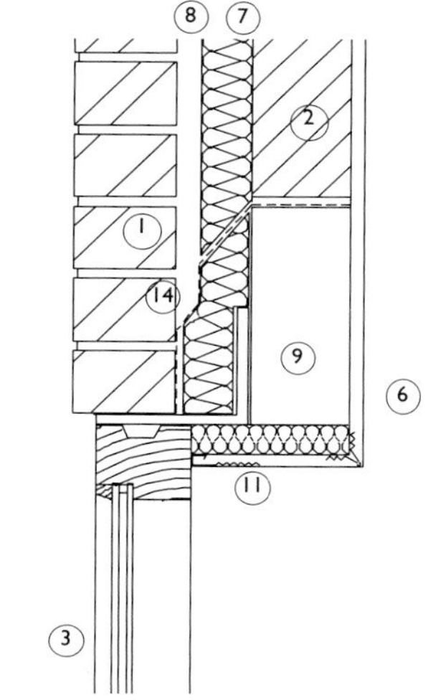

Vertical section

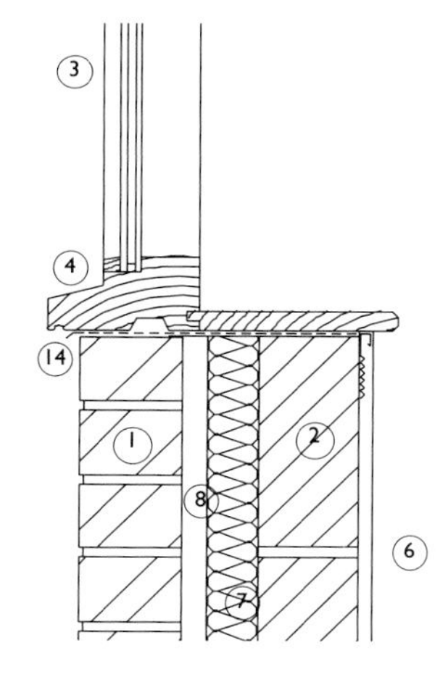


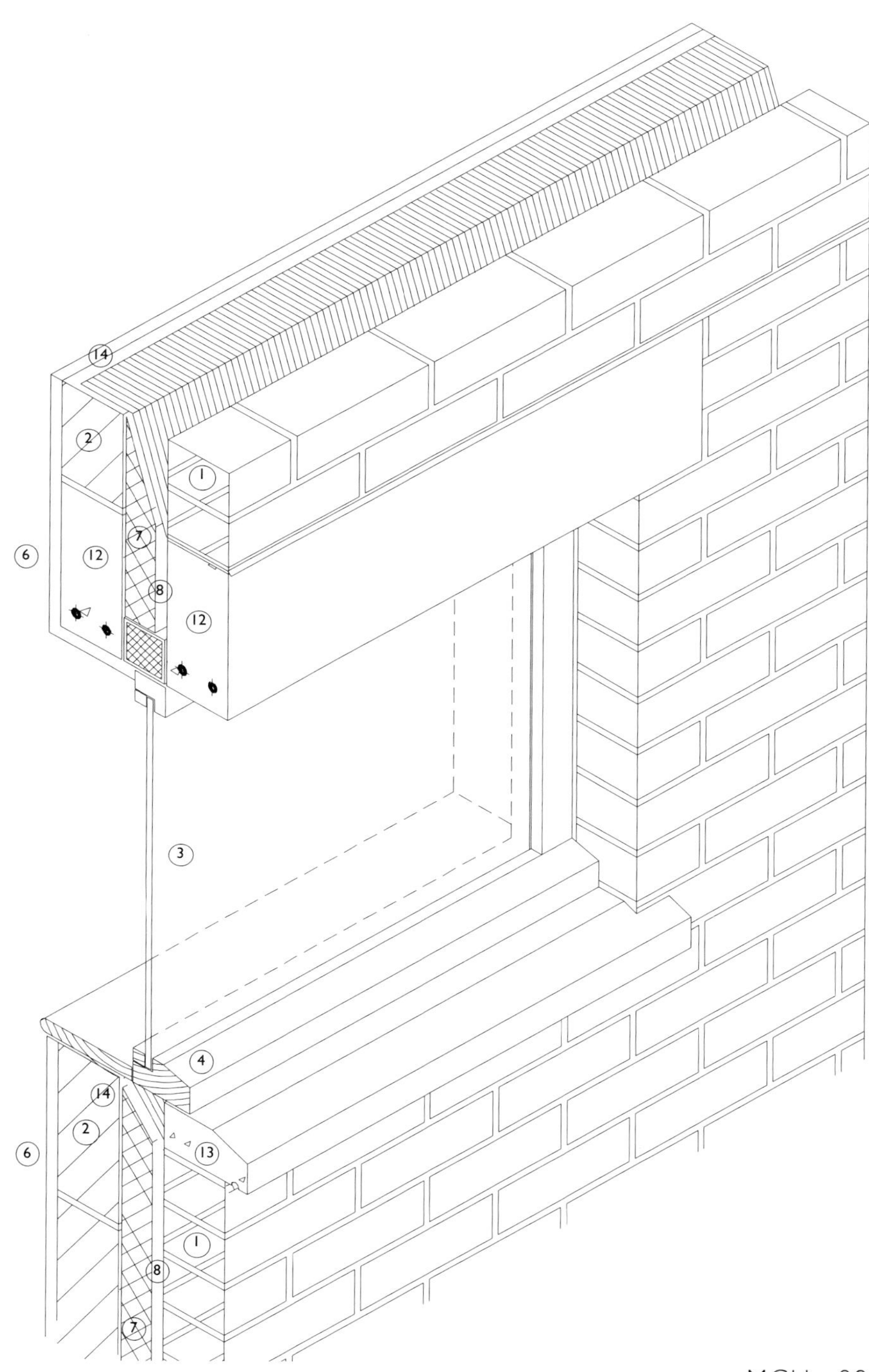

50

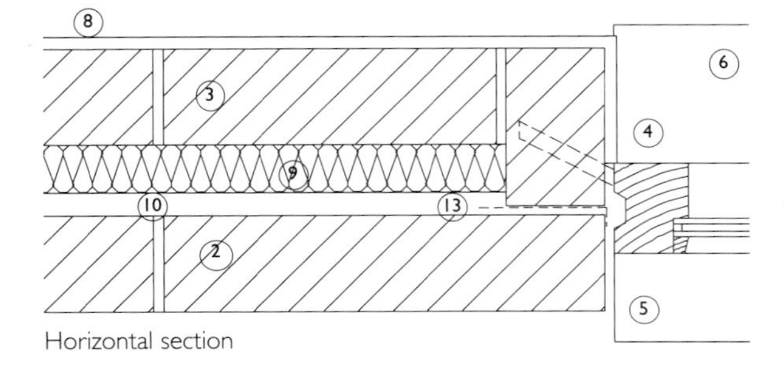


Horizontal section

## Masonry facings and backing walls

Cavity walls can be considered to comprise two forms of construction: an inner wall, and an outer 'facing' wall. The outer facing is commonly used to face a range of backing walls that may be either a loadbearing wall or an infill panel to a framed structure. Inner and outer walls are actually two different generic forms of construction that can be used with other materials. The inner wall can be a variety of constructions and does not generally need an outer leaf other than as a rainscreen. An exception is in small-scale construction, such as in houses, where the outer wall can provide some bracing to the inner wall. The brick cavity wall is mostly used as a rainscreen with complex junctions around openings in order to drain water from what is essentially a closed cavity (which is ventilated at the top and bottom only). Unlike true rainscreens, here the detailing aims to imitate the outward appearance of traditional loadbearing masonry for essentially aesthetic reasons. This situation has come about because the inner and outer leaves cannot be tied together structurally with floor slabs without creating a thermal bridge across the wall. The exposed slab and lintels over openings, so characteristic of 1960's detailing, are difficult to insulate thermally. Interstitial condensation and associated problems resulting from the thermal bridging, together with inadequate ventilation, has led to increased separation of inner wall and outer facing.

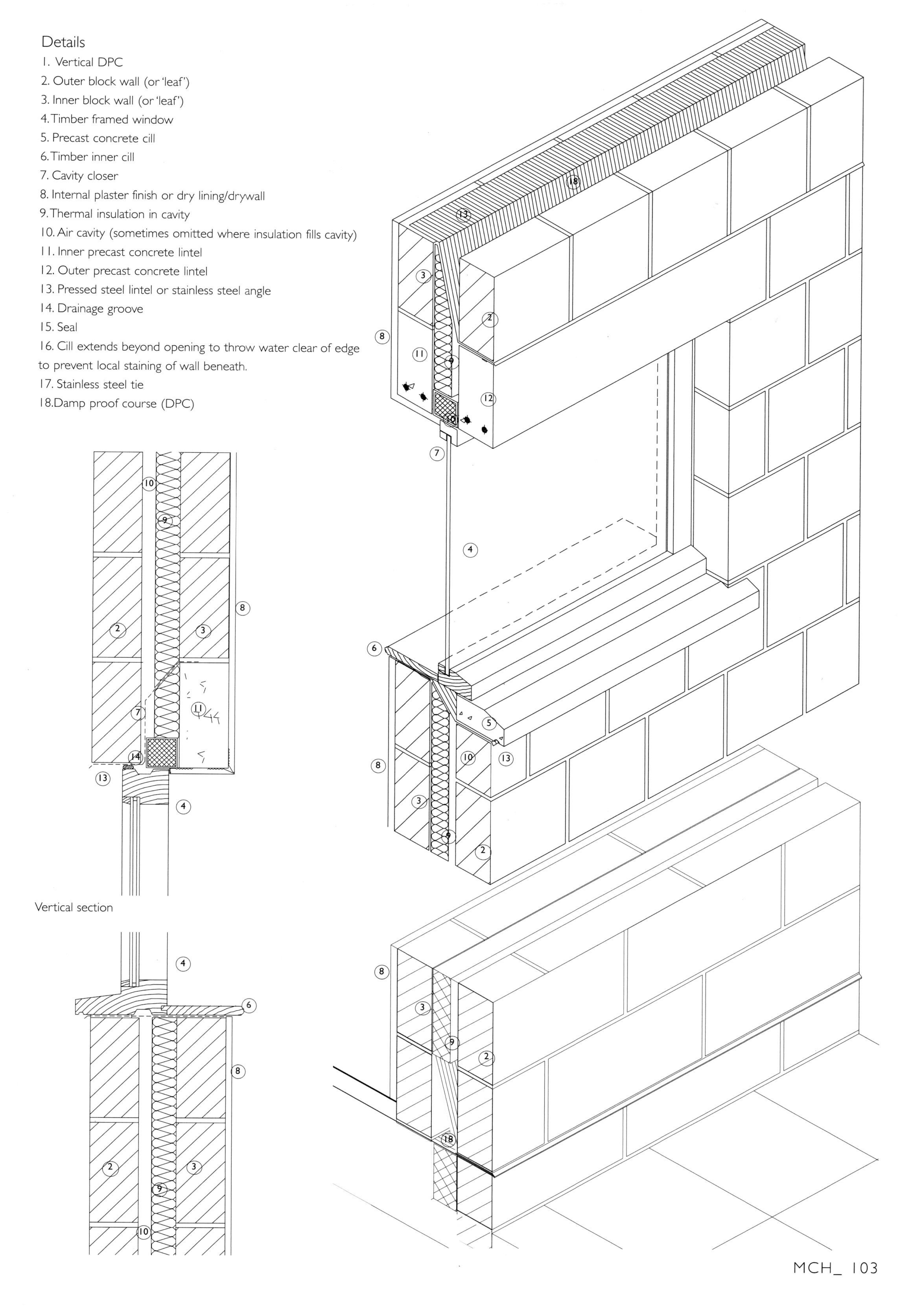
Details
1. Vertical DPC
2. Outer block wall (or 'leaf')
3. Inner block wall (or 'leaf')
4. Timber framed window
5. Precast concrete cill
6. Timber inner cill
7. Cavity closer
8. Internal plaster finish or dry lining/drywall
9. Thermal insulation in cavity
10. Air cavity (sometimes omitted where insulation fills cavity)
11. Inner precast concrete lintel
12. Outer precast concrete lintel
13. Pressed steel lintel or stainless steel angle
14. Drainage groove
15. Seal
16. Cill extends beyond opening to throw water clear of edge to prevent local staining of wall beneath.
17. Stainless steel tie
18. Damp proof course (DPC)
Vertical section

36

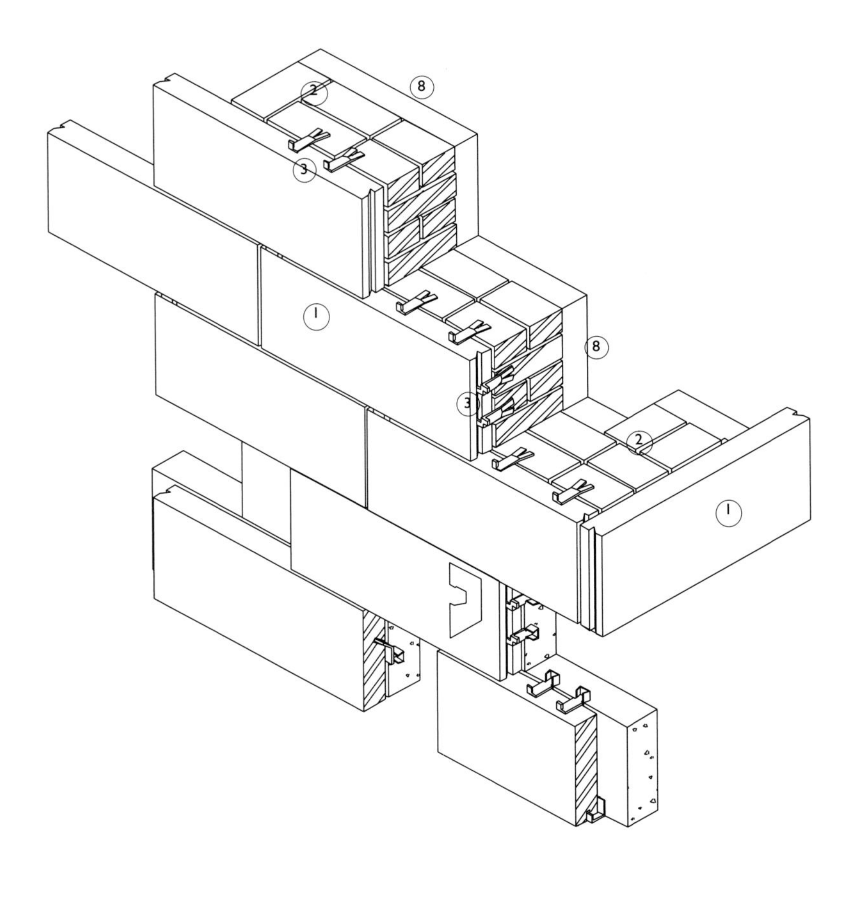


Vertical section

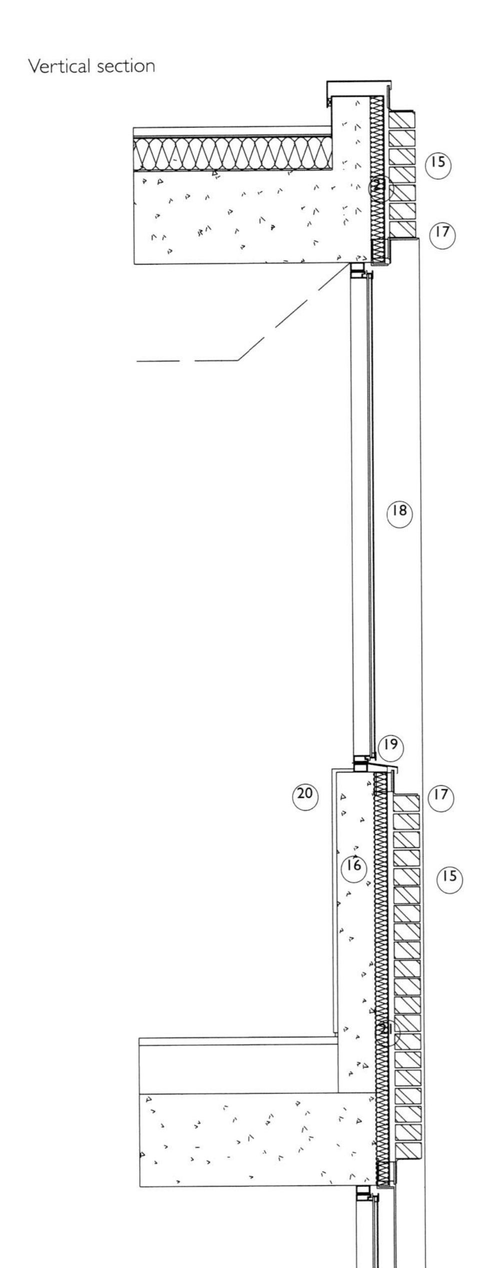


Masonry can be used as part of a cladding system with sealed joints where each piece of masonry is fixed back to a backing wall. Stone panels and prefabricated brick panels are typically used. Panels are supported on brackets fixed either back to a metal frame or to a backing wall. Joints between panels can be open, as in rainscreen construction, or closed.

In the case of stone, the material can be considerably thinner, as little as 30mm (1¼in) if it is fixed to a backing wall rather than built as a cavity wall facing, which makes it more economical as well. The stone is fixed back to the supporting wall or frame using a system of steel cramps or brackets; these are fixed into concealed slots, or rebates, in each stone panel. Because the stone is supporting only its own weight, the design of joints can follow traditional patterns of jointing such as stretcher bond, or it can express its non-structural role with a different arrangement of joints.

Masonry cladding struggles to meet the demands of energy conservation, since the further away the external cladding is from the supporting wall to accommodate a layer of thermal insulation, the more complex and expensive becomes the fixing. The increased dependency on the thermal insulation makes it difficult to ensure that the insulation material will not deteriorate from any leaks in the external wall. Sealed systems have responded to this challenge by improving the drainage ability of the supporting framework. Any rainwater that penetrates the outer seal is directed down the cladding rather than onto the insulation layer. Grooves are introduced into the edges of each stone panel to help prevent any water from reaching the back of the panels.

## Stone details

1. Stone panel
2. Backing wall, typically concrete block
3. Stainless steel fixings (a wide range is available)
4. Timber framed window
5. Stone cill
6. Timber inner cill
7. Internal plaster finish or dry lining/drywall
8. Thermal insulation in cavity
9. Thermal insulation set on internal face of backing wall
10. Air cavity
11. Precast concrete lintel
12. Damp proof course (DPC)
13. Seal
14. Stone coping

## Brick details

15. Prestressed brick panel
16. Backing wall, typically reinforced concrete
17. Stainless steel angles
18. Window
19. Pressed steel or extruded aluminium cill
20. Internal plaster finish or dry lining/drywall
21. Thermal insulation in cavity
22. Air cavity
23. Seal

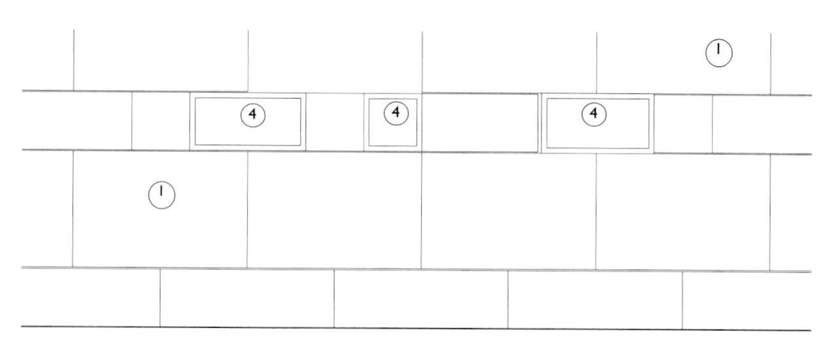

Part elevation

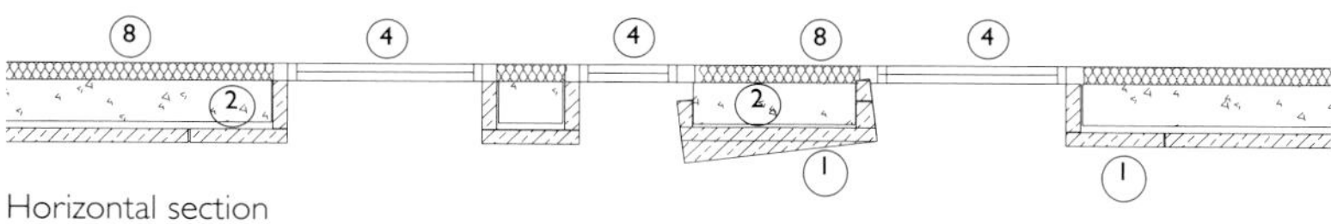

Horizontal section

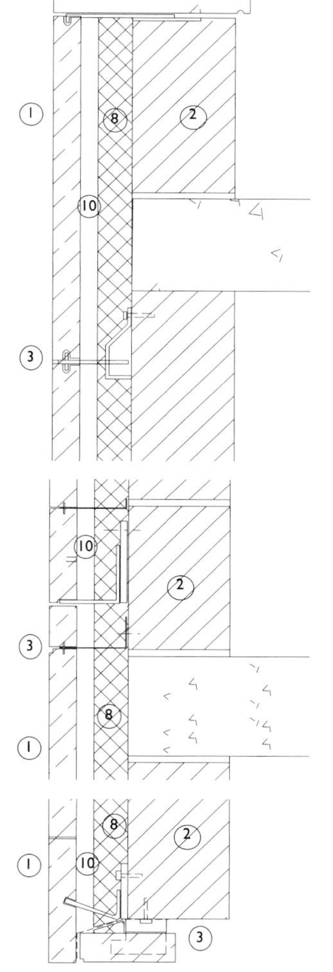

Vertical section

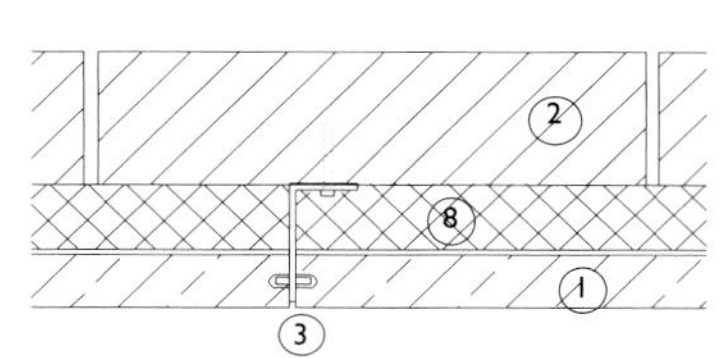

Horizontal section

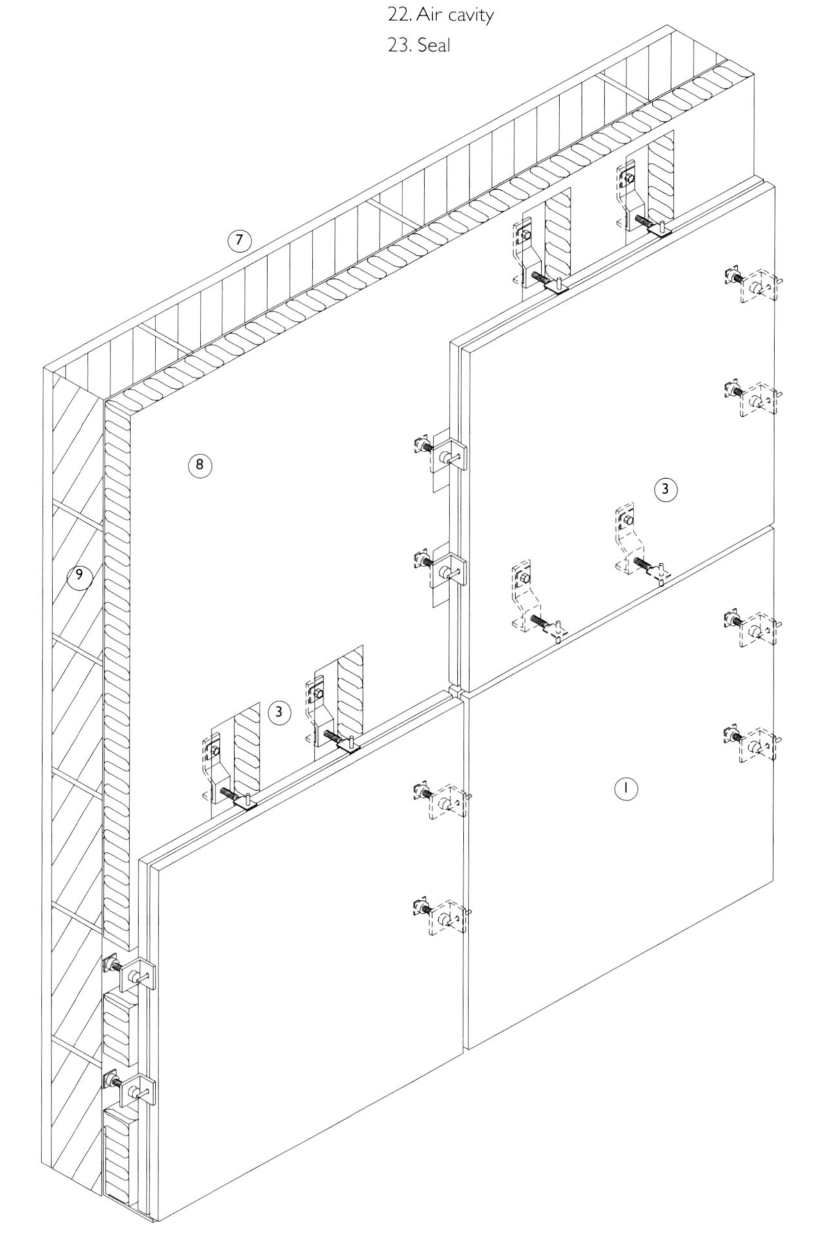

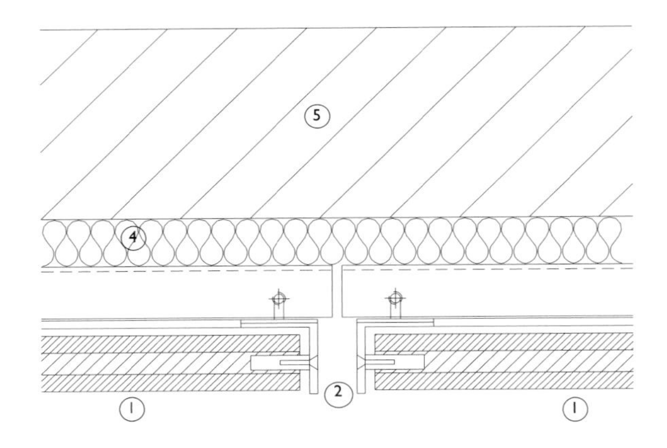

Horizontal section

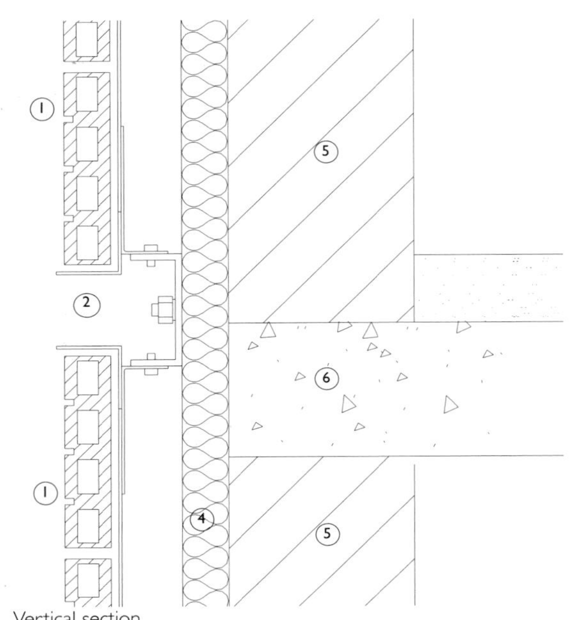


Vertical section

82A

82B

Masonry rainscreen cladding consists of masonry panels mounted in front of a supporting wall that typically consists of a backing wall fixed to a structural frame. The outer screen drains most of the rainwater and shields the backing wall behind. The small amount of rain that passes through, or behind the screen, is drained away by the real waterproof layer on the backing wall behind. The void between the two layers is ventilated to the outside air. The backing wall provides the necessary thermal and acoustic insulation.

The outer layer can also be used as a means of screening a more economical waterproofing material behind, which otherwise would not be suitable for use as wall cladding. It can be used to screen anything from a blockwork wall to a PVC membrane on a supporting surface. Gaps can be left between outer screens, since no seal between outer panels is required.

A system for a terracotta rainscreen has been developed and patented by the Renzo Piano Building Workshop. Similar types have been developed by other manufacturers. The Piano system consists of a series of extruded terracotta panels mounted in a steel frame. The frame is then fitted to the facade protecting the building's waterproof layer. Its use permits brick to be used in demountable units each of which has grooves cut into its face to resemble four separate bricks. The units are fitted into place by means of a series of metal pegs. The equal location of each unit within the frame is ensured by a series of cruciform spacers.

82D

82C

Vertical section

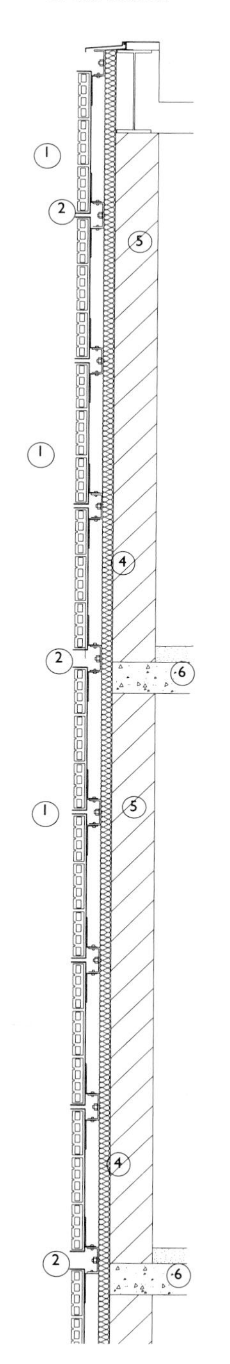


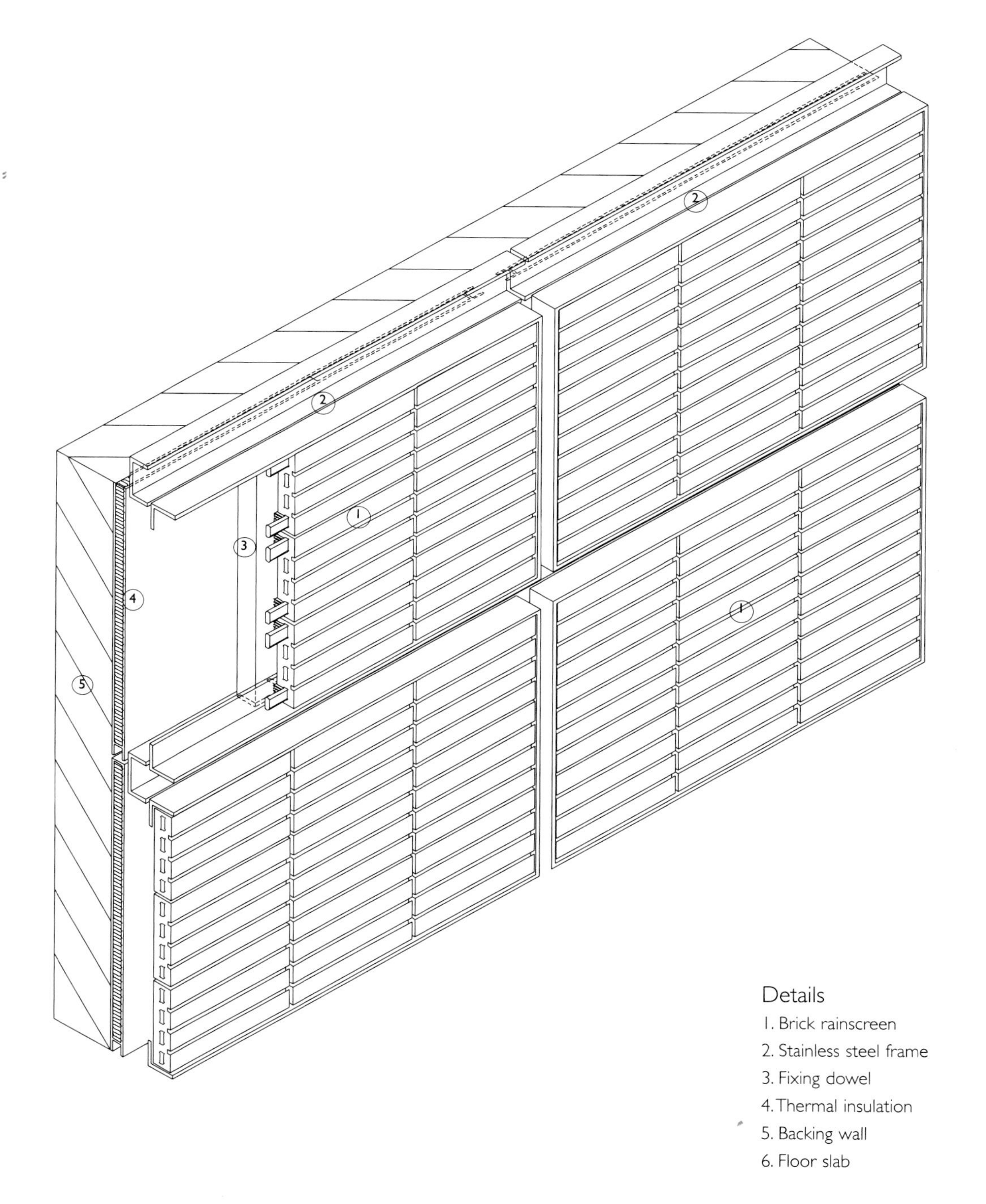


## Details

1. Brick rainscreen
2. Stainless steel frame
3. Fixing dowel
4. Thermal insulation
5. Backing wall
6. Floor slab

38A

Plastic-based cladding is used where either lightweight construction is required or when its high strength, but relative lack of stiffness, can be used to advantage. There are currently problems with long-term fading of the colour finish, particularly the darker shades. For this reason, painted or natural finish metals are often the preferred materials for such applications.

## GRP cladding panels in stick systems

GRP is used as a facing material to cladding panels. Due to the lack of rigidity of GRP, the panels are constructed as composite, or 'sandwich', panels with integral thermal insulation. These panels consist of outer layers of GRP integrally bonded with a rigid core of thermal insulation. The inner and outer faces of the panel provide an integral finish. Polystyrene sheet or polyurethane foam is commonly used to form the insulation core. Whilst heavier insulation helps to make a more rigid panel, the lighter, less rigid types provide better thermal insulation. As a result, the type chosen is a balance between these two requirements. Ribs are sometimes introduced to improve panel rigidity.

GRP has limited fire resistance, but this quality can be enhanced by the addition of additives to the polyester resin during manufacture. GRP panels have a good waterproof self-finish that does not require any further coating and weathers well. A disadvantage is the fading of the finish surface due to ultraviolet (UV) rays. UV stabilisers can be added to the polyester resin but fading will still occur.

GRP panels can be used as part of a glazing-based stick system. Joints between panels resemble those used in metal or glass cladding. Joints in the system must be designed to allow thermal and structural movement, eliminating de-lamination, where the facing becomes separated from the core of insulation.

## Polycarbonate insulated systems

Polycarbonate sheet, in the form of flat sheet, twin wall section and profiled section, has been in constant development over the past few decades. This material provides a translucent, highly insulating and economic cladding system for use in a wide range of building types. Different systems exist, some as proprietary prefabricated systems, others as configurations designed by the design team for a specific project. All types share a common principle of two outer skins of polycarbonate with an inner core of translucent thermal insulation material. Unlike composite metal panels, a support framework of aluminium sections supports the two skins and creates a void. This provides a unitised system that can be assembled inexpensively on site and make use of non-rigid insulation materials such as quilt or loose fibre. A recent development has been the introduction of silica as an insulation material.

Polycarbonate panels with a layer of around 100mm translucent insulation have made a significant impact on cladding. These high insulating, translucent qualities have been used in a range of industrial buildings and storage facilities. Profiled polycarbonate sheet is a relatively inexpensive material that is undergoing a revival in both large and small-scale buildings. Its striking use at the Congrexpo building at Lille, France, has led to a raised awareness of the possibilities of this new material. A series of public buildings in Holland has further explored this material. Its hazy, translucent qualities allow a simple concealed support frame to be used to carry the sheeting.

38B

38C

Details
1. Twin wall polycarbonate sheet
2. Proprietary fixings
3. Supporting structure
4. GRP cladding panel
5. Stick support system

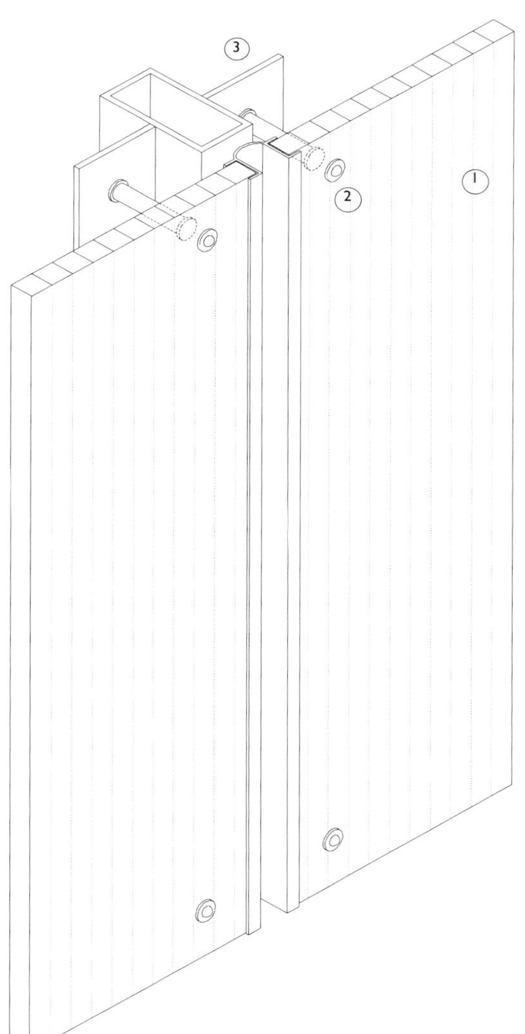


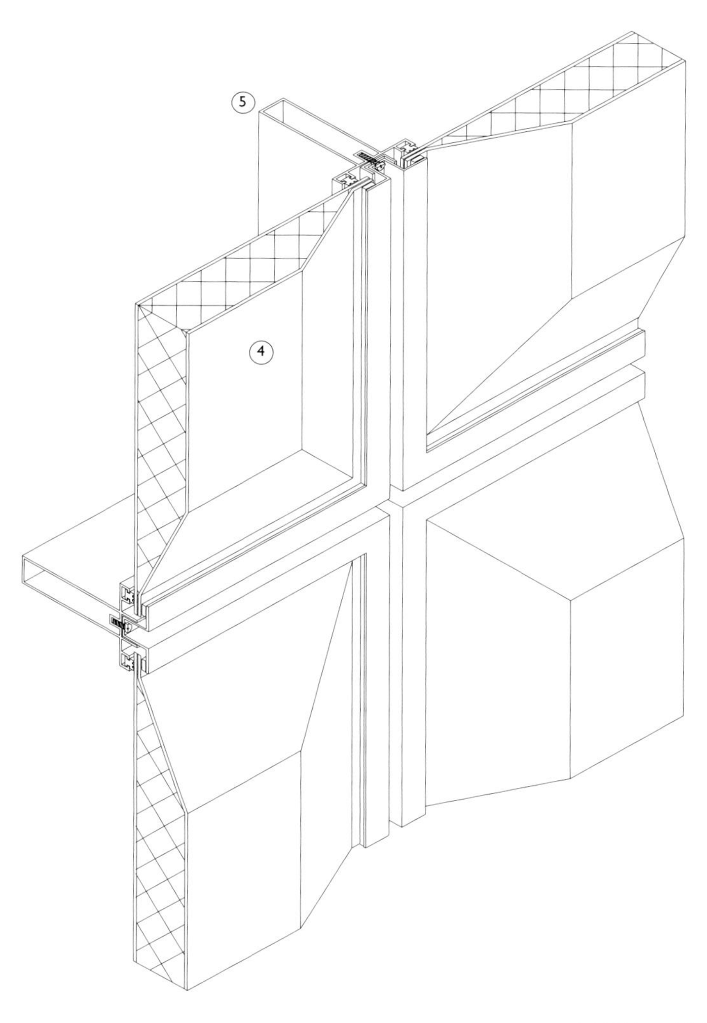

48

Vertical section

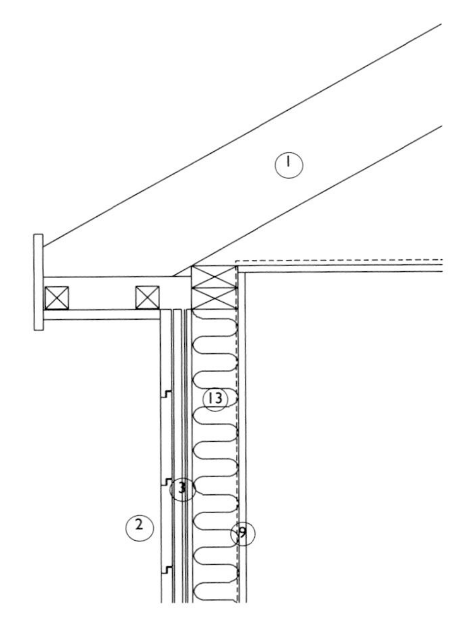


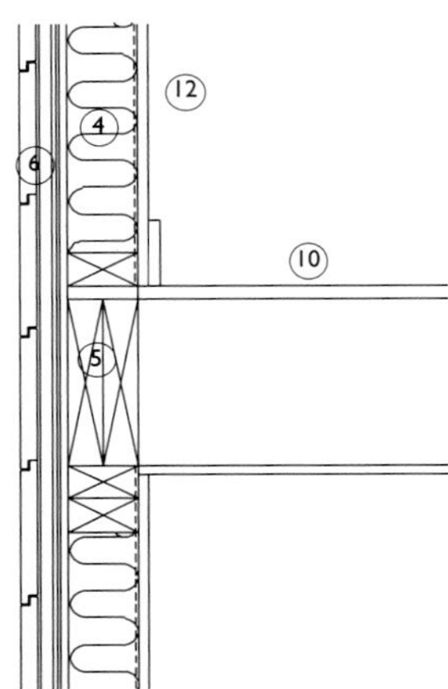


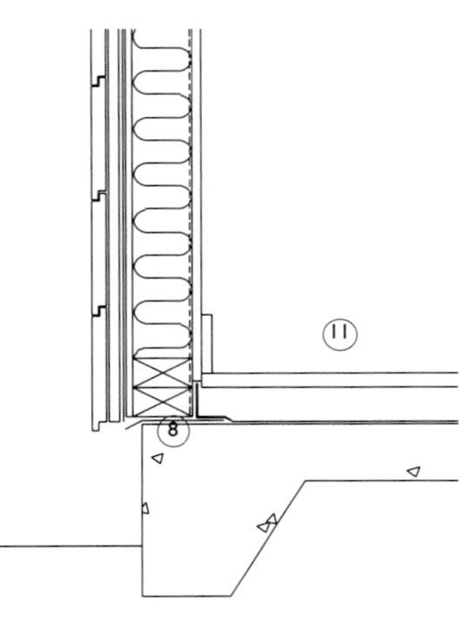


## Wall structure

The walls of the platform frame consist of timber studs (verticals), which are usually set at 400mm (16in) centres. 50x100 mm (nominal 2" by 4") sections are typically used. In addition to being an efficient structural solution, these centres are a multiple of 1200 mm (4ft) wide plywood or lining (drywall) sheets. Studs are arranged vertically from floor-to-floor between timber plates (horizontals). Floor joists are aligned with the studs to transfer the loads efficiently from floor to wall. Timbers are butt jointed and nailed. These joints are not rigid and, as a result, the frame requires lateral bracing. This is commonly done by cladding the outside face of the frame in sheathing grade plywood sheet approximately 10mm (3/8in) thick. In addition to providing lateral rigidity, sheathing gives a stable background for the fixing of the breather membrane and flashings. A more traditional method is to set diagonal bracing into the framework and clad the outside face of the wall with tongue-and-groove timber boards.

## Insulation and waterproofing

Insulation quilt is set into the void between the studs to provide adequate thermal insulation. The inside face of the frame is lined with an internal board such as plasterboard. A vapour-resistant sheet, usually polythene, is fixed on the inside face of the studs. The purpose of the vapour check is to stop the passage of water vapour, generated inside the building, from passing into the depth of the external wall where it might condense and cause damage. The external sheathing is lined with a breather membrane to provide a back-up to the external cladding when complete. It sheds any rainwater that either penetrates or passes behind the timber boarding on the outside. The membrane is also permeable to allow any moisture trapped inside the timber wall to escape to the outside. An alternative to fixing the external boards directly is to fix them on vertical battens that are in turn fixed through the breather paper to the sheathing. This rainscreen method allows all surfaces of the timber to dry thoroughly. Exposed ends of timber boards are protected with trims against checking and splitting through constant exposure to wind and rain.

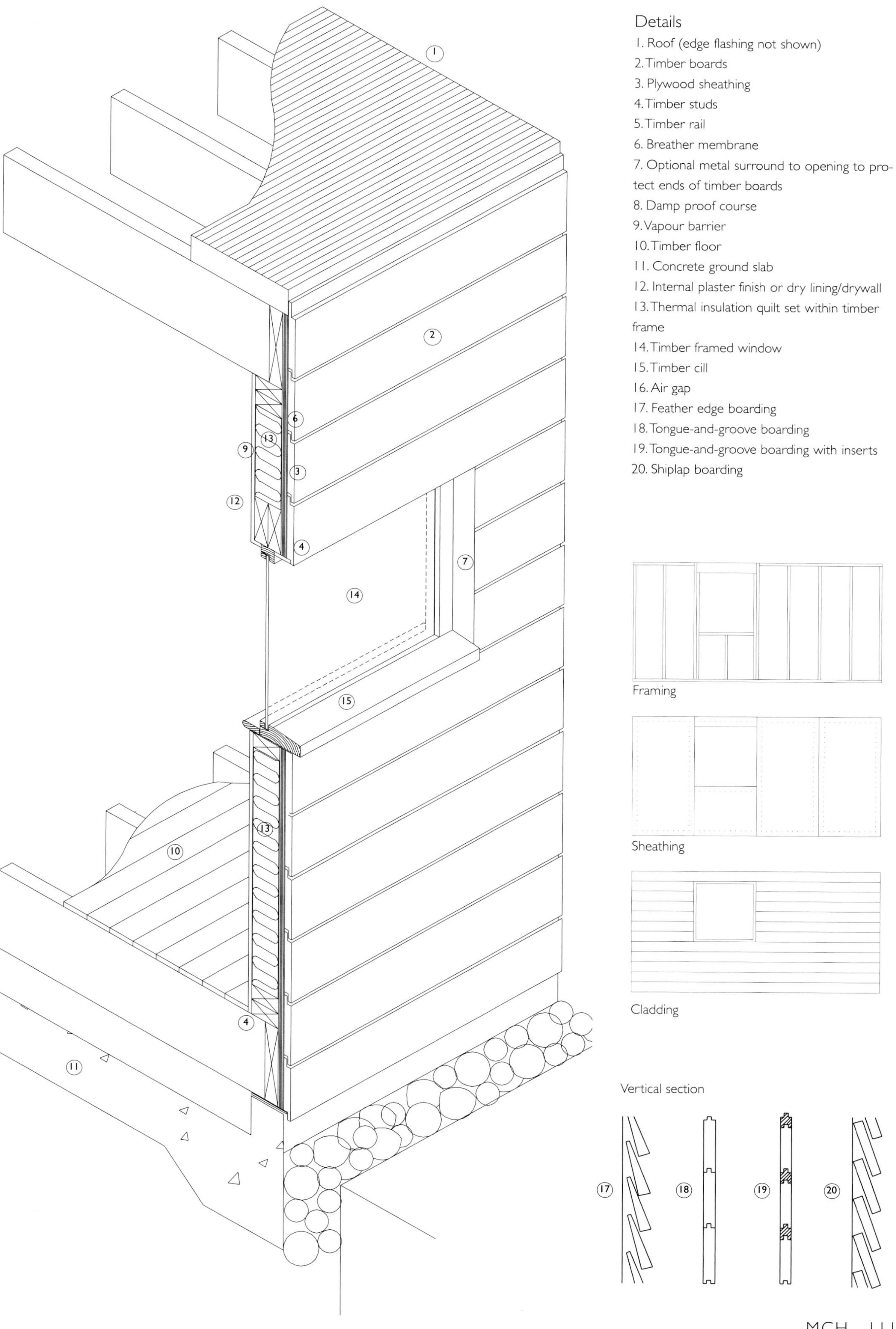
Details
1. Roof (edge flashing not shown)
2. Timber boards
3. Plywood sheathing
4. Timber studs
5. Timber rail
6. Breather membrane
7. Optional metal surround to opening to protect ends of timber boards
8. Damp proof course
9. Vapour barrier
10. Timber floor
11. Concrete ground slab
12. Internal plaster finish or dry lining/drywall
13. Thermal insulation quilt set within timber frame
14. Timber framed window
15. Timber cill
16. Air gap
17. Feather edge boarding
18. Tongue-and-groove boarding
19. Tongue-and-groove boarding with inserts
20. Shiplap boarding
Framing
Sheathing
Cladding
Vertical section

11A

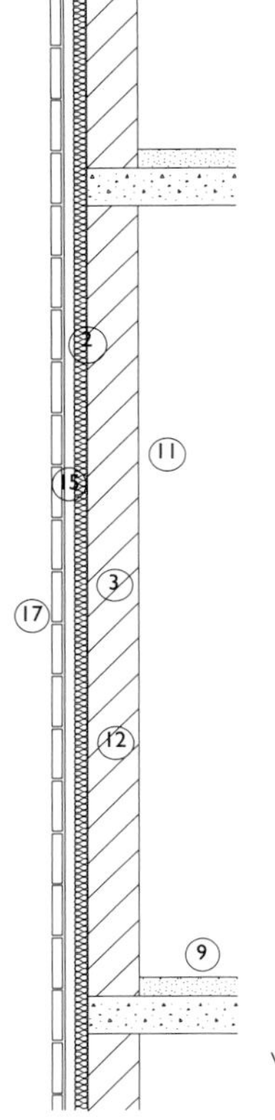


Vertical section

11B

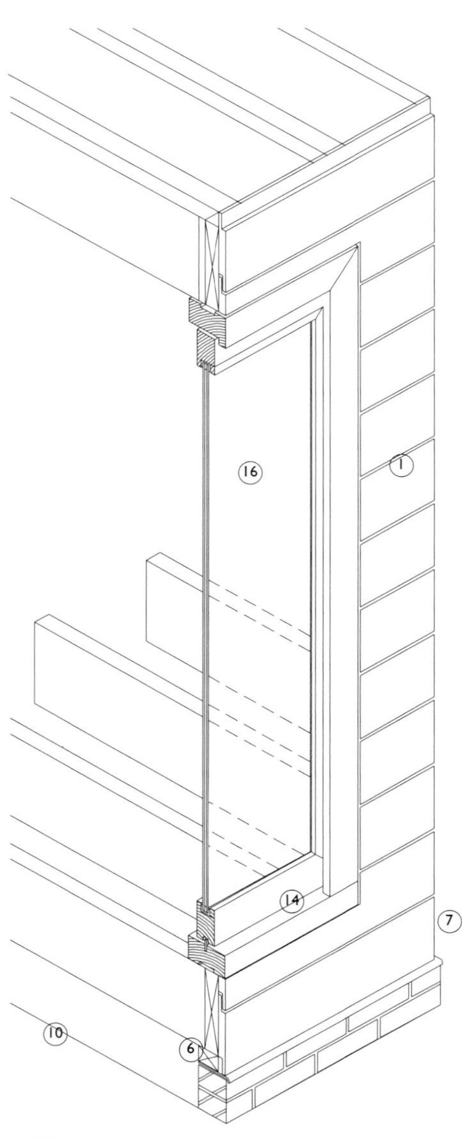


In North America, wood framed buildings are commonly 'stick' built on site as platform frames. In European countries, most timber-framed buildings are built using partial factory prefabrication. Stud wall panels and roof trusses are prefabricated, whilst floors, together with finishes, are typically made on site. In Japan, completely prefabricated house units are delivered to site in sections and lifted into position by crane. Timber cladding panels, that is non-loadbearing panels, are used to enclose building frames, typically in concrete, steel or timber. The construction of panels follows the same principles as walls of platform frames. These have the advantage of being prefabricated off-site where finishes can be closely controlled. Panels may have additional bracing, either as cross-ribbing (cross bracing) or as steel brackets, which allow panels to be transported to site and lifted into place.

Timber cladding has undergone considerable change in recent years. Tongue-and-groove or shiplapped boards were originally used as both sheathing and weather protection to the platform frame. For some time now, this role has been superseded by the use of plywood sheets as sheathing to timber frames. The boarding is then applied to the outside face of the ply as a weatherproof finish with a membrane behind. A layer of breather membrane is set between the timber and the ply to provide this weatherproof layer. The fixing of timber directly to the face of the membrane resulted in problems of ventilating the back surface of the timber boarding. It is now common to fix the timber boarding on battens to allow the back of the cladding to be ventilated. Ends of boards are butted together to protect the endgrain, which is susceptible to deterioration if left exposed. Since the timber no longer has a structural role, its use has become that of a semi-rainscreen. Recent applications have developed its use to that of a full rainscreen. The issue of exposing the endgrain has been resolved either by using a resilient hardwood or by

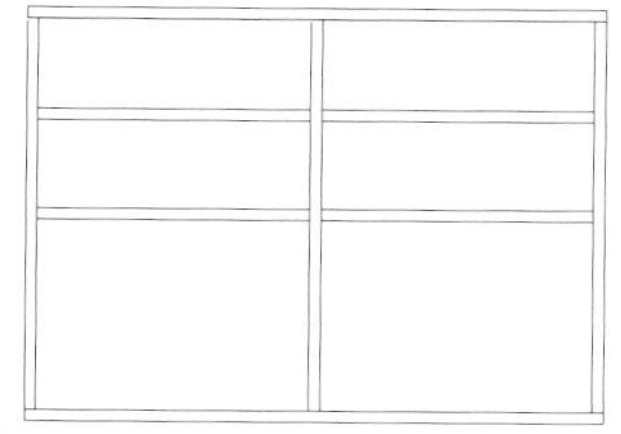

Framing

Sheathing

Cladding

## Details

1. Timber boards
2. Plywood sheathing
3. Timber studs
4. Timber rail
5. Breather membrane
6. Optional metal surround to opening to protect ends of timber boards
7. Damp proof course
8. Vapour barrier
9. Timber floor
10. Concrete ground slab
11. Internal plaster finish or dry lining/drywall
12. Thermal insulation quilt set within timber frame
13. Timber framed window
14. Timber cill
15. Air gap
16. Window or door panel
17. Rainscreen boards in durable timber

34A

inserting the timber boards into a metal frame that protects the end-grain. The adoption of the rainscreen principle for timber allows both faces of individual boards to be ventilated rather than by a single ventilation slot at the top and bottom of the area of cladding to ventilate an entire wall.

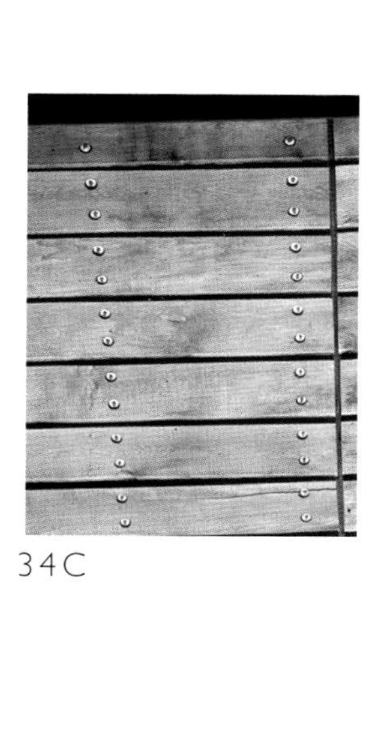

34C

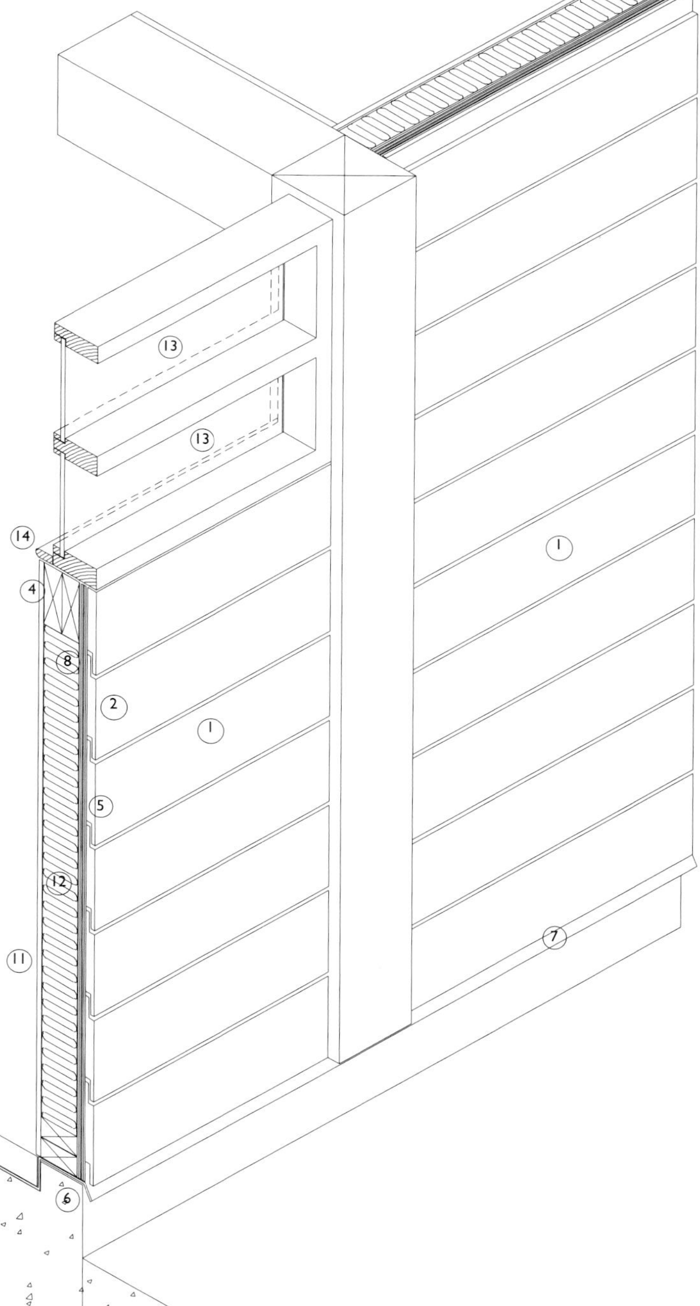


## Rainscreens

The use of timber has increased dramatically in recent years with the emergence of rainscreen construction. This method allows timber to be used free of the constructional constraints associated with using it as a weatherproof skin. Open joints between boards, curved surfaces and the omission of coverstrips in rainscreens have given timber a different, crisper appearance. The use of timber as cladding to the platform frame remains a single generic type. The rainscreen also has a single generic form of construction.

# 3

## ROOFS

112D

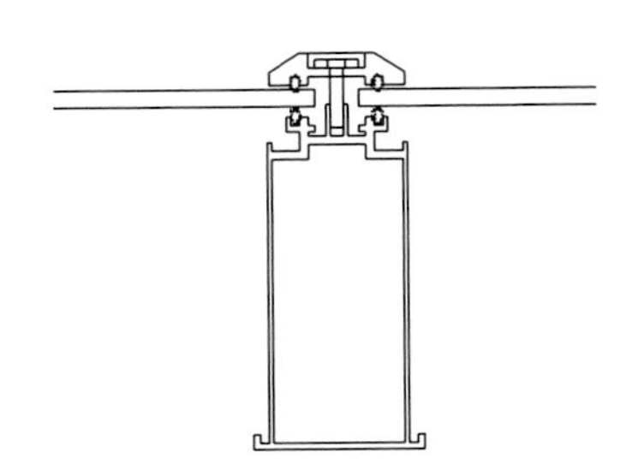

Vertical section.
Rooflight with single glazing.
Double glazing is becoming the norm.

## Increased treatment of roof as 'non accessible' facade

There is a slow but steady increase in the cladding of roofs to form a complete building envelope. This is part of a trend to bring the roof into the 'visible' design of the building, and create a fifth facade as it were, rather than relegating the top of a building to a concealed space or a traditionally-based inclined element. Generally, though, roofs are distinct from wall cladding in their detailing, even though very similar techniques are used for glazed walls and glazed roofs. As a result, cladding techniques are being adapted for use on roofs, particularly where the roof becomes a visual continuation of the wall. However, the drainage of rainwater from roofs is fundamentally different from walls in that some roof surfaces are often submerged beneath water during rain. Where water drains down facades and can be directed down joints, roofs often remain submerged beneath water for long periods. For this reason, the use of single membranes has increased.

When used in conjunction with single membranes, rainscreen panels provide a new economic method of enhancing surface appearance and making the roof visually a part of the design, especially on non-accessible roofs. Fully supported sheet metal roofing is an alternative to rainscreen panels, with its characteristic continuous standing seams that can create unusual roofscapes with their ability to cover complex shapes.

Bolt fixed glazing is being used as a roof glazing system, also in conjunction with solar shading, but often shading is used internally, where it is protected from the effects of weather, thus allowing the bolt fixed glazing to provide a smooth external roof surface.

## Increased use of flat roofs as usable external space.

Single membranes can be either exposed as a roof finish, or concealed beneath a covering in an 'inverted' configuration in flat roofs, where insulation is set above the waterproofing layer, with another material, such as paving, which protects both the insulation and the waterproofing layer beneath. This has led to the new use of a large number of different finishes on flat roofs from timber to concrete to metal. A durable walking surface can be provided without use of the inverted roof configuration. The introduction of plastic supports which resemble those used to support simpler types of raised floors used in office buildings, allow wearing materials such as concrete paving, steel grating and slatted timber panels to be supported by a waterproof membrane without the need for fixing through the material, which would present an additional risk of water penetration. The supports also serve as spacers to set out open joints between panels forming the wearing surface, allowing rainwater to pass through open joints between the paving and drain away beneath. They provide a walking surface that protects the membrane beneath.

Planted roofs have developed both as accessible gardens and as 'visible' roofs that are non-accessible. When accessible, the roofs usually have larger plants, which require more soil. The water required for irrigation can increasingly be provided by rainwater, which is collected and stored, partly in tanks and partly in the roof itself. The blocks of rigid thermal insulation, when set on top of the waterproof membrane, can be shaped to hold or drain water as required by the irrigation.

The use of glazed walkways and stair treads externally has led to the development of proprietary systems for rooflights that can be walked across. These are laid to a modest fall of around 3° and comprise laminated glass sheets supported on all edges by a metal support frame, usually steel. The glass can be treated with a range of fritted or silk-screened finishes both partially to obscure views through and control solar gain and glare.

## Generic roof types

Until recently, roofs were described as being generically 'flat' or 'pitched'. Pitched roofs used mainly traditional techniques of lapping materials such as tiles, slates or metal sheet, to allow rainwater to run over the lap joints. Flat roofs had very shallow pitches that were covered

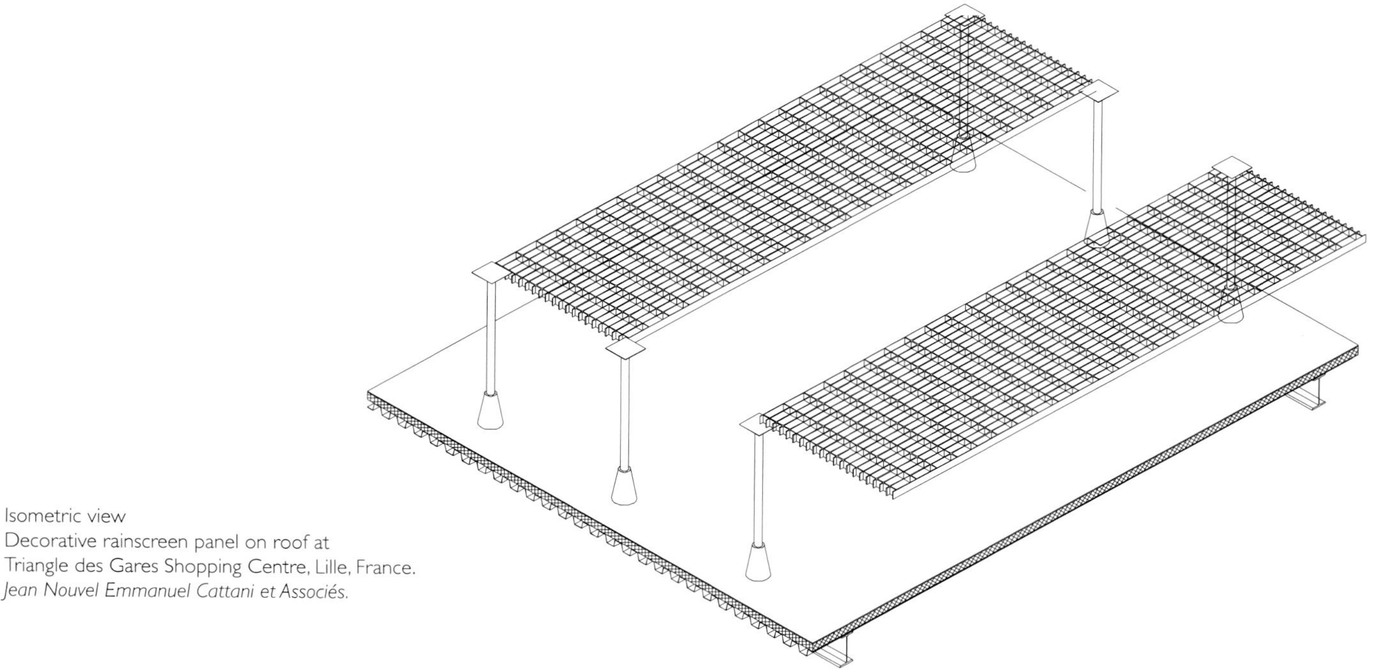

Isometric view
Decorative rainscreen panel on roof at Triangle des Gares Shopping Centre, Lille, France.
*Jean Nouvel Emmanuel Cattani et Associés.*

mostly in a continuous single material. This classification is becoming ever more blurred as roofing materials can increasingly be used in either flat or pitched configurations. The increased use of inverted roofs to form both accessible decks, and rainscreens providing 'visible' roofs, has resulted in an ever-increasing mix of substrates and finishes. There is now very little correlation between substrates and finishes. Most types of finish can be adapted to suit a particular substrate. Some materials that form a combined substrate and finish can be said to be generic roof 'types', including glazed, fabric and composite panel types. Others have combined substrate and supporting substrate such as glazed roofs and composite metal panels. The main generic types of roof finish used are as follows:

1. Membranes: sheet metal, elastomeric, asphalt.
2. Rainscreens: slates, tiles, shingles, metal panels.
3. Planted roofs: heavy vegetation, light vegetation.

The most commonly used generic roof substrates are as follows:
1. Metal 1: Profiled decking
2. Metal 2: Composite panels
3. Glass 1: Adapted glazed walls
4. Glass 2: Bolt fixed glazing
5. Concrete 1: Concealed membrane
6. Concrete 2: Exposed membrane
7. Timber 1: Flat roof
8. Timber 2: Pitched roof
9. Plastics: Polycarbonate and acrylic
10. Fabric 1: Tensile supported
11. Fabric 2: Air supported

The issue of ventilation within roof build-ups is still important, particularly in timber, with the emphasis on providing increased thermal insulation to reduce energy consumption within buildings. The specific issues of ventilation are discussed at the beginning of the Walls chapter. Even these generic types are being mixed as part of a bigger roof build-up where different roof types are combined to form a 'layered' roof.

## Layered roofs

Where a roof build-up may vary across its surface, either as a result of differing design requirements or as a result of budget constraints, this often results in different roof types being used together, for different functions. A recent high-profile example is the Triangle des Gares Shopping Centre in Lille, France. Metal mesh decking, normally used for walkways decking, is used primarily as a visual screen to conceal roof mounted mechanical plant for mechanical ventilation, economic rooflights and smoke extractors. It provides limited wind protection to the roof beneath. Panels are set approximately 1.5 metres (5ft) above the roof deck to provide access for the regular maintenance of equipment, as well as clearing wind-blown debris, without disturbing the rainscreen. The outer layer of this highly visible roof is reflected inside the building with the use of a suspended ceiling in the same material. Using the same metal mesh both inside and outside makes the roof appear to be made from a single material. This 'virtual mesh roof' allows the real decking, insulation and waterproofing layer to be concealed within the construction. During the day, natural daylight is diffused through the mesh suspended ceiling without the low cost rooflights being visible. At night, lighting within the building is reflected off the outer rainscreen mesh to create a glowing roofscape.

Another recent example, this time of a fully glazed roof, is an office building at Bedfont Lakes in England. The roof structure is formed from steel trusses from which is suspended a bolt fixed glazed roof. A layer of fabric-based solar shading is fixed to the top of the trusses. This configuration forms a layered build-up where functions of weatherproofing, structural support and solar control are separated. With ever more stringent design requirements for water-tightness, solar control and thermal insulation with modest budget constraints, the use of layered roofs for opaque, translucent and transparent applications is set to increase significantly.

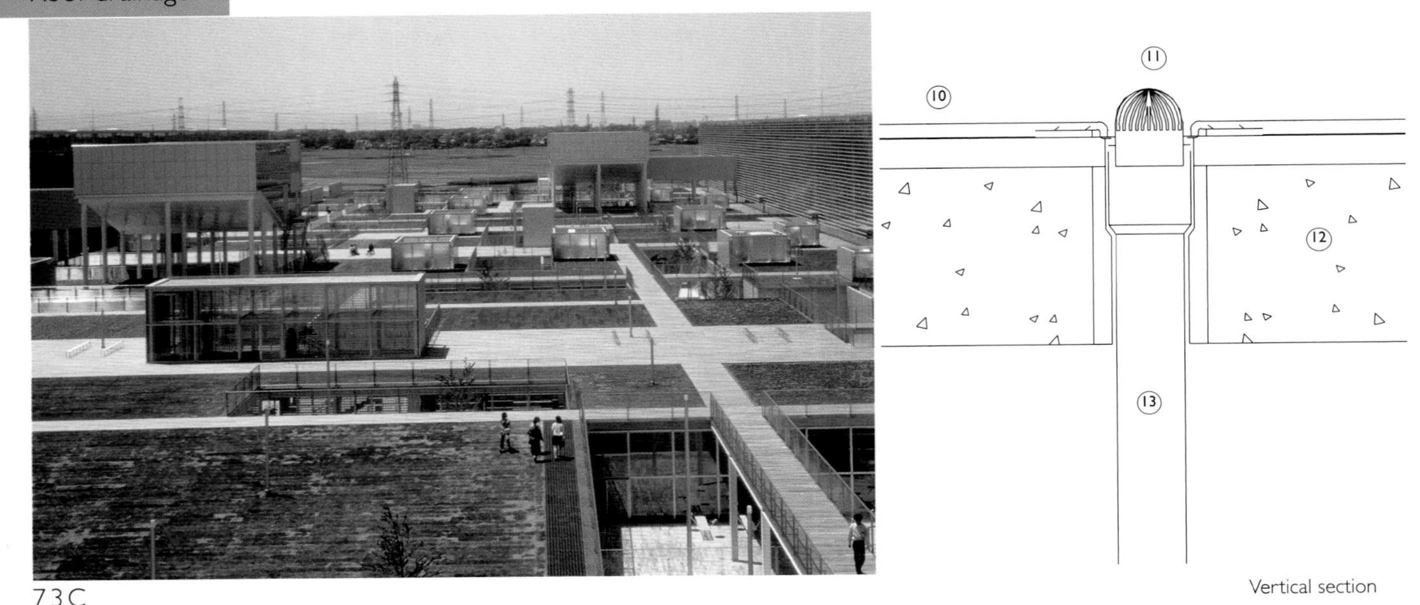


73C

Vertical section

Drainage systems are designed to remove rainwater from roofs by directing water to rainwater outlets in a controlled manner. Typically, water is directed into gutters or downspouts directly to an outlet. The traditional gravity system relies on a sufficient slope across a roof for the rainwater to travel across it and into down-pipes. The recent introduction of the siphonic system has eliminated the need for gutters. Instead, it relies on rainwater outlets connected to small diameter tubes that literally suck the water off the roof. Both gravity and siphonic systems are used for pitched roofs and flat roofs with shallow slopes as low as 3 degrees. The siphonic system can also be used with essentially 'flat' areas of roof without slopes, but larger structural deflections due to the load from standing water should be taken into account. In practice, flat roofs are designed with falls that mimic the geometry of pitched roof systems

## Gravity drainage

Gravity systems drain away rain-water at outlets located either within a gutter or through a roof drain located at the lowest point of a flat roof. Gutters are laid to a minimum fall of around 3 degrees and down-pipes are sized to be large enough to cope with the conditions that can be predicted . Similarly, the number of outlets must be able to cope with the runoff during the worst storm conditions.

## Siphonic drainage

The siphonic system was developed to drain large roofs without slopes in the gutter. Instead of sloping gutters, flat channels are provided which contain rainwater outlets that are connected together beneath the roof by a horizontal collector pipe, (sized according to need). From the collector pipe, water runs first into a single vertical downpipe and then directly into a belowground drainage system. Downpipes can be of small diameter, down to around 25mm (1 in.), but can be as large as 200mm (8 in.) for a large roof in conditions where heavy rainfall is expected.

Siphonic systems utilize a gravity-induced vacuum process. Water is drained away by filling the downpipes with rainwater that backs up from the base of the downpipe, creating a vacuum that then draws the water off by siphonage. Rainwater is literally sucked down from the roof. An advantage of the siphonic system over the gravity system is that the number and size of downpipes is drastically reduced. A disadvantage is that the collector pipe is set beneath the roof deck, usually within the building, which may not suit the interior, particularly if an exposed roof deck is desired. In addition, when the pipe is filled with rainwater, it increases the possibility of leaks. Conversely, if exposed, it may be easier to maintain than external gutter systems.

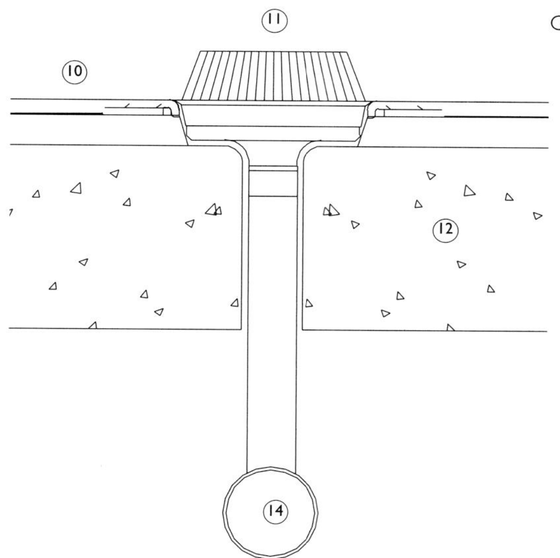

Vertical section

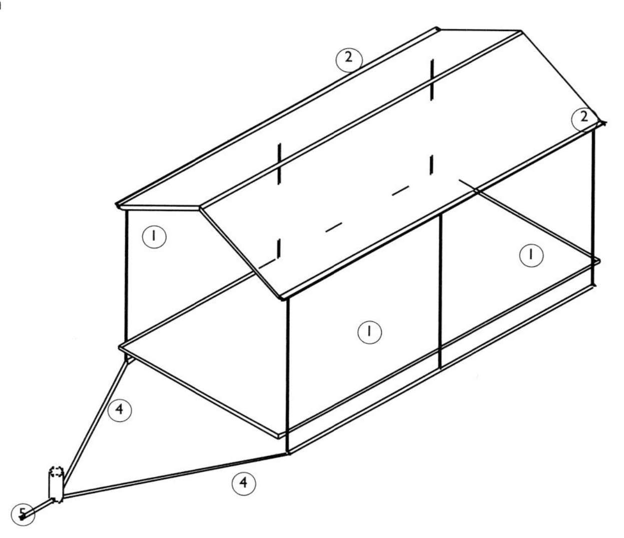

Gravity system

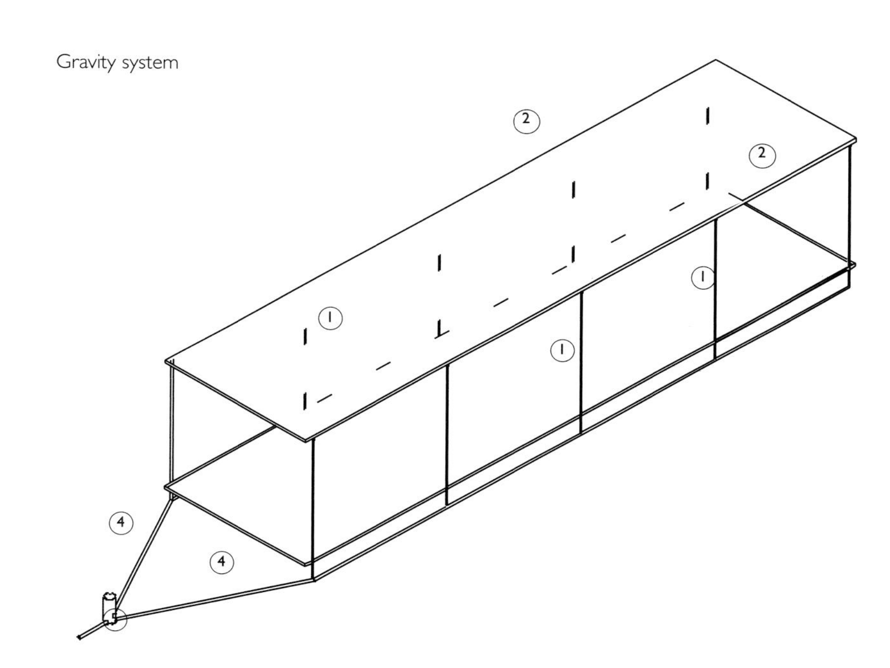

Gravity system

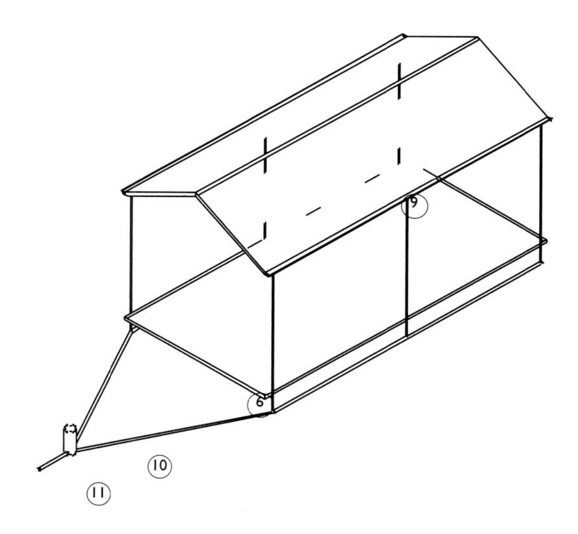

Siphonic system

## Isometric view of gravity system

1. Downpipe
2. Gutter
3. Gravity type rainwater outlet
4. Pipe below ground
5. Inspection chamber

## Isometric view of siphonic system

6. Downpipe
7. Gutter
8. Siphonic type rainwater outlet
9. Collector pipe
10. Pipe below ground
11. Inspection chamber

## Rainwater outlet details

12. Waterproofing layer
13. Outlet
14. Roof deck
15. Downpipe
16. Collector pipe

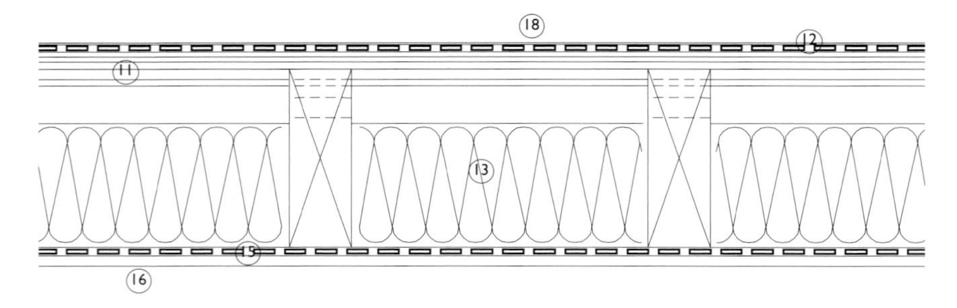


Vertical section

7B

The most common materials used are painted steel, painted aluminium, galvanised steel, lead, copper, zinc and stainless steel. Detailing of roofs in most common types of metal are very similar, with the exception of zinc, which requires ventilation on the underside to avoid corrosion occurring within the construction.

Sheet metal fixed to an unprotected surface is the traditional method of using this medium as a waterproof layer and external finish. It is used with a substrate or structural deck in any of the typical forms of construction in concrete, steel and timber. Sheets are laid onto a supporting surface such as plywood, usually incorporating an insulating layer and which is itself fixed back to the supporting structure. It is often used to weatherproof a material that provides excellent sculptural qualities, such as timber.

Sheet metal roofs are a site-intensive form of construction. Metal is rolled out in place with the edges of adjoining sheets lipped to form a

4B

Isometric view of sheet metal assembly

1. Metal sheet
2. Standing seam joint
3. Breather membrane
4. Thermal insulation
5. Substrate, typically timber/metal rafters with plywood facing
6. Vapour barrier
7. Drywall/dry lining if required

Details

8. Metal sheet 1
9. Metal sheet 2
10. Clips at centres
11. Plywood substrate
12. Breather membrane
13. Thermal insulation
14. Substrate
15. Vapour barrier
16. Drywall/dry lining if required
17. Standing seam joint
18. Metal sheet

85B

Vertical section

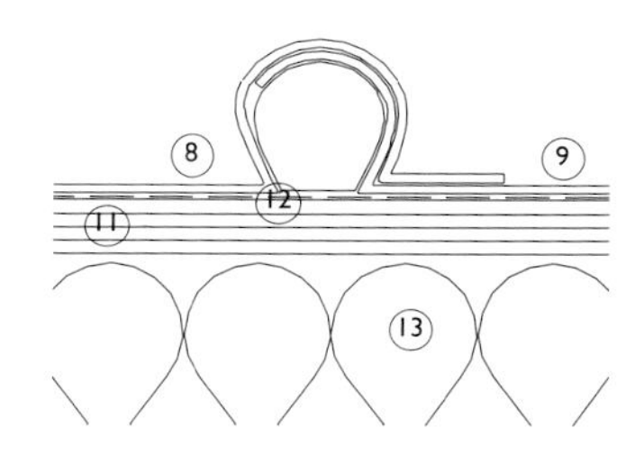


Vertical section

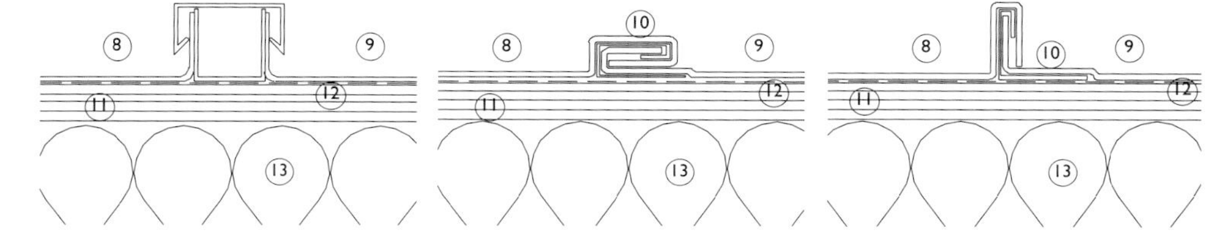


standing seam. A metal or timber roll is often used if the operation is done by hand. The roll helps both to form and align the joint. Larger roofs are installed mechanically. Metal sheets are rolled through a machine that sets them in place. A second machine rolls across the roof forming a continuous standing seam.

The metal covering must allow for thermal movement. A layer of building paper or felt, laid between the metal and the substrate, allows the metal skin to move independently of the supporting structure. Rolls, drips and welts are used to accommodate thermal expansion.

Rolls are laid parallel to the slope of the roof and run from top to bottom. The top is covered with a metal capping strip of the same material, held in place by cleats. Drips are positioned across roofs where the pitch is below 15°. The joint can be welted or beaded. Where the roof pitch is greater than 15 degrees, metal sheets are welted together in a flattened joint and held in place by cleats.

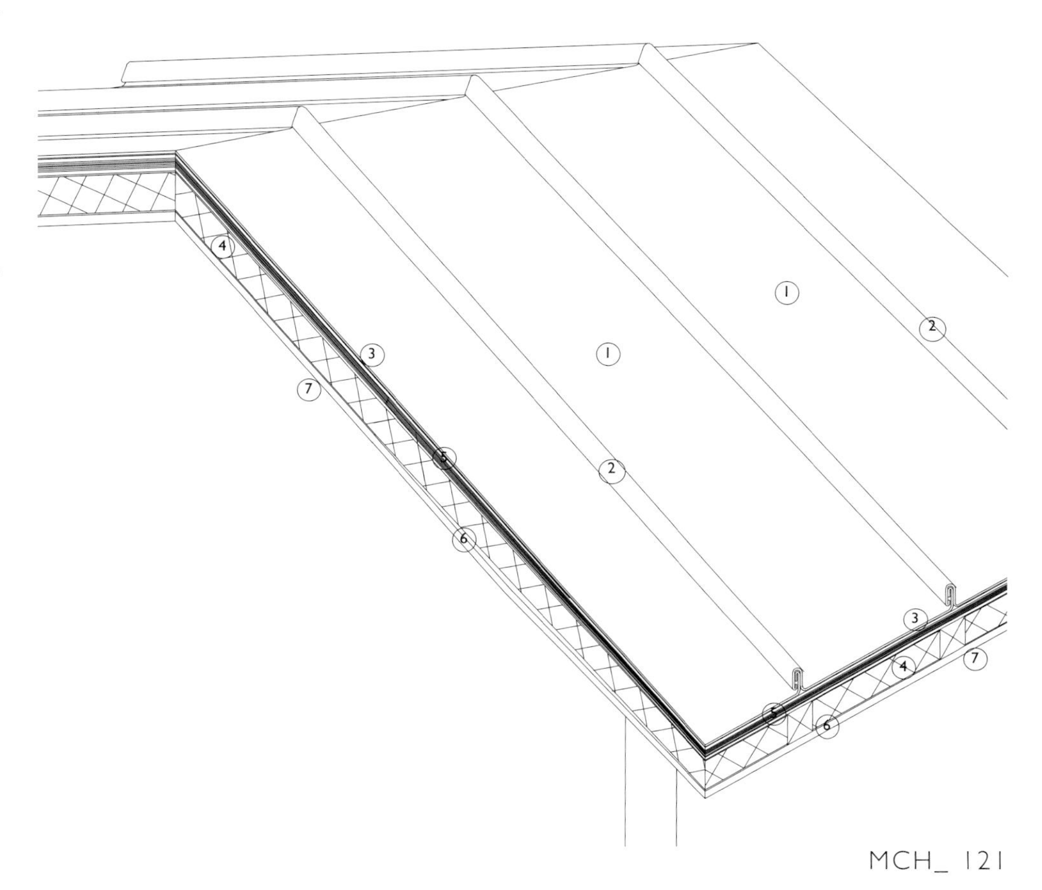

Single and multi-layer membranes are continuous materials that are laid across a complete section of roof.

## Asphalt

Asphalt is used essentially as a waterproofing layer for concrete decks. In an exposed configuration, it can, with regular maintenance, last up to 25 years. In an inverted configuration, the material can, with regular maintenance, last up to 50 years. However, asphalt performs less satisfactorily over timber and steel decks, due to the greater amount of thermal movement associated with these types of construction.

Roofing asphalt consists of coarse and fine aggregate bonded together with bitumen. Some types have polymers added to reduce the tendency of the material to move at high temperatures. It is usually delivered to site as blocks that require melting before use. The material is poured hot and is applied by hand using a float, or large trowel, in two equal coats totalling a minimum of 20mm (3/4in). The asphalt is laid onto a bitumen-based sheathing felt, a slip material that allows the asphalt waterproofing to move independently of the structural deck. A 50mm (2in) overlap of the sheets is required to accommodate movement within the substrate. On vertical surfaces such as parapets, penetrations, rooflights or drainage pipes, the asphalt is keyed into a supporting material to protect the edges and prevent separation. Specific detailing depends on the expected movement between the vertical and horizontal elements.

Asphalt is a continuous material that is broken only at structural movement joints. Large areas may require daywork joints (construction joints) if a complete area of asphalt cannot be completed in a single day. The liquid application allows complex areas to be covered. The material is vulnerable to both foot traffic and solar radiation and, as a result, requires surface protection. Damage from sunlight can be countered by the use of stone chippings (ballast) or solar reflective paint, while an upside down roof configuration provides both types of protection. Asphalt is laid to a minimum fall of 3 degrees. Where insulation is placed over the membrane, the ballast (stone chips) stabilizes both. The insulation also greatly increases the performance and life of the membrane by cushioning foot traffic and reducing thermal stress.

## Bitumen-based roofing felt

Like asphalt, bitumen-based roofing felt forms a single membrane across a complete roof. It is welded together with a lap joint to form a continuous surface. Minimum slopes of 3 degrees are necessary. Breaks in the material usually occur only at structural movement joints. In common with asphalt, sharp angles created by changes in direction of the material are to be avoided as they cause cracking.

Felt is laid on a slip material, such as bitumen-based sheathing felt, which allows the waterproofing to move independently of the structural deck. The felt is then bedded in a layer of hot poured bitumen, and is applied in three to five layers, depending on the system used. Some felts have polymers added and are called 'high performance' felts, while some manufacturers supply the material with the bitumen pre-applied,

66B

requiring the rolled-out sheet to be heated on site in order to melt the bitumen. Joints between sheets are staggered to assure complete coverage and thus avoid excessive stresses being formed at any single point. Felt is usually manufactured with a plain or mineral colour finish. Stone chippings are applied to resist effects of solar radiation and UV light.

Felt is vulnerable to damage where foot traffic is necessary, such as for maintenance of roof mounted equipment. Protection is provided with either cementitious stick-on tiles or through the use of paving slabs in an upside-down configuration. In common with asphalt, sharp angles created by changes in direction of the material are to be avoided as they cause cracking. Gutters and flashings are made from either the same material itself or a metal roofing material such as lead, copper or zinc. Minimum slopes of 3 degrees are necessary.

## Single layer membranes

These thin membranes consist of a range of relatively new materials developed over the past 20 years. They are mostly made from elastomers such as neoprene, or PVC-based products, and are used as proprietary systems with details designed by the manufacturer. Single-layer membranes are used in a similar way to roofing felts but are more expensive. Usually they are employed over lightweight decks in metal or timber, which experience higher amounts of movement than concrete decks, where asphalt is most commonly used.

The membrane is loose laid, which makes it easier to deal with structural movement in the supporting deck. The material can be restrained with local fixings, enabling it to resist wind uplift. Like asphalt, single layer membranes form a single skin laid across a complete section of roof. Sheets are welded together with either heat or solvent to form a lap joint that creates a continuous sheet. Breaks are introduced only at structural movement joints. The main advantage of the single layer membranes is their ability to provide a continuous roof covering that requires little maintenance. In common with roofing felts, these membranes must be tested for watertightness.

Single layer membranes are vulnerable to foot traffic, and are difficult to cover with another protective material without risk of puncturing the membrane. As a result, they are best used on roofs that are accessed for maintenance purposes only. Some types can resist solar gain and UV light without additional surface protection. Other types require a protective finish such as smooth stone ballast to prevent damage from the sun.

Rainscreen cladding can be used on roofs where most of the principles of rainscreen walls apply, as described in the Walls chapter. In this situation, roof rainscreens are often used as the continuation of a rainscreen wall where the roof forms a visible part of the design, but they are beginning to be used independently of adjacent wall construction. Rainscreens on roofs consist of panels that check the passage of wind-blown rain, mounted in front of a sealed waterproof substrate that supports the panels. The outer screen drains away most of the rainwater and shields the roof membrane behind from the worst effects of wind-blown rain. Rain passing through the screen is drained away on the membrane that is the real waterproof layer. The void between the two layers is ventilated to the outside air.

Where they differ from walls is that most of the rainwater is drained onto the weatherproof layer beneath the panels, rather than down the panels themselves. This is primarily because dirt collects much faster on inclined roof surfaces than on wall panels. Dirt that accumulates on the panels is washed off by rainwater directly onto the layer beneath rather than running from panel to panel. This makes some applications on roofs less effective from the weatherproofing point of view than rainscreen walls. However, the rainscreen method provides the possibilities of using less conventional roof substrates such as composite panels, which are more vulnerable to wind-blown rain than a single membrane roof and which can be protected both from the effects of windblown rain and UV light on rubber-based seals. In addition, panels may have to support foot traffic when accessed for maintenance purposes. In order to gain access to the roof membrane beneath, panels may also have to be easily removable with fixings that avoid noise associated with the movement of panels. Plastic-based fixings, typically in PVC or nylon, are used to avoid rattling.

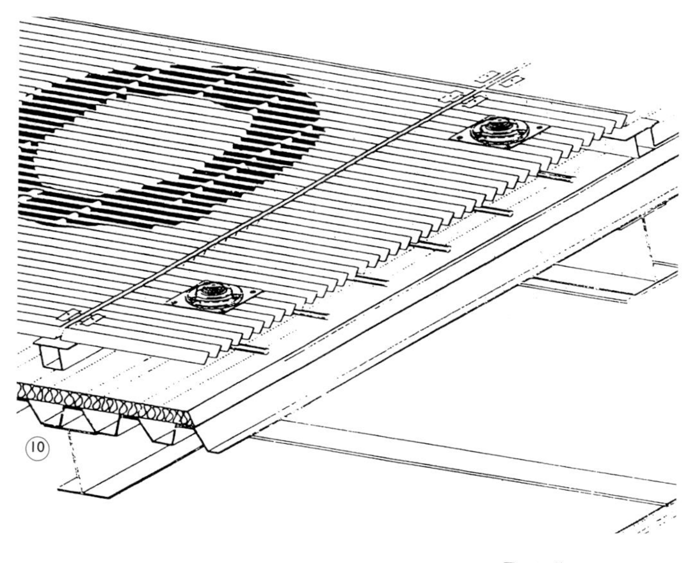


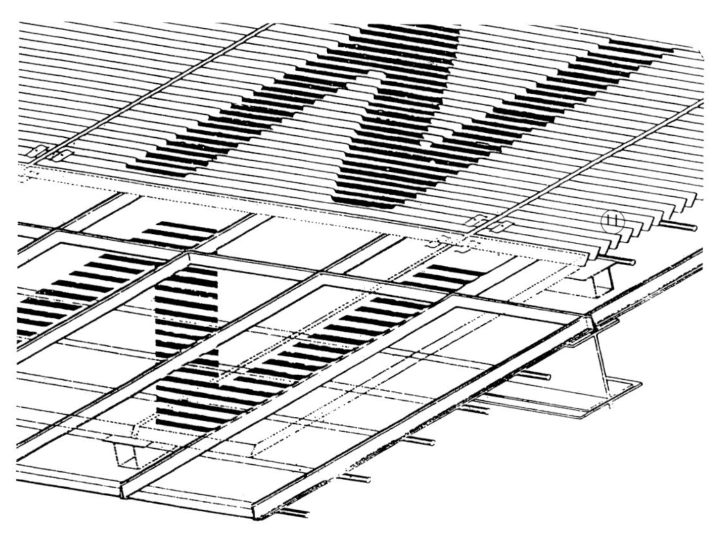


Details
1. Metal rainscreen panel
2. Support brackets
3. Tubes
4. Rigid thermal insulation
5. Profiled metal cladding
6. Purlins
7. Laminated timber portal frame
8. PVC-based waterproof layer
9. Nylon clips
10. Profiled metal roof deck
11. Mill finish aluminium grating

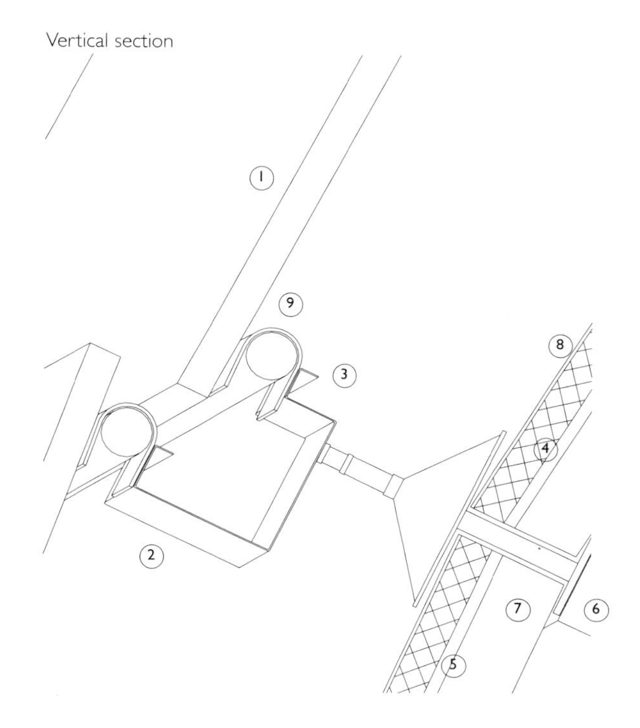


99A

99B

Bercy Shopping Centre, Paris, France and Kansai Airport, Japan.
*Renzo Piano Building Workshop.*

Both buildings use a stainless steel cladding panel fixed with plastic hooks that are attached to a series of steel tubes set forward of the roof covering. Plastic is used to reduce noise associated with thermal movement. The tubes are fixed back to the roof by posts that project forward of the roof. The roof is waterproofed with a PVC membrane. Whereas the rainscreen at Bercy forms a continuity of wall and roof in a single shell-like form, that of the Kansai roof is a separate element that is distinct from the glazed walls.

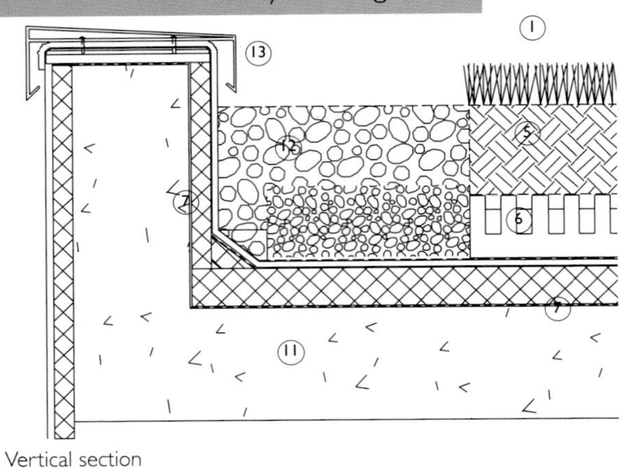


Vertical section

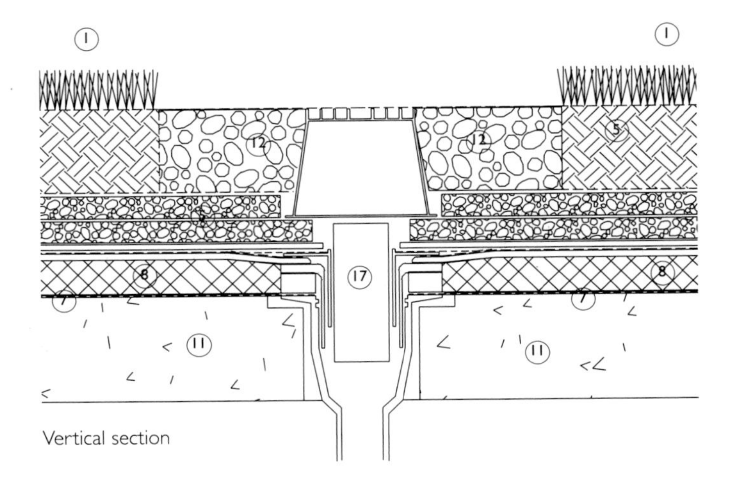


Vertical section

Vertical section

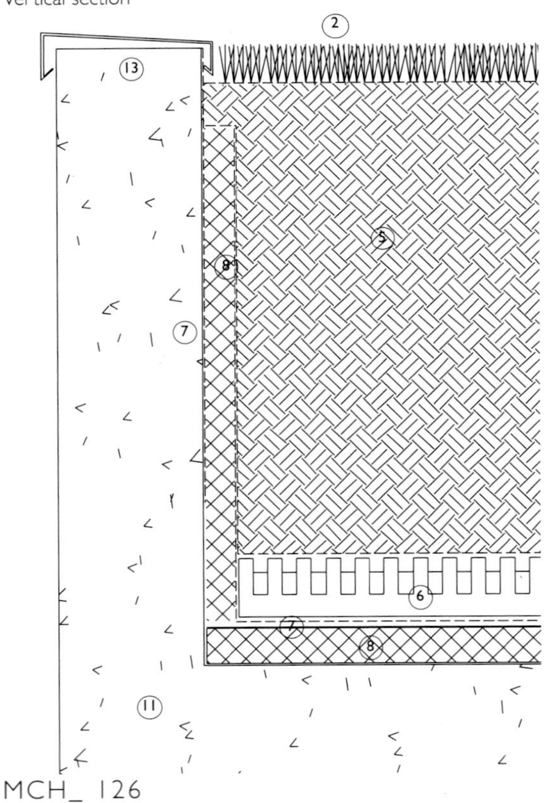


## Details

1. Light vegetation
2. Heavy vegetation
3. Filter layer
4. Lightweight organic substrate
5. Soil
6. Polystyrene drainage boards
7. Waterproofing layer
8. Thermal insulation
9. Vapour barrier
10. Lightweight structural deck
11. Reinforced concrete deck
12. Edge gravel
13. Coping
14. Top of cladding
15. Sprinkler supply pipe
16. Supply pipe
17. Rainwater outlet
18. Wall cladding

Plants and shrubs require a greater depth of soil than grass and light vegetation. Soil is heavy and is usually supported on a reinforced concrete deck. Planted roofs have the advantage of retaining rainwater, as well as providing protection from excessive heat and noise and a reduction in the effects of temperature extremes to the building.

In planted roofs, the thermal insulation must be protected both from above and below. In addition to being waterproof, the system must be decay resistant. Insulation is placed immediately above the waterproof layer, on which the planting is set. The insulation must have adequate compressive strength to withstand the total load of the planted roof. A geotextile sheet is laid directly on top of the waterproof membrane. This geotextile has the dual function of preventing root penetration to the membrane and preventing the soil fines from being carried away by the flow of water through the system.

The depth of soil required will depend on the type of plants and shrubs used. Tall shrubs can require up to one metre depth of soil.

Drainage boards beneath the geotextile sheet allow water to pass through to rainwater outlets. To keep the soil moist some incorporate a continuous series of pockets to collect and store rainwater when more water is required. Supports to shrubs and other fixings must not penetrate the waterproofing layer. Separating slip layers are introduced between the upper waterproofing layer and the supporting structure above in order to allow independent movement. Plant irrigation is carried out by sprinkler or drip feed from a perforated pipe and may above the surface of the soil or sub-surface. Water supply and drainage piping should avoid penetrations in the waterproofing membrane. The substrate or structural deck is laid to a gentle fall towards the roof drains.

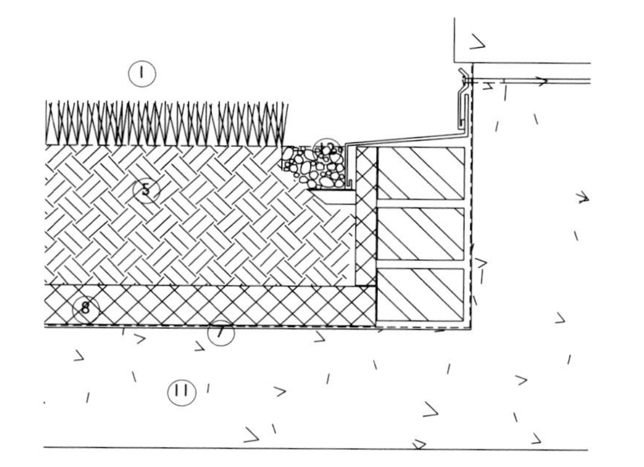


Vertical section

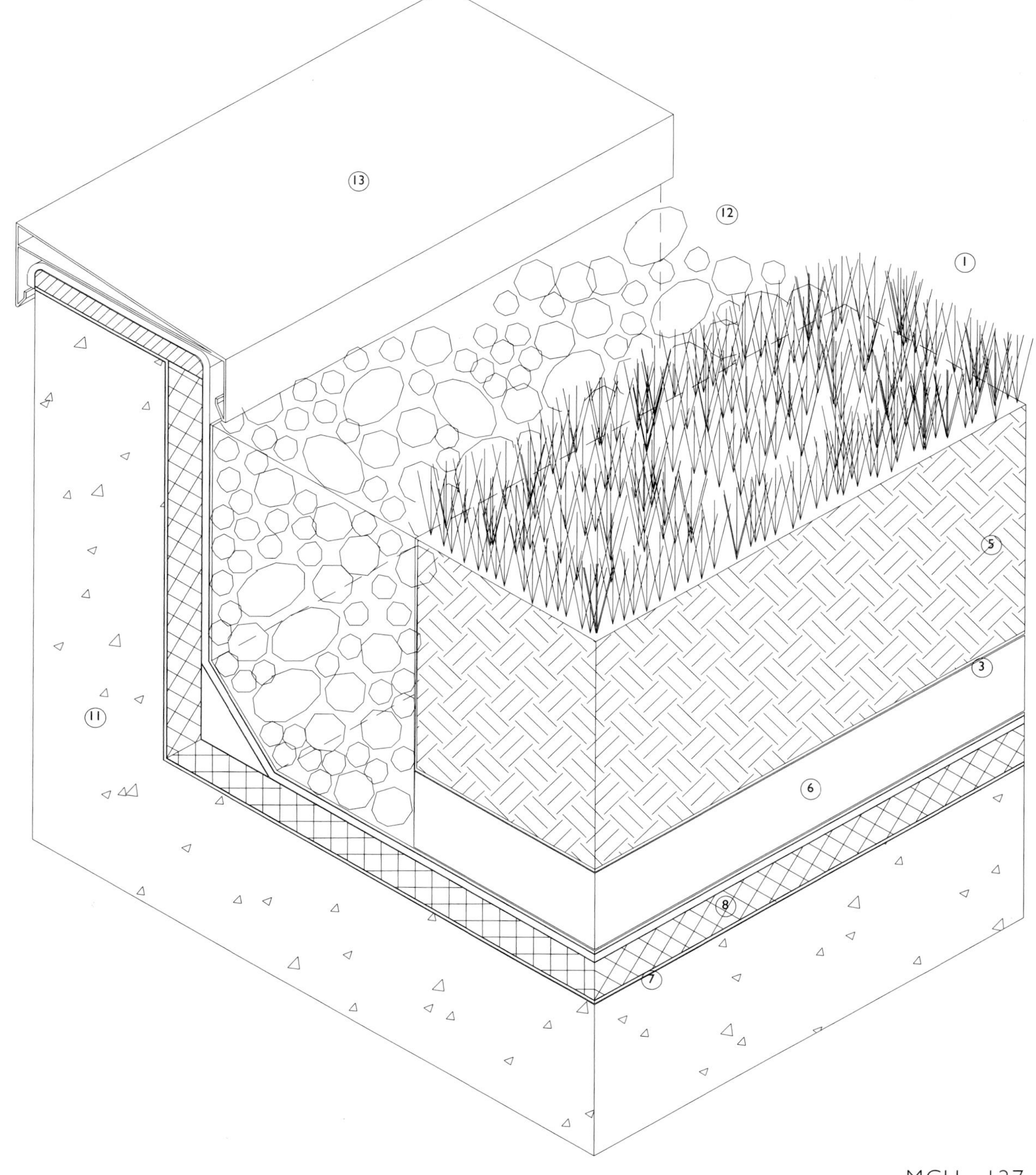

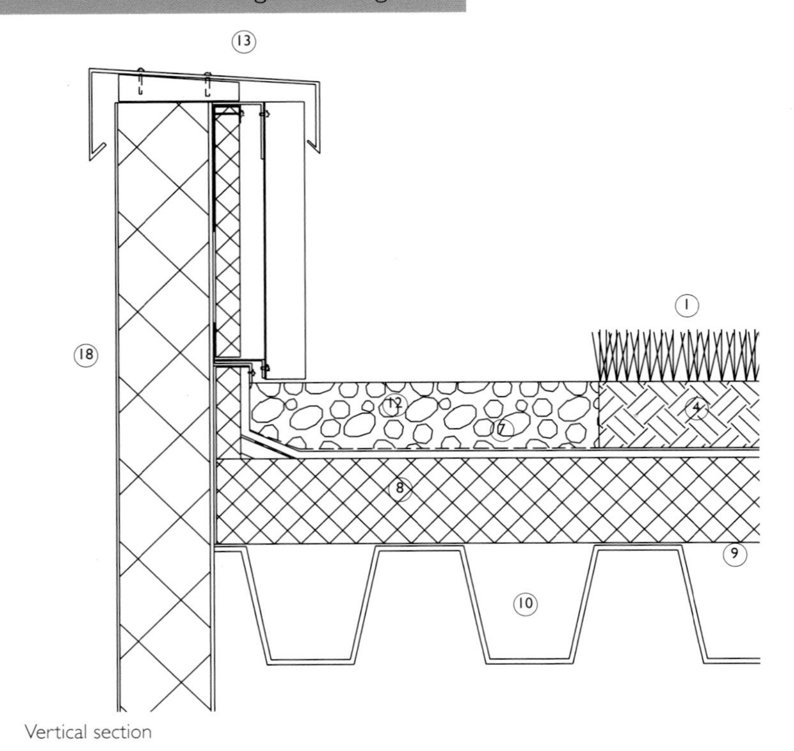


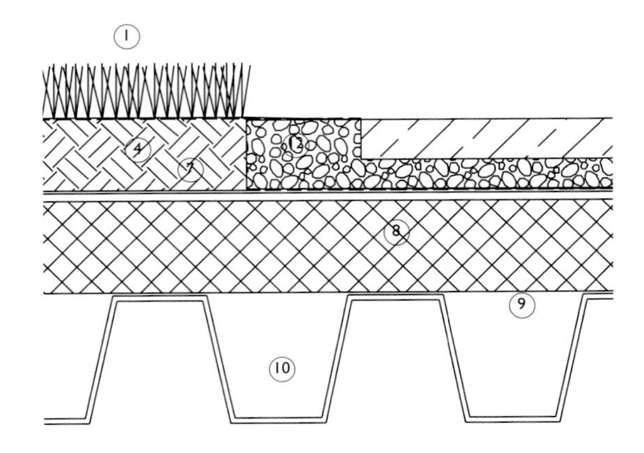


Vertical section

Vertical section

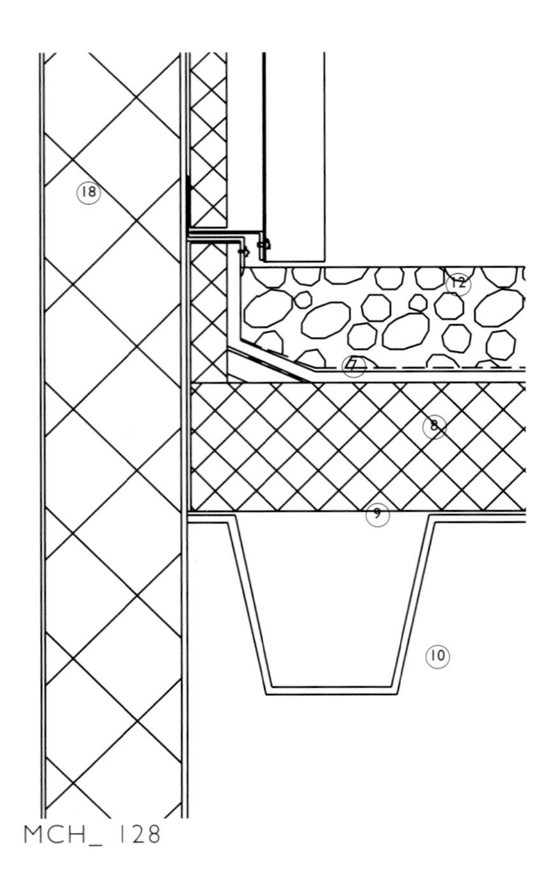


Grass and light planting require much less soil than heavy vegetation, making them suitable for lightweight forms of construction such as a profiled metal deck. A soil depth of 30cm (12") is recommended. Manufacturers' proprietary systems are most commonly used, consisting of turf laid on a geotextile sheet, resting on rigid insulation in an inverted roof configuration. They usually have an integral method of irrigation. Light vegetation roofs retain less rainwater than the heavily planted ones. They also offer less protection from solar heat gain and noise, together with a diminution of the reduction of temperature extremes within the building.

In a conventional membrane configuration, the thermal insulation is placed immediately beneath the waterproof layer, on which the planting is set, and has a root barrier laid above it to prevent root penetration. Supports to shrubs and other fixings must not penetrate the waterproof membrane. There is a substrate mat instead of a filter layer to prevent soil from being carried away by rainwater. Separating slip layers are introduced between the waterproof membrane and the supporting structure to allow the planted roof construction and roof deck to move independently.

Plant irrigation is carried out by sprinkler or drip feed from a perforated pipe and may be above the surface of the soil or sub-surface. Water supply and drainage piping should avoid penetrations in the waterproofing membrane.

## Details

1. Light vegetation
2. Heavy vegetation
3. Filter layer
4. Lightweight organic substrate
5. Soil
6. Polystyrene drainage boards
7. Waterproofing layer
8. Thermal insulation
9. Vapour barrier
10. Lightweight structural deck
11. Reinforced concrete deck
12. Edge gravel
13. Coping
14. Top of cladding
15. Sprinkler supply pipe
16. Supply pipe
17. Rainwater outlet
18. Wall cladding

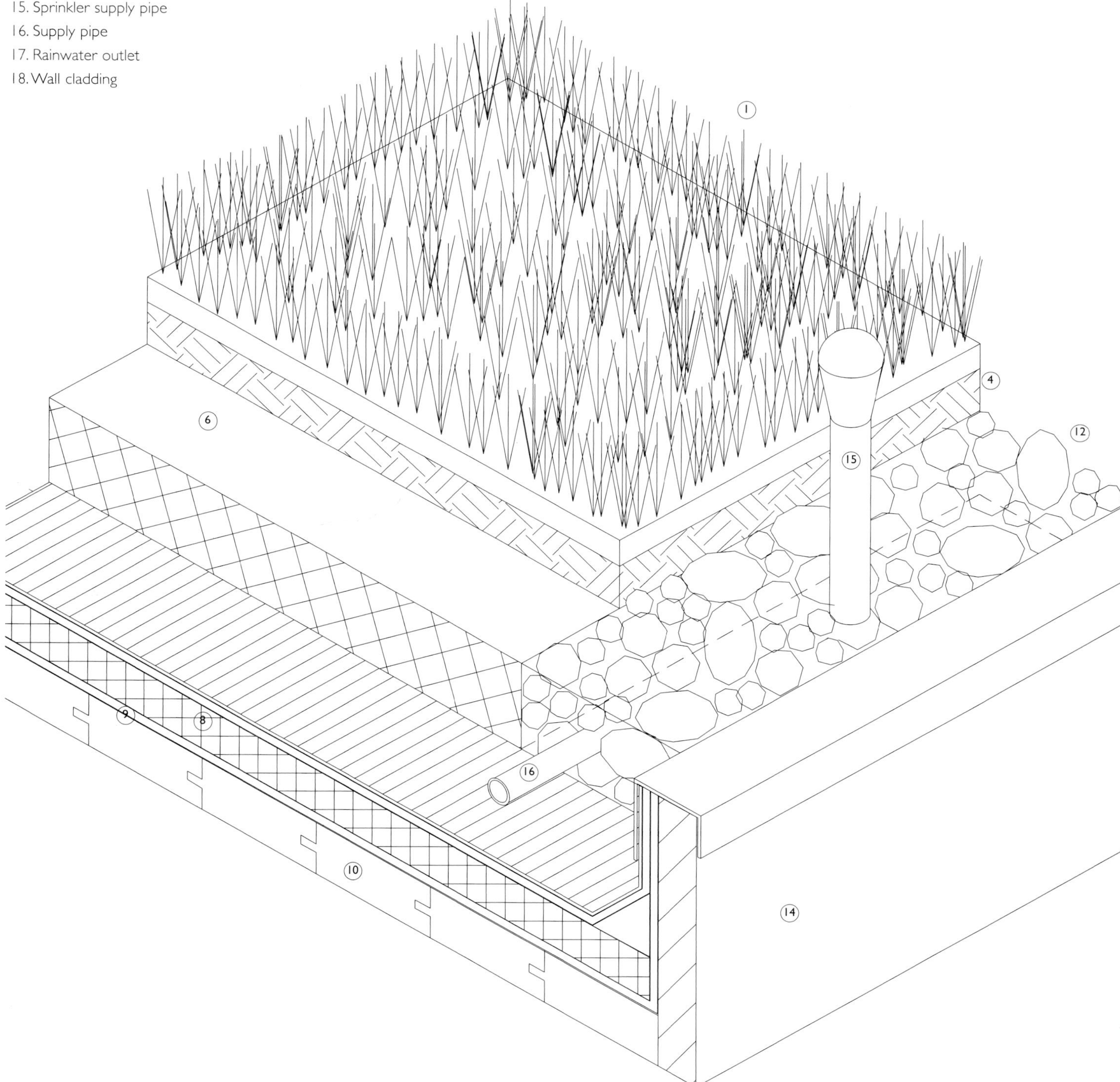

55A

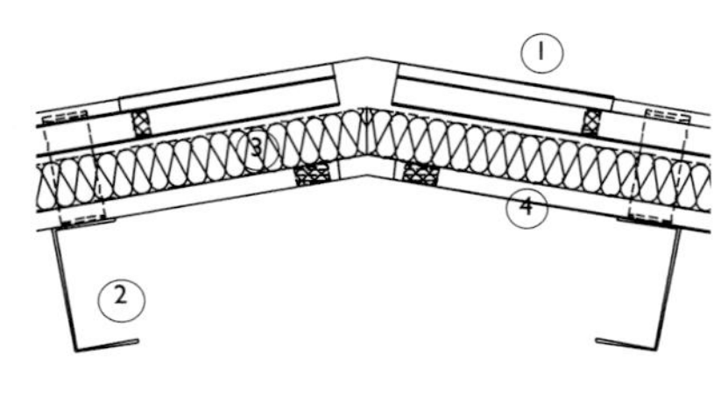


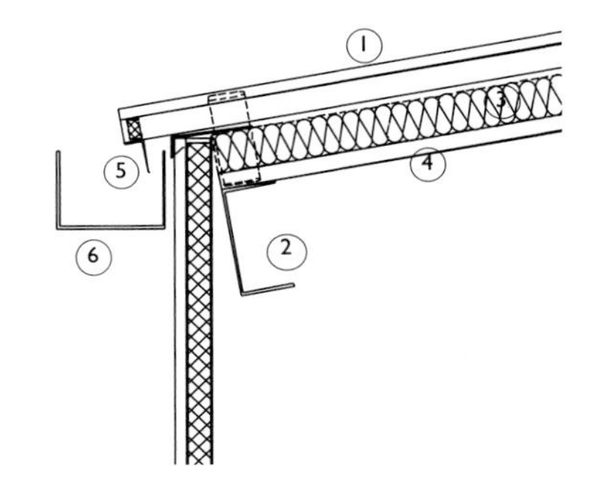


Vertical section
Unventilated roof

55B

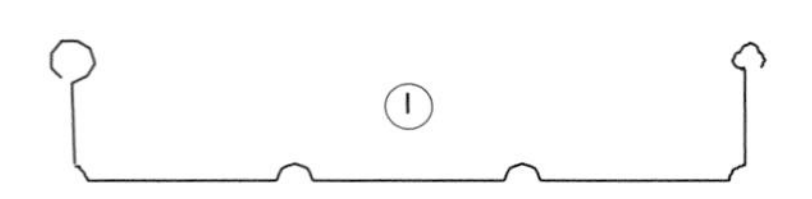


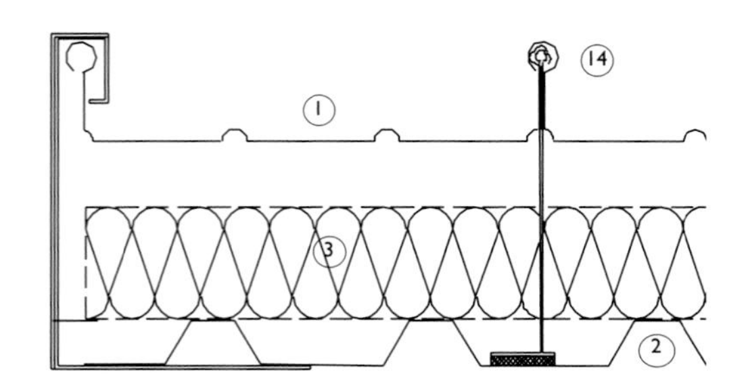


Vertical section

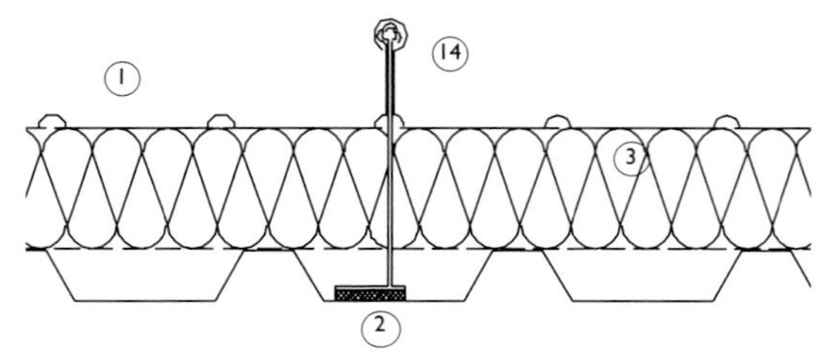


Steel and aluminium profiled sheet can be used as roof decking to span economically up to around 3.5 metres. Spans up to 10 metres can be achieved for lightly loaded roofs, but the depth of profile increases to around 200mm (8") thick, making it considerably more expensive. Decking is most commonly fixed back to purlins spanning between primary structural supports. This construction method uses two layers of profiled sheet separated by metal sheeting rails, with high-density mineral fibreboard or lower density quilt filling the void. Sheeting rails are typically made from pressed metal to form channels, zeds or top hat sections. The upper profile forms both the structural deck and the waterproofing layer. The lower layer is a lighter gauge sheet that supports only the insulation, and provides a decorative or acoustic lining. A vapour barrier is placed between the inner sheet and the insulation. A breather membrane is usually set between the insulation and outer sheet to ensure the one-way passage of water vapour, preventing any moisture that finds its way under the sheet from reaching the insulation before it evaporates. An alternative method is to provide a through-ventilated, or 'cross ventilated', cavity between the outer sheet and the insulation. The inner profile can act as an acoustic lining when it is perforated to allow the insulation behind to absorb sound. The upper profile on top is set at a minimum three-degree fall to allow rainwater to drain away. An alternative method of construction is to lay rigid insulation onto the decking with another profiled sheet as a waterproof layer installed on top, with a vapour barrier between the insulation and the metal profiled deck. In this configuration, the structural deck is perforated to provide an acoustic performance similar to that of an internal lining sheet. The external surface of the profiled sheeting is typically finished in a proprietary coating as set out in the Materials chapter.

A recent development has been in the introduction of simple thermally broken purlins that both reduces heat loss and reduces the condensation risk. These purlins are typically made from rigid insulation with a metal reinforcement on each side.

## Jointing

Joints between profiles are made by overlapping the sheets. The amount of lap varies with the system used, ranging from around 100-200mm (4in-8in). They are fixed using self-tapping screws back to a structural support. Since profiled metal

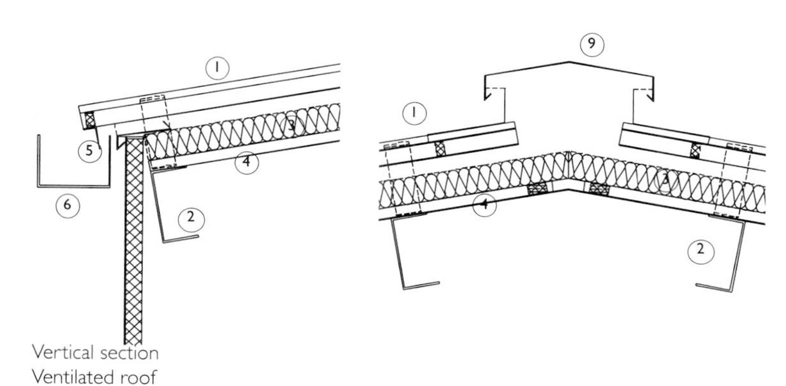


Vertical section
Ventilated roof

sheeting is formed from factory-coated coil, a range of pressed metal accessories is made from the same coil material in matching colours. These are typically pressed metal gutter linings, ridge capping and covers for eaves and gables. Profiled filler and flashing pieces are made to close the gaps between profiled sheet and flat sheet when they are fixed perpendicular to one another across the ridges of the profile. Gutters are formed by lapping a metal lining under the insulation layer and covering the joint with a metal flashing that is tucked under the outer profiled metal sheet. Ridges are formed by lapping pressed metal sheet over the gap between the two adjacent profiles and sealing the gap between sheet and flashing, along the length of the roof, with proprietary filler pieces.

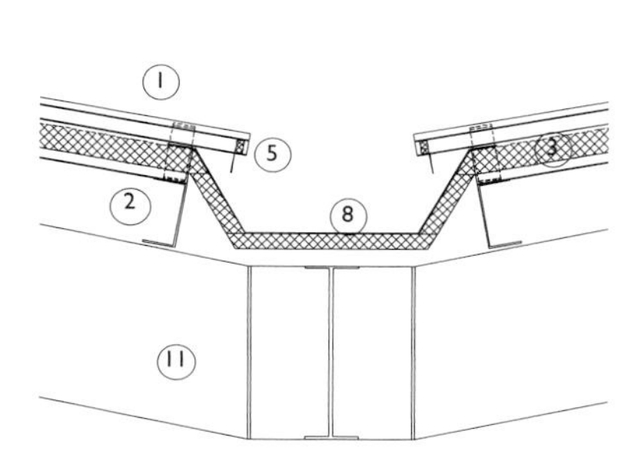


Vertical section

Canopy at Parc de La Villette, Paris, France.
*Bernard Tschumi Architects.*
The wave profile generated by the canopy allows a complex structural form to be used without having to depend on other elements, such as walls to provide rigidity to the system. The higher structural deflections permissible in a canopy allow smaller, elegant components to be used.

Details
1. Profiled metal sheet
2. Purlin
3. Thermal insulation
4. Internal liner tray
5. Metal drip
6. Exposed gutter
7. Preformed and insulated concealed gutter
8. Preformed and insulated valley gutter
9. Pressed metal capping to ridge vent
10. Metal flashing
11. Structural support
12. Wall cladding panel
13. Purlin
14. Standing seam joint
15. Ventilated air gap
16. Tubular steel structure

87A

87B

Horizontal section

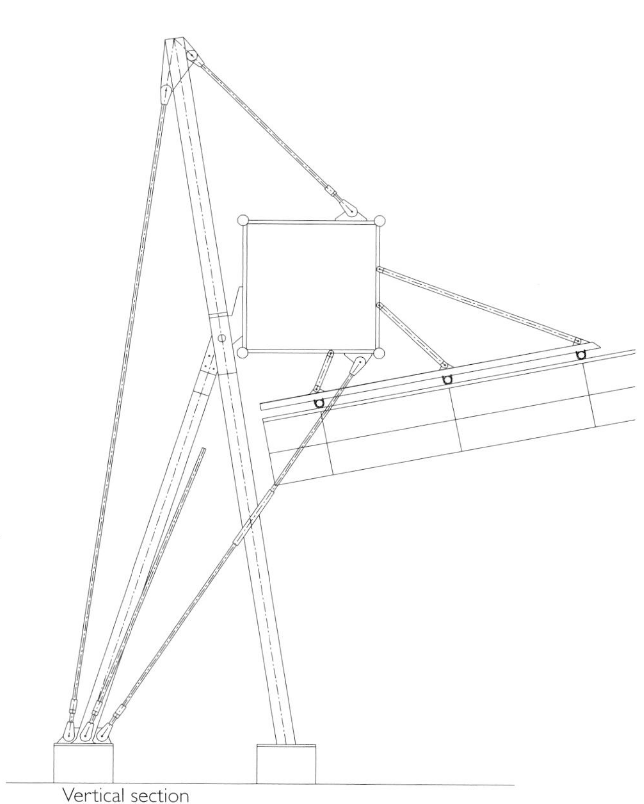

Vertical section

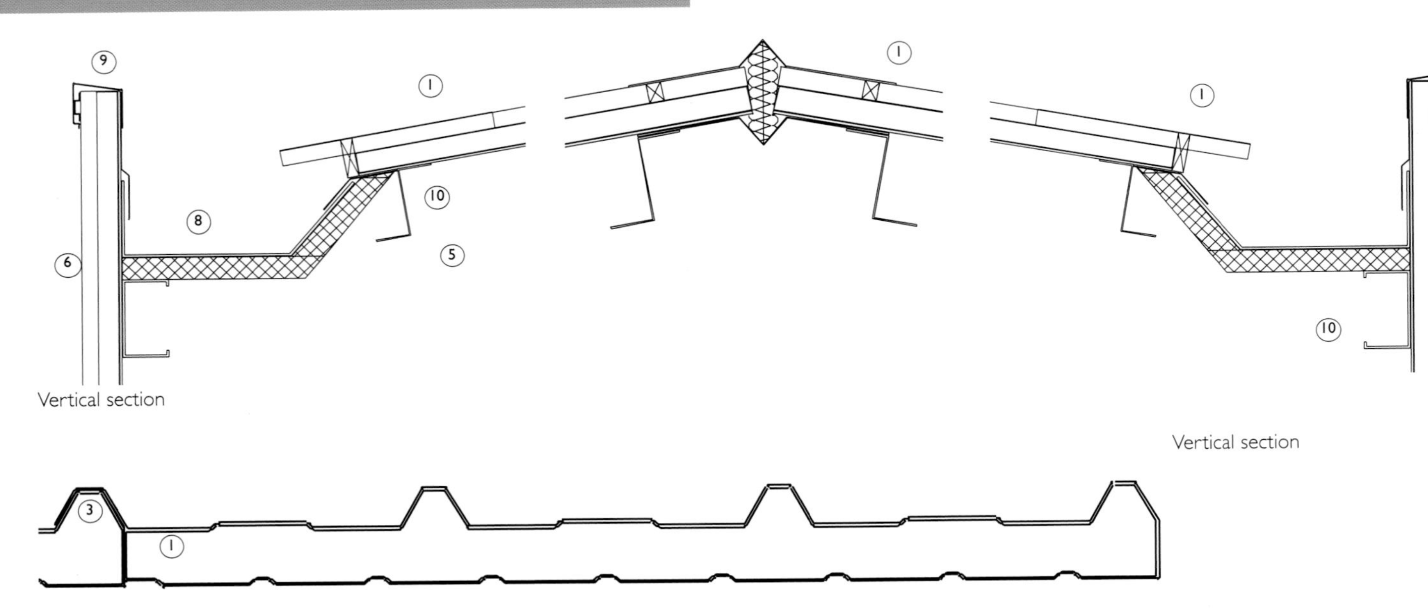


Vertical section

Vertical section

## Details

1. Roof panel
2. Rubber-based gasket
3. Rubber-based seal
4. Structural support
5. Extruded aluminium frame
6. Wall panel
7. Exposed gutter
8. Concealed gutter
9. Parapet coping

Vertical section

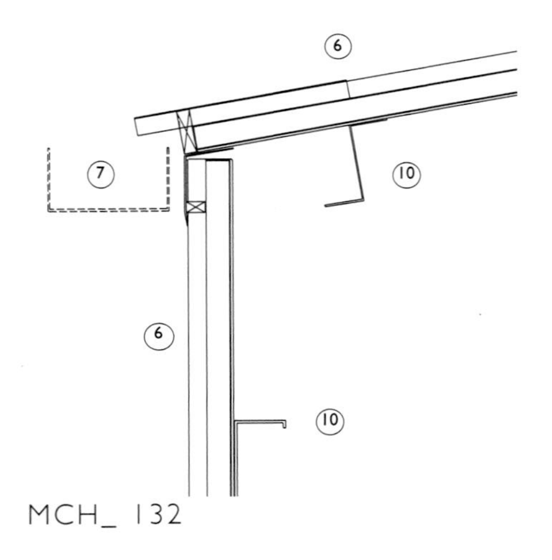


Self-supporting composite roof panels are similar in concept to wall panels, and are usually used in conjunction with wall panels. Like wall panels, they consist of outer layers of metal bonded together integrally with a rigid core of thermal insulation, forming a complete roof construction where substrate and waterproof finish are combined. Their advantage lies in both speed of construction, which can be as little as half the time used by profiled metal sheet, and in their thermal insulation, which can avoid thermal bridging with greater ease than can profiled sheet.

Where composite roof panels differ from wall panels is both in the profile shape and in the junctions. Where wall panels tend to be flatter and have joints that have a flush appearance from the outside, roof panels use a variation of the standing seam on joints running parallel to the slope of the roof. Since panels can be manufactured up to approximately 25 metres long (80ft.), horizontal joints between panels, (between the ridge and the eave), can be avoided. Where panels are joined at the top and bottom of the panels, a lap joint is used, resembling that of a profiled metal sheet, but with an additional rubber-based seal. The use of the standing seam principle ensures that water running down the roof is shed from the joint onto the panel, rather than down the joint itself. The principle of the flush joint, used in wall panels, is difficult to apply on roofs, since water will always have a tendency to run into the joint, particularly when the surface of the panel is effectively submerged in water during a heavy rainfall.

There are two generic solutions to this standing seam joint between panels. Panels can be manufactured with lapped sides with an integral sealing cover strip on one side, or panels are made with flush sides, where a separate sealing cover strip is fixed across the joint. The type with an integral joint is used for steeper roofs, with a minimum pitch of approximately 4 degrees. The separate cover strip is used for shallower pitches down to around 2 degrees, depending on the proprietary system being used. In both cases, the additional structural deflections of panels must be taken into account when working with low pitches to avoid ponding, where water cannot be drained away.

100E

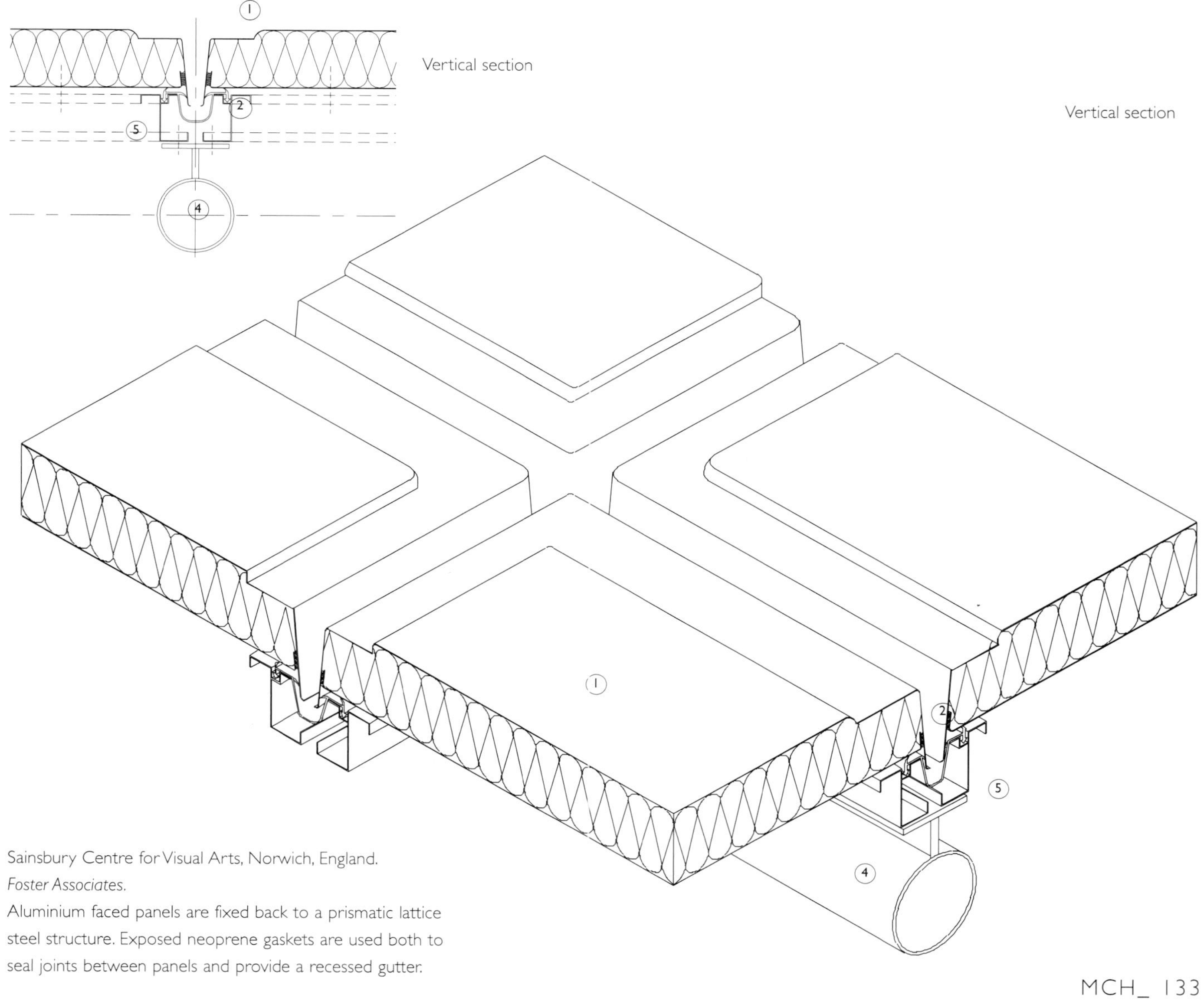


Sainsbury Centre for Visual Arts, Norwich, England.
*Foster Associates.*
Aluminium faced panels are fixed back to a prismatic lattice steel structure. Exposed neoprene gaskets are used both to seal joints between panels and provide a recessed gutter.

Details
1. Structural steel support
2. Connector plate
3. Bolt fixing
4. Silicone seal between glass panels
5. Single glazed or double glazed unit

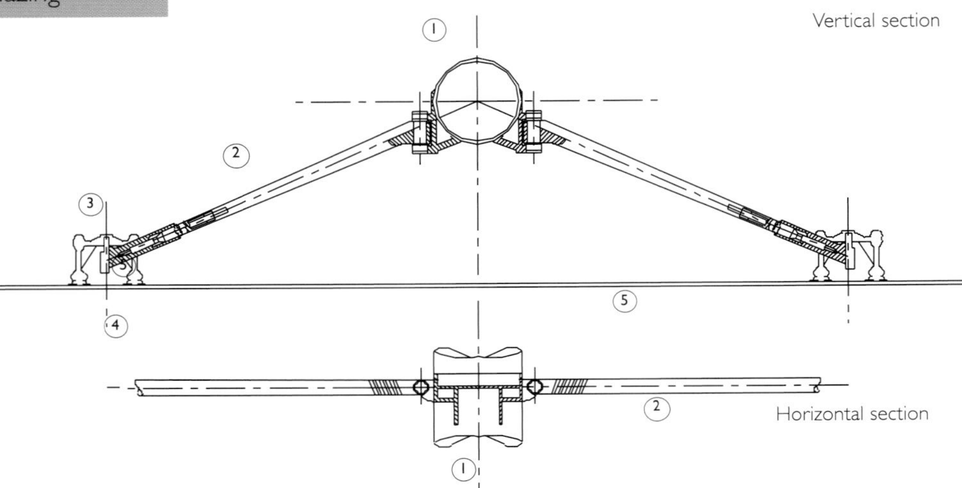


47E

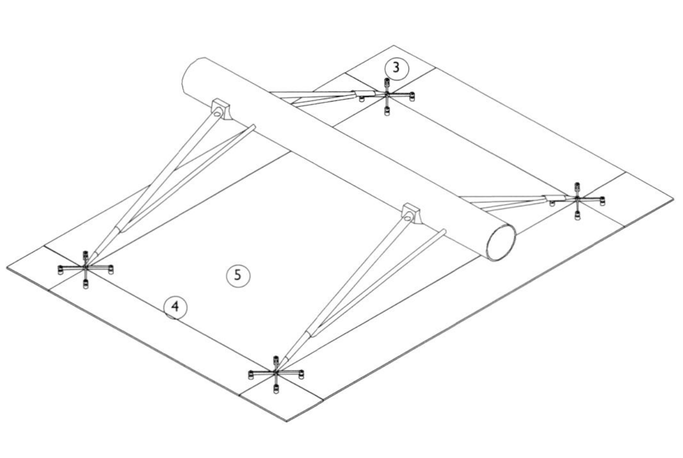


This system was originally conceived as a glazed wall system with benefits of high visual transparency. Used as a glazed roof, the techniques are very similar but panels are seated onto the corner bolts rather than being hung from the bolts. As with bolt fixed glazed walls, single glazing or double glazed units are fixed back to supports with a bolt in each corner.

Bolt fixed glazed roofs vary widely in the fixing system from an externally fixed steel disc or plate bolted through the joints between sheets to tiny bolt connectors countersunk directly into the glass. The supporting structure varies enormously depending largely on whether a secondary layer of solar shading is to be used, usually mounted externally, above the glazing to reduce solar gain. As with bolt fixed glazed walling, the amount of tolerance and adjustment possible with each connector varies according to its design. In addition, the bolt fixing can be fixed from below the glass, putting the waterproofing seal on the opposite side to that used in walls. Whether the glass is supported internally or externally, there must be sufficient adjustment in the supports to provide the correct fall for rainwater drainage.

Bolt-fixed glazing can be laid with modest falls of 3 degrees, but it requires frequent access for cleaning the outside of the glass. Access is usually provided by walkways set above the glass. The roof is drained into gutters that are made from an opaque material, usually metal. Forming gutters in glass can lead to deposits of dirt down the length of the gutter which are left after every rainfall. Low slope allows water to collect on the surface and as the water evaporates, dirt is left behind. In addition, dirt accumulation can accelerate the degradation of the seals.

Connections between the bolt fixed glazing can be achieved either with silicone seals or bonds to related items such as gutters, or by introducing glazing bars fixed with pressure plates. A limited mixing of glazing systems can reduce the complexity of detailing without the loss of transparency that is the main advantage of bolt fixed glazing.

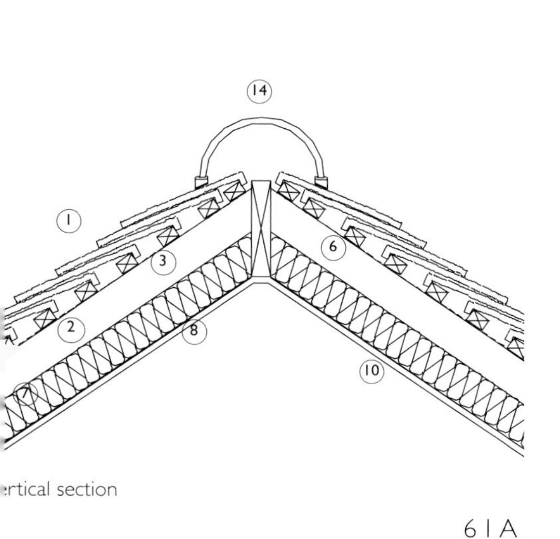


ertical section

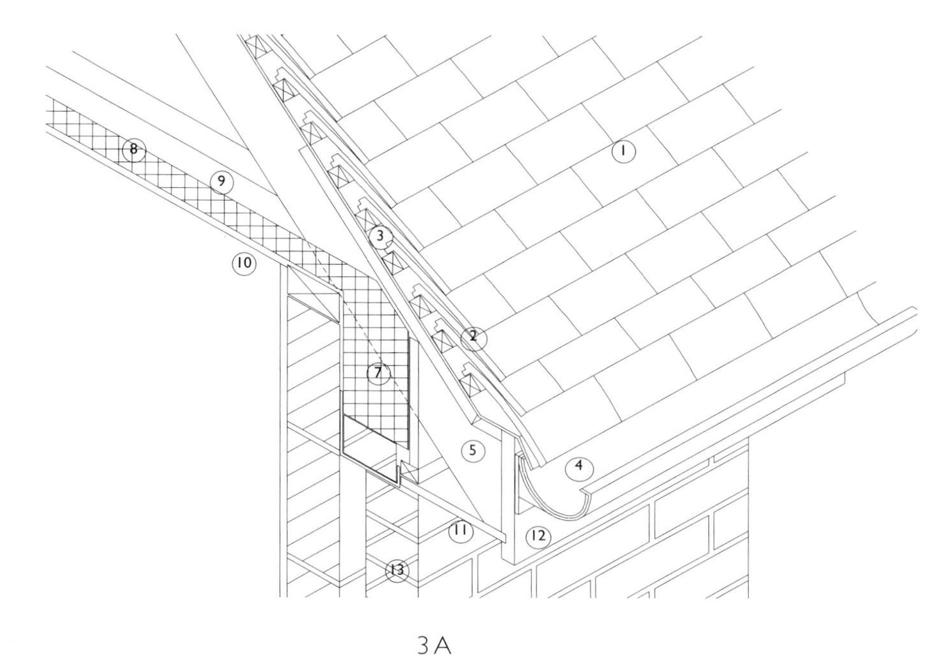


61A

3A

there are always at least two layers of protection. For this reason they are laid in staggered rows so that the gap between adjacent tiles is always covered. Slates and tiles are fixed to battens that span between rafters or trusses and additional protection may be provided by a layer of bituminous felt set beneath the battens. Slates and tiles impose heavier loads on the structure than most other roofing materials. They can provide a long life but are difficult and expensive to maintain when this is required.

Junctions at ridges, hips and chimneys are waterproofed with metal flashings, usually made in lead. This is a soft metal that can be moulded to the shape of the junction. Flashings are dressed over tiles in a lapped joint. Valleys are sealed with soakers (valley flashing) that are interleaved with the tiles in order to maintain the waterproof layer. These are also usually made from lead. Gutters can be of a variety of types. External gutters are fixed to the wall at the eaves. Internal gutters are concealed within the depth of the tiles above the eaves line . Parapet gutters are concealed behind the external wall rising above the eaves line.

3B

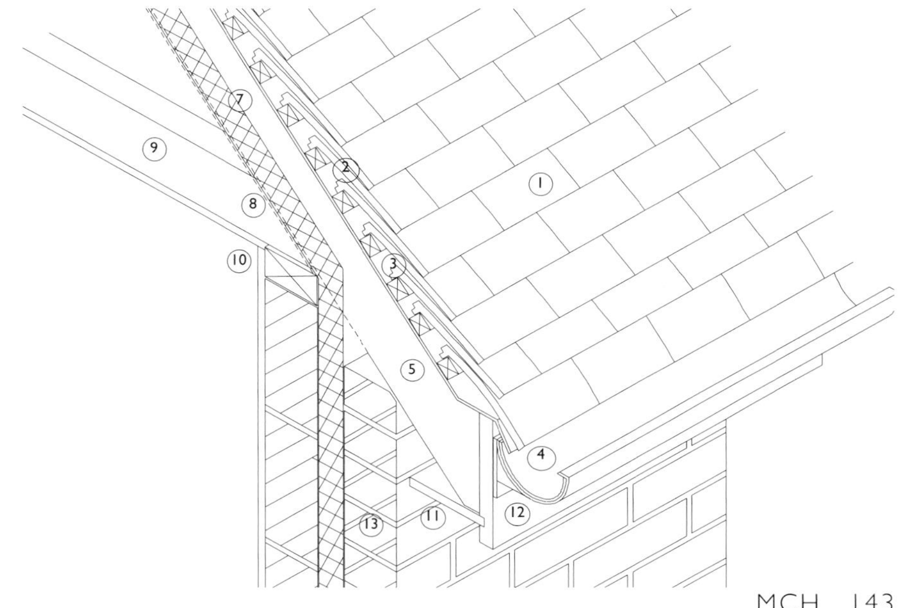

92C

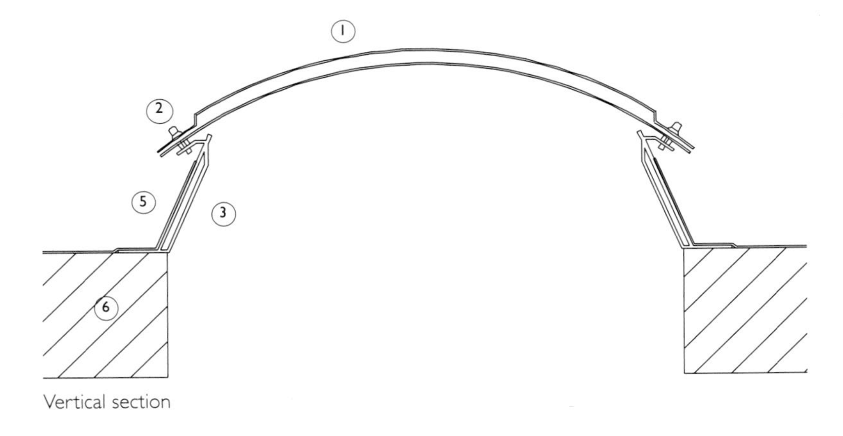


Vertical section

92D

92A

92B

Rooflights can be made more economically in plastics than in glass, though U-values are much poorer than are comparable double glazed units in glass. A typical double skin plastic-based rooflight has a U-value of 3.3 W/m²K compared with a double glazed unit with a value of around 2.0 W/m²K. Small rooflights, manufactured to around one metre square (3ft x 3 ft) in size, are economic and are fitted into openings in a roof. Some are now made in varying sizes up to around 2400mm x 1200mm (4ft x 8ft). These are made in either regular polycarbonate sheet or acrylic as 'shells' that are fitted onto an extruded aluminium or PVC-U frame. Until recently, before glass roofs were more commonly used, these small rooflights with their characteristic pyramidal or curved profile were set out in arrays on space frames to provide a transparent, but visually textured, roof.

More recently, longer, rectangular rooflights are being used with openings from 1.0 to around 3.5 metres (3ft to 10ft). Twin-wall polycarbonate sheeting is used, glazed into aluminium framing similar to that used for patent glazing, allowing them to be built in long lengths. New systems use translucent composite panels of fibreglass skins infilled with an inner core of translucent insulation. These are fixed using an aluminium frame secured with a pressure plate or glazing clip, but will no doubt be developed with lap joints similar to those used for metal composite roof panels. Plastic rooflights are often used as smoke vents when fitted with electrically operated arms that open automatically in the event of fire.

All plastic rooflight types are fitted into openings formed in a roof. They are sealed against the waterproofing layer of the roof by dressing the roofing material up the upstand to the rooflight around its base. Flashings extending down from the top of the upstand cover the joint. Rooflight upstands usually have an insulated frame but also this can be formed on site as a continuous part of the thermal insulation used for the roof deck.

## Details

1. Double skinned acrylic rooflight
2. Bolt fixing to upstand
3. Preformed and insulated upstand and frame to rooflight
4. Waterproofing layer secured to upstand forming part of rooflight
5. Waterproofing layer secured to upstand forming part of adjacent roof construction
6. Adjacent roof construction

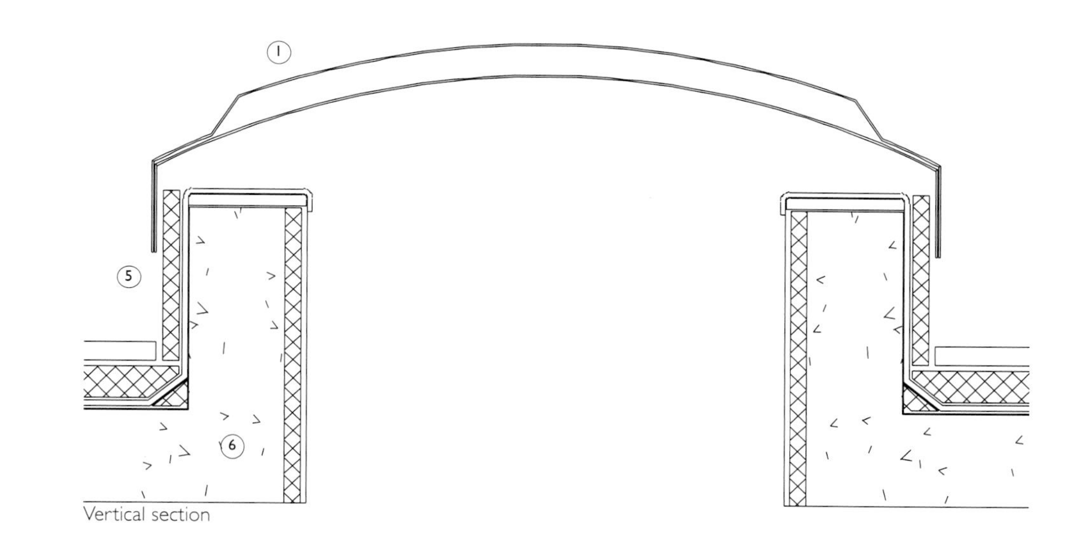


Vertical section

89A

89C

Fabric roofs are extremely lightweight and can enclose large spans with a minimum amount of structure. Poor thermal and acoustic performance can be improved by using a double skin construction with insulation introduced into the cavity. Coating the outer surface with a reflective finish controls solar gain.

There are two generic types of fabric-based roof in general use. These are the type supported by a structure of masts, rods and cables that keep the material in tension, or air supported types made as twin-wall inflated membranes. Until recently, fabric membranes were not available as proprietary systems. Even with these systems, fabric structures are custom-designed by an architect and structural engineer for each specific project.

Typically, a fabric covering of translucent Teflon-coated glass fibre membrane, is used which achieves a light transmission of around 20 per cent. An alternative material is Teflon-coated polyester, which achieves around 15 per cent light transmission. The Teflon (PTFE) coating is an inert plastic, with high resistance to chemical attack and discoloration by UV light. This coating is applied to the woven fabric using a dipping process. The material is joined together as a series of strips, which are lapped and heat-sealed during manufacture. The edge is typically wrapped over a steel strand cable. The completed roof membrane is then transported to site where it is attached to the main structural cables using clamps, which tension the fabric when tightened. The supporting structure of masts, rods and cables incorporates adjustable pinned joints that also rotate in response to changing live loads.

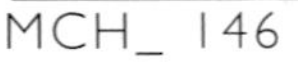

89B

89E

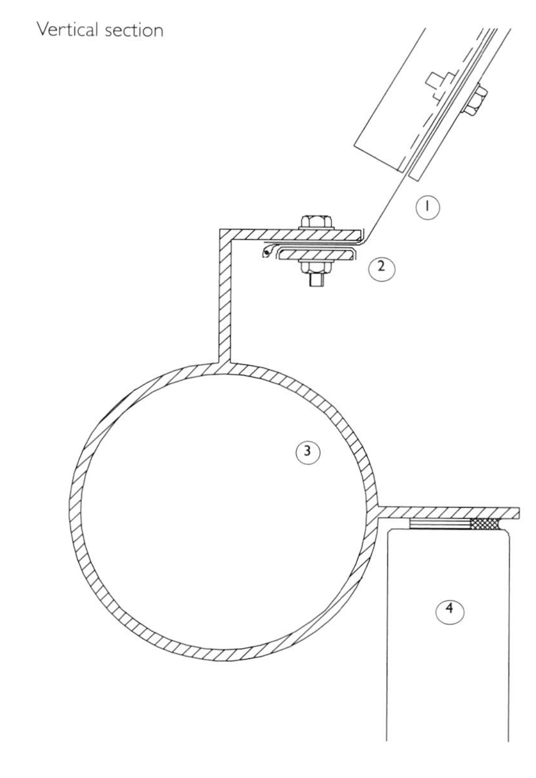


A recent example is the Millennium Experience in London, England. The 'Dome' covers an area of 10,000m$^2$ and encloses an exhibition. It has a circumference of one kilometre (1100 yards) with a diameter of 365 metres (1199ft 6in) and 50 metres (164ft) at its highest point. It is suspended from 12 steel masts that are 100 metres (328ft) high. The roof is made from Teflon-coated glass fibre.

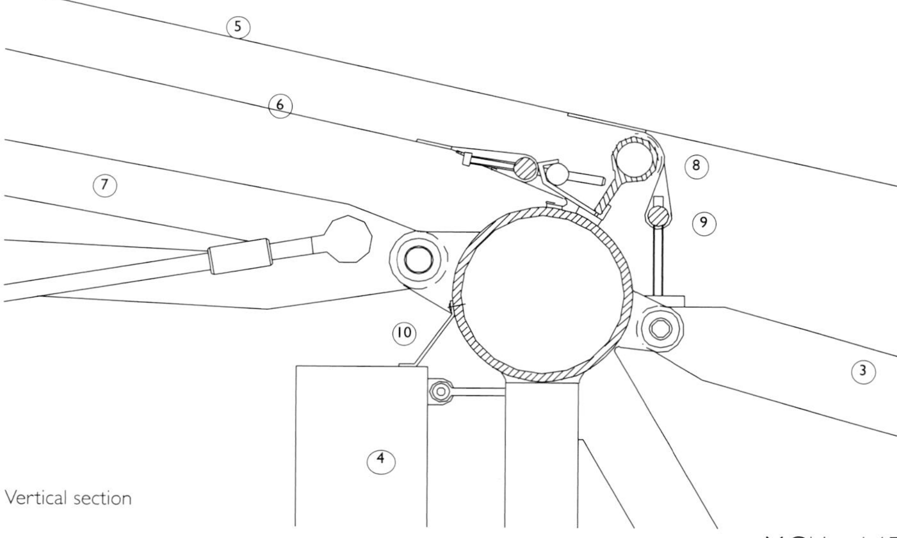


Details

1. Fabric membrane
2. Edge seal clamped to continuous plate
3. Structural support
4. Wall panel
5. Outer membrane
6. Liner membrane
7. Steel supporting structure
8. PTFE film or similar slip layer
9. Continuous stainless steel or aluminium bar
10. Pressed metal flashing

6B

6C

6D

Details
1. Upper fabric membrane
2. Lower fabric membrane
3. Edge seal clamped to continuous plate
4. Steel hollow section ring beam
5. Polycarbonate sliding panels set in extruded aluminium frame
6. Steel supporting structure
7. PVC reinforced membrane
8. Continuous stainless steel or aluminium bar
9. Pressed metal flashing

Air-supported fabric roofs are used to cover large areas that require a degree of transparency together with high thermal insulation. They are available commercially as permanently inflated small 'pillows' set into a metal frame. These supporting frames are set in a grid and are supported as required by the design. The pillows are inflated using an air pump. In the event of collapse, the steel frame provides support. Its curved outer surface allows rainwater to run off. Proprietary systems have gutters in a PVC fabric that are secured between metal frames. Both pillows and gutters are held in the metal frame in a similar way to glazed roofs, with a pressure plate and glazing bar. Maximum sizes of pillows are currently about 8 metres x 2 metres (26ft 3in x 6ft 6in).

Pillows can be formed in two or three layers to provide higher thermal insulation. A very light wire net is often incorporated on the inner face of the material to withstand very high snow loads. The fabric most commonly used is a fluoro-polymer such as ETFE foil (ethylene tetrafluorelthylene). This material has a design life of 10 to 15 years and is available as either translucent or transparent film. The latter type provides around 95% light transmission. This material is non-combustible and does not absorb dirt, which is washed away in heavy rain. The internal face of the pillow, within the building, can become slightly dirty due to small amounts of static electricity that can be present in the material.

## Removable roof covering, Nimes Roman Arena, France. Geipel and Michelin

This roof is used in the winter months only, and is demounted and stored during the summer. The structure, completed in 1988, encloses a Roman amphitheatre with space for 24,000 spectators. The roof has been designed to withstand the high wind load from the 'Mistral' that blows in that part of France. It consists of a single fabric pillow filled with air. The fabric membrane is supported by a set of 32 steel posts that rest on the massive masonry foundations of the arena. The steel posts are restrained by the existing structure at the base and are spaced apart by a continuous elliptical ring beam, formed in rectangular hollow steel sections, around

6F

6E

6A

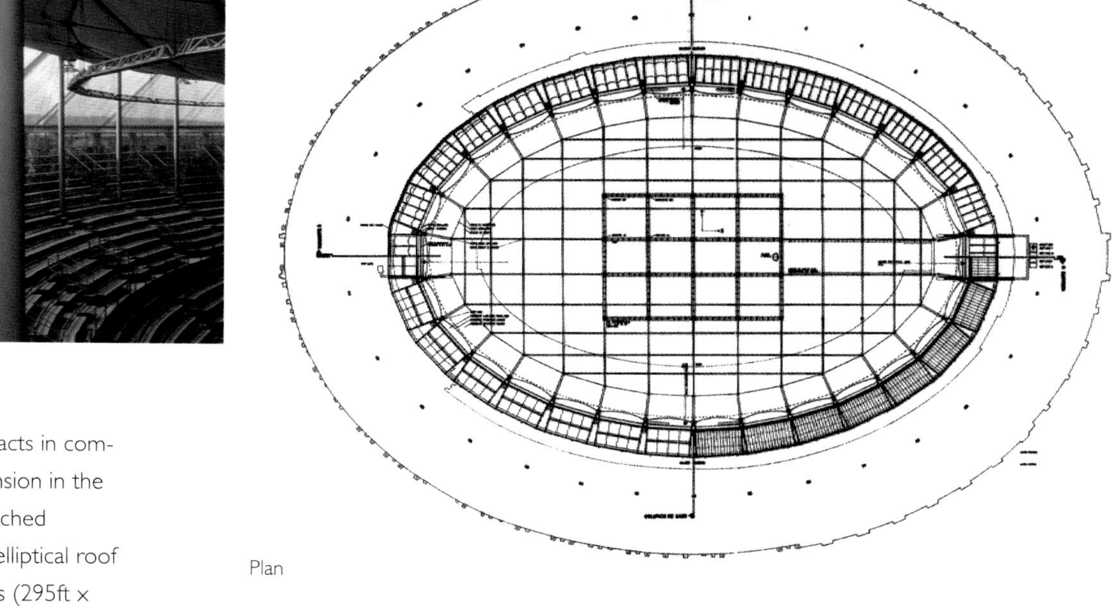

Plan

the roof. The ring beam acts in compression to resist the tension in the fabric roof, which is stretched between the posts. The elliptical roof is 90 metres × 30 metres (295ft × 98ft) wide, covers 4200m² (45,000 sq ft) and weighs 17.5 tonnes (18 tons). The roof structure is first assembled on the ground, and then lifted in place by balloon or crane. The fabric pillow is inflated only when the structure is in place and fully secured. The gap between the tent and the rear wall of the arena is closed with a separate sloping roof in polycarbonate sheet. This polycarbonate roof has areas of opening louvres to provide natural through-ventilation across the arena.

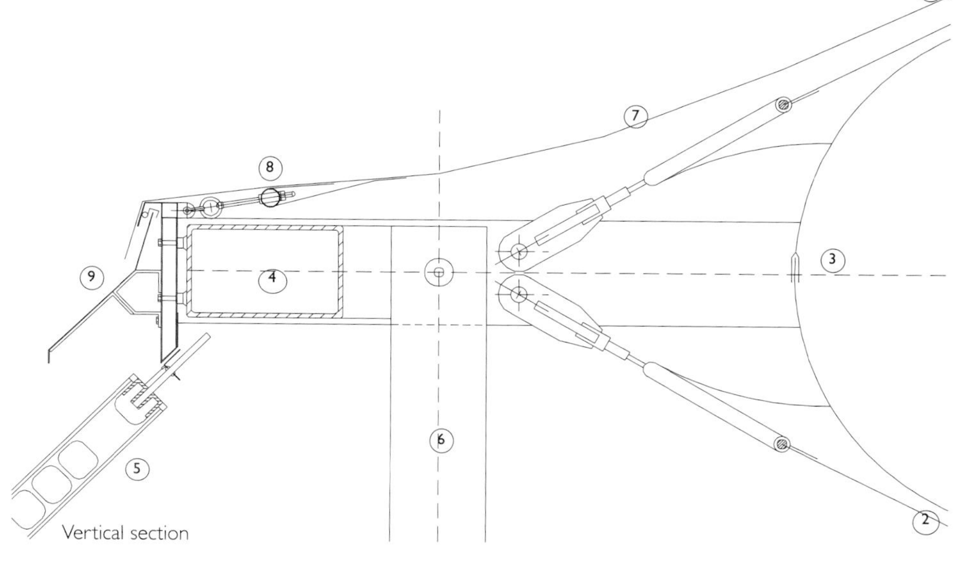


Vertical section

## STRUCTURE

Elements of structure
1. Braced frames
2. Portal frames
3. Loadbearing boxes
4. Trusses
5. Arches and shells
6. Space grids

Floor structures
1. Cast in situ / Cast-in-place concrete
2. Precast concrete
3. Steel and steel mesh
4. Timber
5. Glass

Foundation types
1. Strip and Raft
2. Pad and Piles

A future for building structures: The use of technology from other fields in construction

58B

58A

104A

## Structural stability

This chapter sets out the most commonly used elements of structure and the relationship between structure and enclosure, with typical details for each element of structure and their interface with walls and roofs. Structural elements are described in this chapter as components used to make complete building structures. They can be used as a single structural system for a complete building or, more commonly, be combined to form a mixed building structure. Building structures are described as being either braced or unbraced types. Braced types use devices such as internal walls or service shafts, or alternatively use cross bracing to stabilise the structure. This vertical bracing is usually provided in two vertical planes at right angles to one another in order to stabilise the structure both along its length and across its width. Horizontal bracing is provided by the floor structures which act as horizontal girders. The roof, if sufficiently rigid in the horizontal direction as in a reinforced concrete slab, can also be used as horizontal bracing. Unbraced structures, in contrast, are allowed to sway as a result of their being unbraced in one direction. Their stability is provided by rigid connections that provide stiffness within the structural elements, such as in portal frames. Like braced structures, bracing in the horizontal direction is provided by floors and, in the case of single storey structures, by bracing in the plane of the roof. The most common elements used to make building structures are braced frames, portal frames, loadbearing boxes, trusses, arches and shells, and space grids. Reinforced concrete and masonry structures, despite their essentially monolithic nature, are considered braced structures. Some of the walls provide specifically bracing elements within the overall monolithic structure.

## Structural movement

The term 'structural movement' covers a wide range of effects of deformation and movement in building structures. The dead and imposed loads, together with wind and snow loads and also forces specific to use or location, such as dynamic and seismic loads, can bring about movements seen as bending and shear deformations which are a normal part of the building in use. Temperature changes, and differences in temperature between different parts of a structure, can cause expansion and contraction of structural members,

104C

with possibly some local bending of those members. The detailing of both the components and any associated weatherproofing or internal finishes must accommodate this movement without causing damage to these finishes. Moisture penetration can cause expansion and contraction in concrete, masonry and timber structures. Frost action, caused by the freeze/thaw cycle of moisture within structural materials, can eventually cause damage if the detailing is not sufficiently robust. Frost action in the supporting ground is of particular importance in the design of foundations.

## Trends in the use of building structures

Much effort in the design of building structures is focused on economy combined with safety, in terms of both the quantity of material used and the amount of fabrication needed to assemble the structure. The appropriate use of structure can often be seen in the 'fine tuning' of the balance of material used and fabrication undertaken. This approach is being increasingly paralleled in architectural design in the external envelope and internal fittings where there is a growing tendency to use the least number of components tempered with the needs of their fabrication. This has led to an important trend in structures of their increased visibility and as self-finished components within buildings. This brings the aims of structural design and architectural design much closer together where the relationship between structure, enclosure and internal finishes must be closely co-ordinated in the design at an early stage and throughout the entire design and construction phases. The greater integration of structure and envelope is discussed at the end of this chapter.

In steel structures, frames have become a full part of the language of architectural design. Assemblies in steel, ranging from small lattice trusses to cable-stayed structures that resemble 'kits of parts', has led to their widespread use as highly visible structures, set either outside or inside the building enclosure. The necessity to join steel sections with either welds or bolts has led to a visual richness in both the design of structural members and the joints between them. This has been helped enormously by the development of intumescent paints, which protect the structure during a fire by expanding to form a heat-insulating layer that protects the integrity of the structure

104B

39B

27D

27A

for a limited period, usually one hour. These finishes, once only roughly textured or trowel applied, can now achieve a smooth finish once associated only with the regular paint finishes used in unprotected structures such as those used for single-storey roof structures. Issues of thermal bridging in steel structures and the penetration of the waterproofing layer with steel components, are now easier to overcome with thermal breaks within joints and across weatherproofing membranes such as with the use of a set of bolted connections that are difficult to achieve in either concrete or timber.

Concrete structures are now being exposed as self-finishes within buildings, both on the external walls and in the soffit (underside) of structural slabs, with a quality of finish traditionally associated with only plastered walls and ceilings. Exposed finishes in concrete are now less common due to their poor weathering, from a visual point of view, but are enjoying a revival in predominantly dry climates where thermal insulation is set within the depth of the wall. The renewed interest in precast concrete structural systems, other than those used for parking garages, is set to continue, with ever-greater expression of jointing forming part of the visible language of detailing.

Timber structures are becoming ever more complex both in terms of construction techniques and geometries used in their design. The development of pinned jointing systems as well as the interest generated in the material by its low levels of embodied energy have resulted in the revival of exposed timber structures. The new generation of pinned joints helps to overcome the traditional problem of creating reliable and elegant joints that perform well in tension, particularly in large-scale glue-laminated structures.

The use of CAD/CAM techniques, or computer aided design/computer aided manufacturing, for both timber and steel structures as well as the use of these materials for concrete formwork, is set to transform the geometries of structures built from all these materials. The use of CAD/CAM permits much greater accuracy and closer tolerances in assembly with joints that are increasingly visually elegant. CAD/CAM is set to push further the drive towards prefabrication. Timber and concrete, considered traditionally to be materials that are worked on site, are now more frequently fabricated at the factory, for example in the pre-cutting of timber compo-

88A

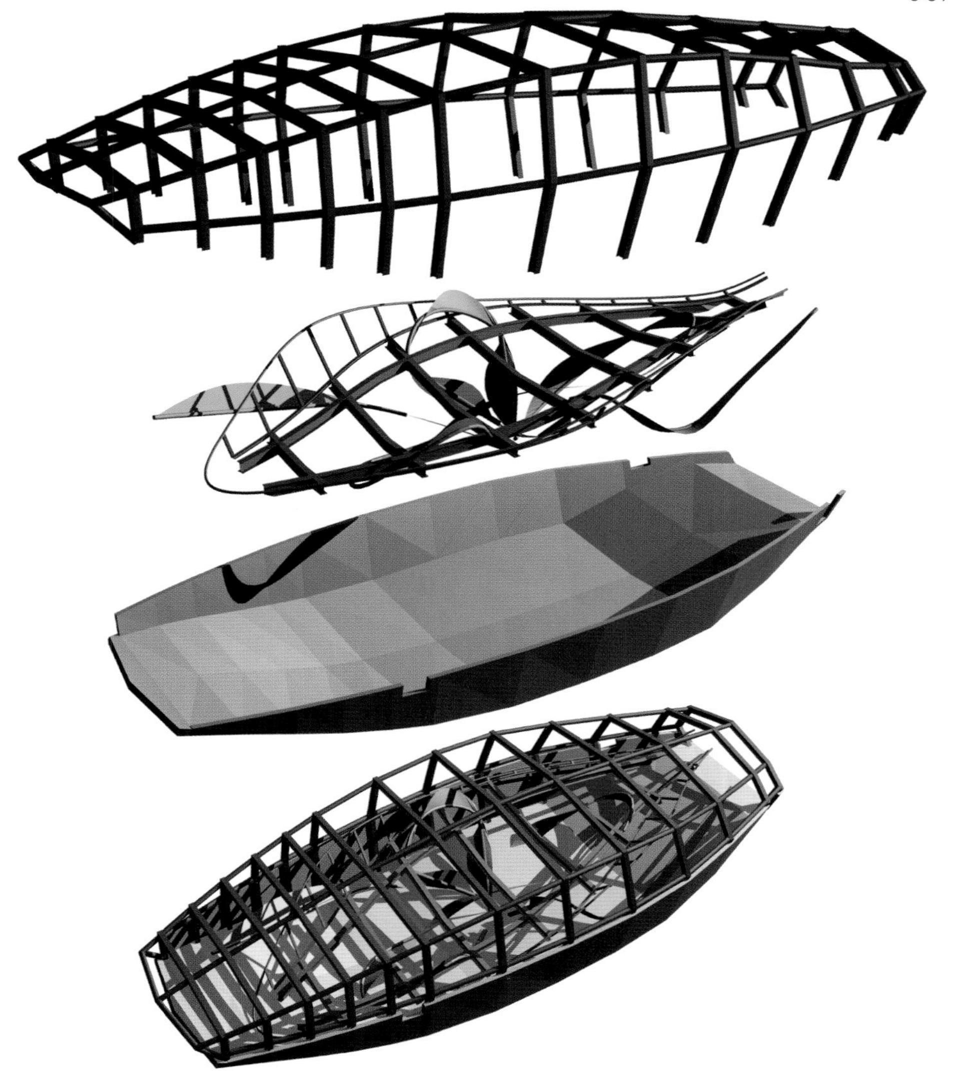

nents and in the precasting of concrete components instead of their casting in place. The building site will inevitably become a place where assemblies are fixed together rather than being the temporary workshop to which we are accustomed.

Thermal bridging, and the related effects of condensation, associated with the integration of building structure with walls and roofs, continues to be a big issue in building design in all but predominantly hot and dry climates. This has led to the increased use of vapour barriers, and the careful placing of thermal insulation, which need to be co-ordinated with the structural design in order to avoid interstitial condensation. The thermal mass of exposed concrete structures is being used for the nighttime cooling of buildings in conjunction with ventilated facades. The increased prefabrication of both structures and facades has led to a greater awareness of issues of co-ordination at an early stage of the project to ensure that highly visible finished components, which in earlier buildings would have been covered over by successive building trades, can be successfully exposed as high quality finished components.

88D

1 A

8 A

The use of precast and cast-in-place concrete in frames provides a homogeneous structure where joints are made as rigid connections. These frames require additional stiffness from bracing that can be provided typically by either concrete shear walls or steel cross bracing. Alternatively, lateral stability can be introduced into concrete frames by stiffening other elements such as cores, shafts or stair enclosures. The inherent fire resistance of concrete provides structures that require no further fire protection measures. With cast-in-place construction, the ability to re-use formwork is important in keeping its use economic. Each 'lift' of concrete, typically a floor with its supporting walls and columns, takes longer to construct than an equivalent steelwork structure, since the concrete takes time to reach an adequate strength to allow another floor to be built on top. Although the construction of a cast-in-place reinforced concrete frame is considered slower than erecting a steel frame, the construction times can be matched if the concrete work is well organised. Frames in concrete combine easily with ground structures such as retaining walls and foundations because they are in the same materials and can be continuous.

## Joints and Connections

Stiffness at the junctions of the frame is important both to provide

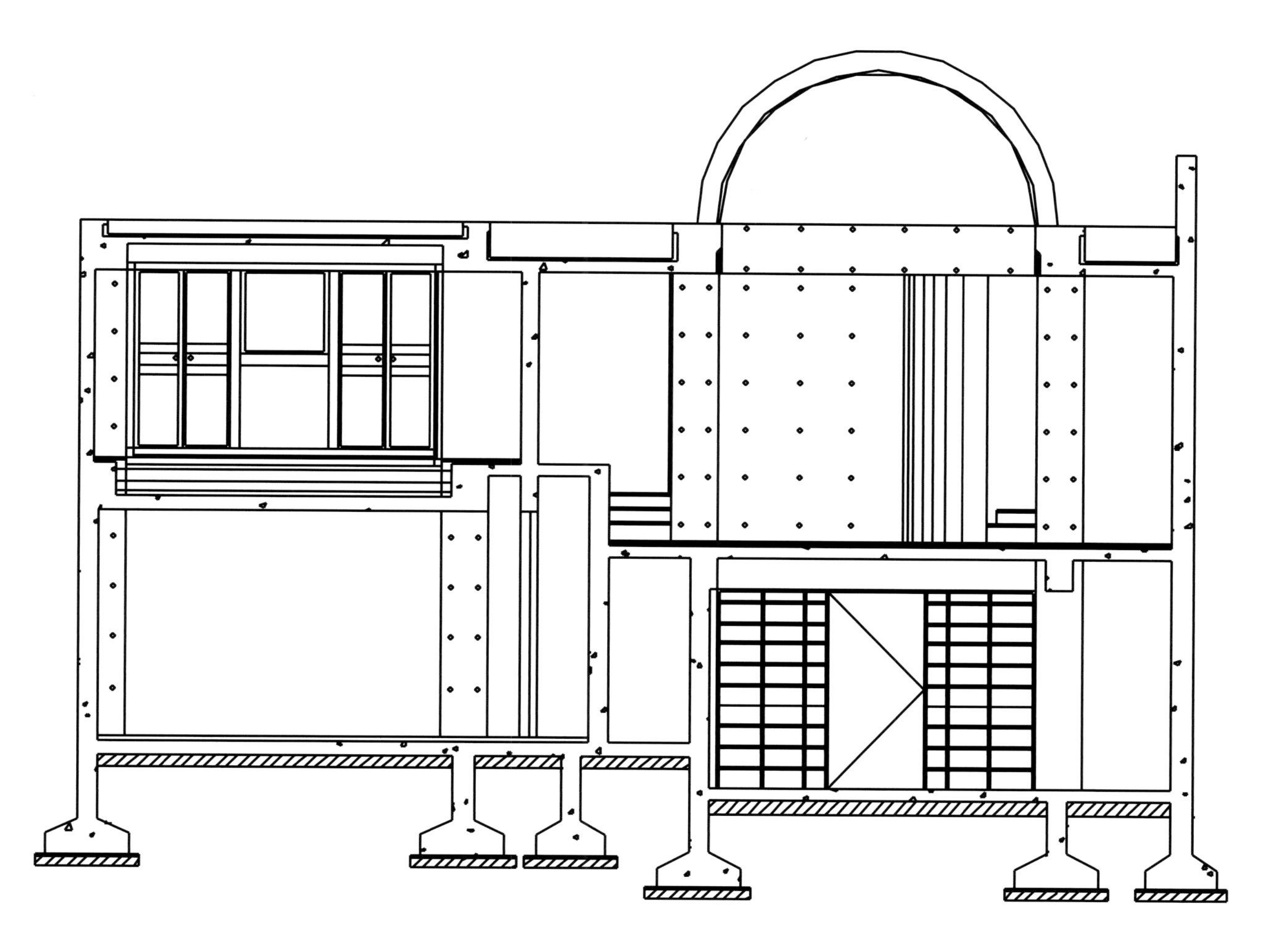

Vertical section

sufficient rigidity in the frame itself and to make its deflections compatible with the cladding and internal elements such as staircases. This can be achieved either within the joints forming the frame, by including extra reinforcement, or by using additional stiffening walls such as those used to enclose elevator (lift) shafts and service cores. Typically, deflections in concrete beams or slabs are designed to be compatible with the external cladding, partitions and fittings within the building.

Movement joints are introduced to allow for thermal movement. These junctions are effectively breaks in the structure, resulting in the overall structural frame comprising a set of smaller linked frames, each an independent stable structure. Movement joints are usually taken through the external walls and roof, requiring envelope details that allow for structural movement while remaining weatherproof.

## Interface with external envelope

Concrete frames can be either exposed outside the weather line of the cladding, that is, the non-loadbearing walls, or be completely enclosed by the external envelope. When exposed externally, the penetration of the frame through the walls and roof requires careful detailing to overcome weatherproofing and thermal bridging around penetrations of structural members through the cladding. Grooves are often cast around penetrating beams to allow flashing pieces to be inserted to assist in forming a weather seal. Cladding is fixed either by a continuous stainless steel channel that is cast into the edge of the concrete slab, or by expansion bolts secured into the slab when the cladding is fixed. The bolts may also secure stainless steel angles, brackets or continuous rails depending on the nature of the lightweight cladding. Reinforced concrete panels can be secured with additional metal pins cast into the top surface of the concrete slab.

88B

88C

Steel frames are made from columns and beams that are fabricated off-site and assembled together on site. Unlike concrete, they can be assembled easily with either rigid moment connections or with pin connections. Lateral stability can be provided with either cross bracing of some bays, or with staircase enclosures, cores or shafts. Steel frames are usually shot blasted, cleaned and primed prior to erection. Frames that are not to be left exposed visually require no further painting. Most steel structures require fire protection, which is provided with either concrete encasement, an intumescent paint or by enclosing the frame with fire resistant board. Most steel frames are painted to avoid corrosion during construction and after the frames are assembled. Members which are to remain visible when assembled, such as in visible roof structures, are often delivered to site as pre-painted components that can be assembled on site and lifted into place by crane, avoiding the need for substantial scaffolding. Care must be taken not to damage the surface finish, as touching up of galvanizing and paintwork on site is slow, laborious and more difficult to achieve where a good quality of finish is needed.

## Joints and Connections

Joints in steel frames are bolted or welded. Bolted connections are the most commonly used, with either cleats or plates used as connectors. The use of steel frames allows partial steel floor decks to be used instead of concrete slabs. This allows the floors themselves to be built at a faster rate. Concrete is used in composite action with profiled steel sheet to form a composite deck structure that spans between steel beams within the frame. In addition, the concrete provides essential properties of fire protection and acoustic separation.

Movement joints are introduced to allow for thermal movement in the steelwork, which can be greater than that of concrete. These joints are breaks in the structure resulting in a series of smaller linked frames, each of which must be independently stable. Movement joints on external walls and roofs require details that provide weatherproofing as well as the various movements occurring along its length.

## Interface with external envelope

Like concrete, steel frames can be either exposed outside the external skin or be set inside the building. When steel sections are exposed externally, the penetration of the frame through the external wall can be complicated by the difficulty of forming a seal around the profile and the effects of thermal bridging. The profile, typically an I-section, is boxed-in with a casing to provide a simpler profile around which to seal. The casing may be partially insulated to avoid the thermal bridging.

7A

88E

Details

1. I-section column
2. I-section beam
3. Floor deck
4. Base plates at foundation not indicated
5. Cleats formed from steel angle
6. T-section or flat sections commonly used for cross bracing

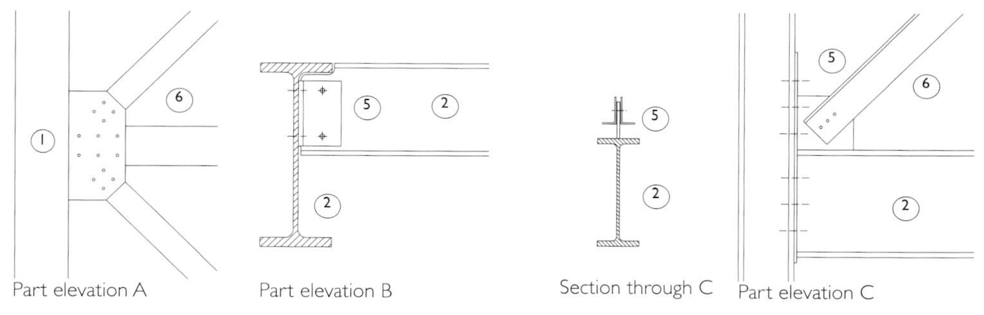


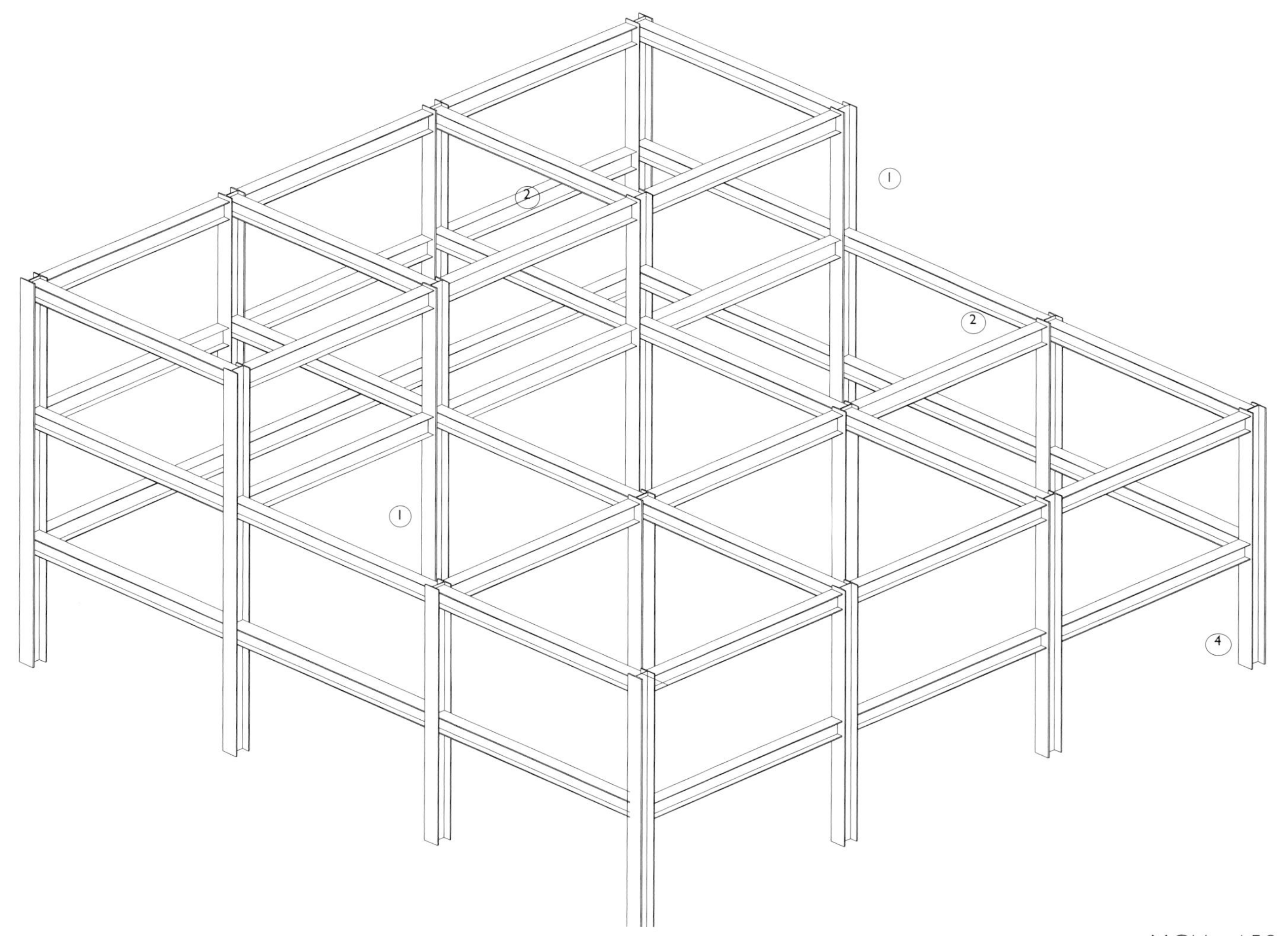

45A

45B

Details
1. Portal frame
2. Cladding
3. Concrete slab

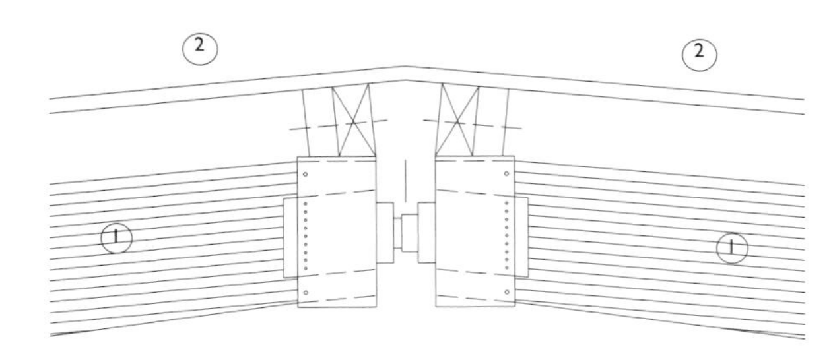


Pin detail. Vertical section

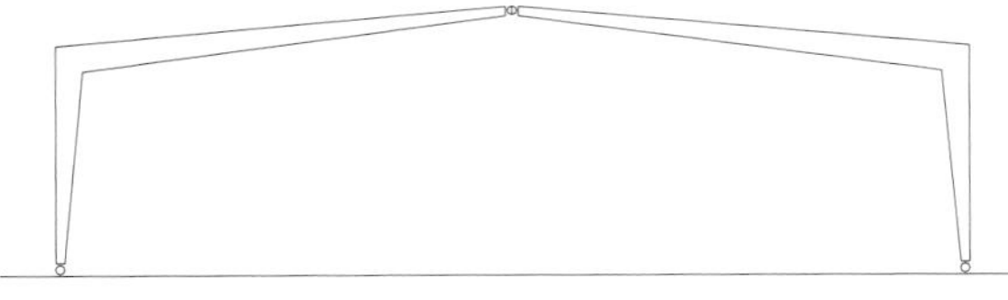

3-pin frame

2-pin frame

Rigid frame

The portal frame supports loads by bending. The frame is stable in the plane of the frame through the rigid moment connection between the column and the rafter, which can be either a beam or truss and resists bending by the use of rigid connections. The principle of the portal frame can be applied to multi-storey applications, and can be found within a wide range of structures. The main advantage of this type of structure is in its economic use of material, combined with a simple construction technique. A wide range of structural components can be used, from trusses and trussed columns to I-sections. Portal frames are typically used to provide economic structures for long-span single storey buildings through their low weight.

Portal frames are made from steel, laminated timber and reinforced concrete. As is the case with arches, there are three types: the rigid frame, the two-pin frame and the three-pin frame. In the rigid portal frame, the structure uses the least material, since the rigidity of the joints is extended to ground level. The pinned types act in a similar way to arches but the loads in each member are resisted in bending. Portal frames are linked together with purlins to form linear arrangements at 3 to 6 metre centres. Since the purlins alone do not provide sufficient rigidity in the transverse direction, the frames are stiffened with lateral bracing in a few of the bays, usually near the ends of the structure.

## Joints and Connections

· Moment connections are either welded or bolted with plates to form haunches.
· Pinned connections use types illustrated here.

## Interface with external envelope

Purlins are used to support the external cladding. Where profiled metal cladding is used they are called sheeting rails. In common with arches, cladding is fixed to these purlins rather than directly to the portals in order to keep the connection type simple. The stiffness of junctions between walls and roof makes the transition in the external envelope straightforward and economic.

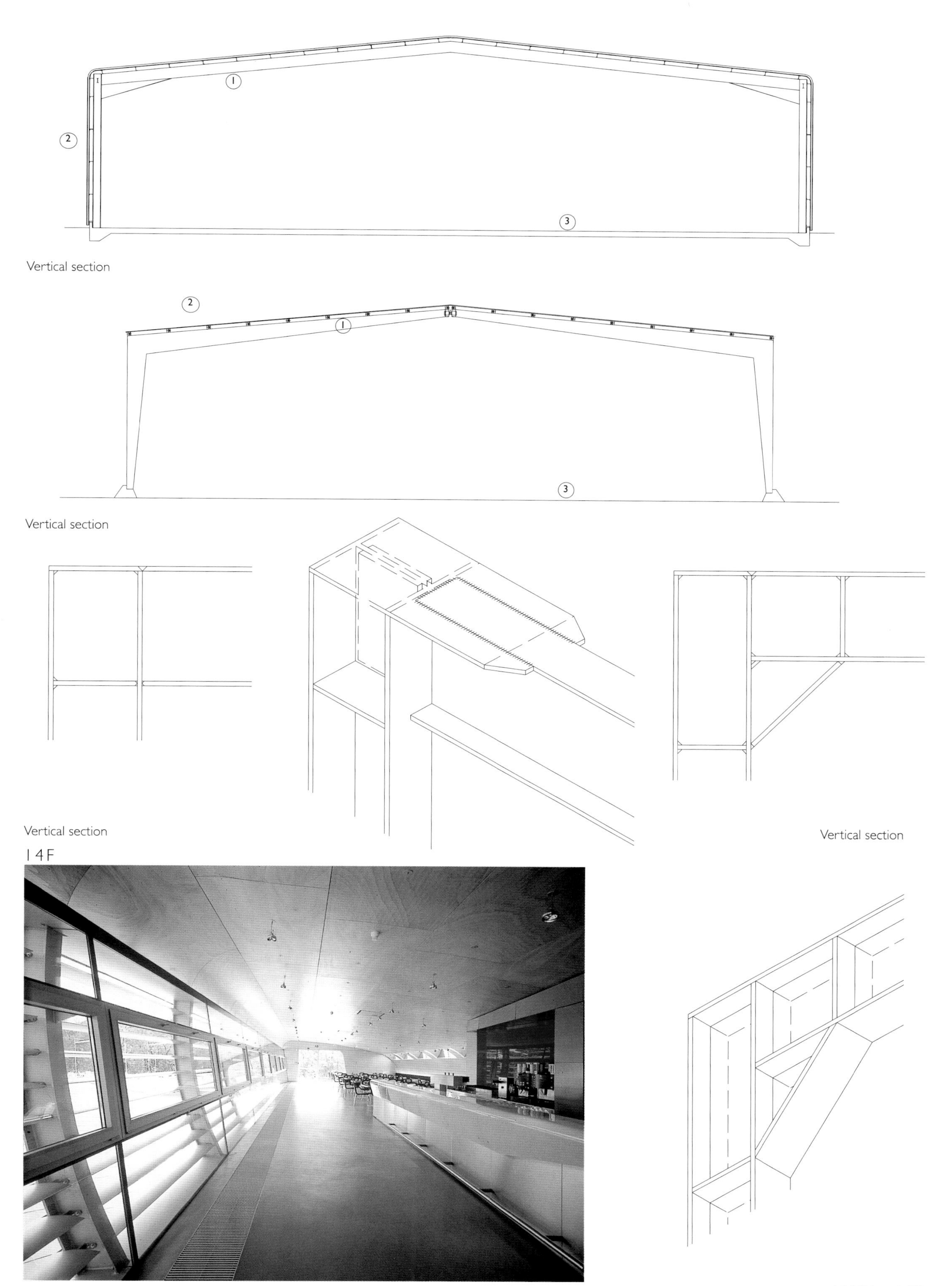


Vertical section

Vertical section

Vertical section

Vertical section

14F

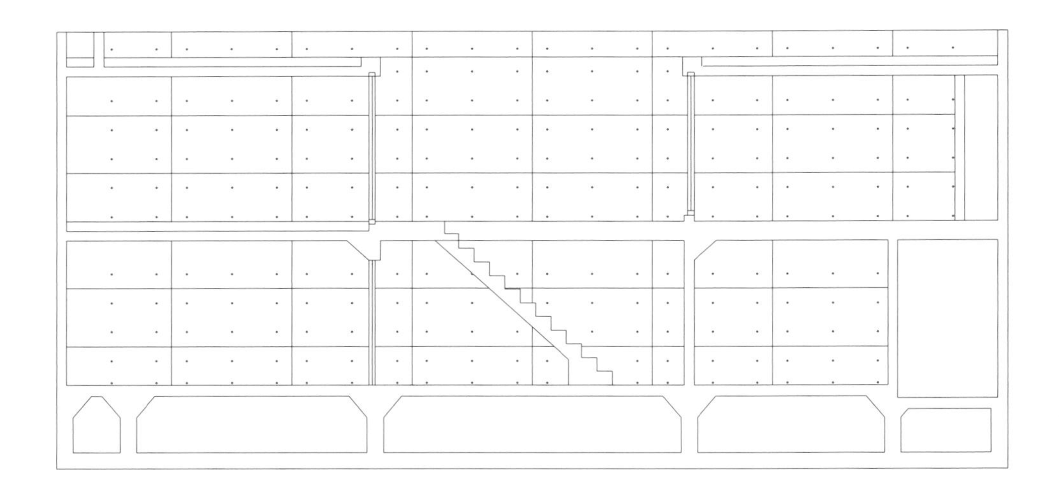

Vertical section

52A

44B

44A

In this section, are considered structures where loadbearing walls and floors are linked together with some degree of interdependency. Although not strictly regarded as a structural element, techniques of loadbearing concrete wall and floor construction are very common in Europe for large-scale housing projects. Most European houses are built with loadbearing masonry walls with either timber or concrete floors that provide stability to the external walls.

## Concrete Loadbearing Boxes

Loadbearing walls and floors made in reinforced concrete can be combined to make a complete monolithic structure. Both cast-in-place and precast concrete techniques can be used, though precast methods suit projects where a high degree of repetition occurs in panel types. Concrete loadbearing boxes have the advantage of good fire resistance combined with good sound insulation from both airborne and impact sound. Concrete can be used in conjunction with both steel and concrete structures. Cast-in-place techniques generate a monolithic connection between floor and walls. Precast components are stitched together to form a similar connection. Since concrete is not regarded as vapour-proof, the outside of the concrete is either rendered or covered with a cladding system, typically a rainscreen.

15B

15A

15C

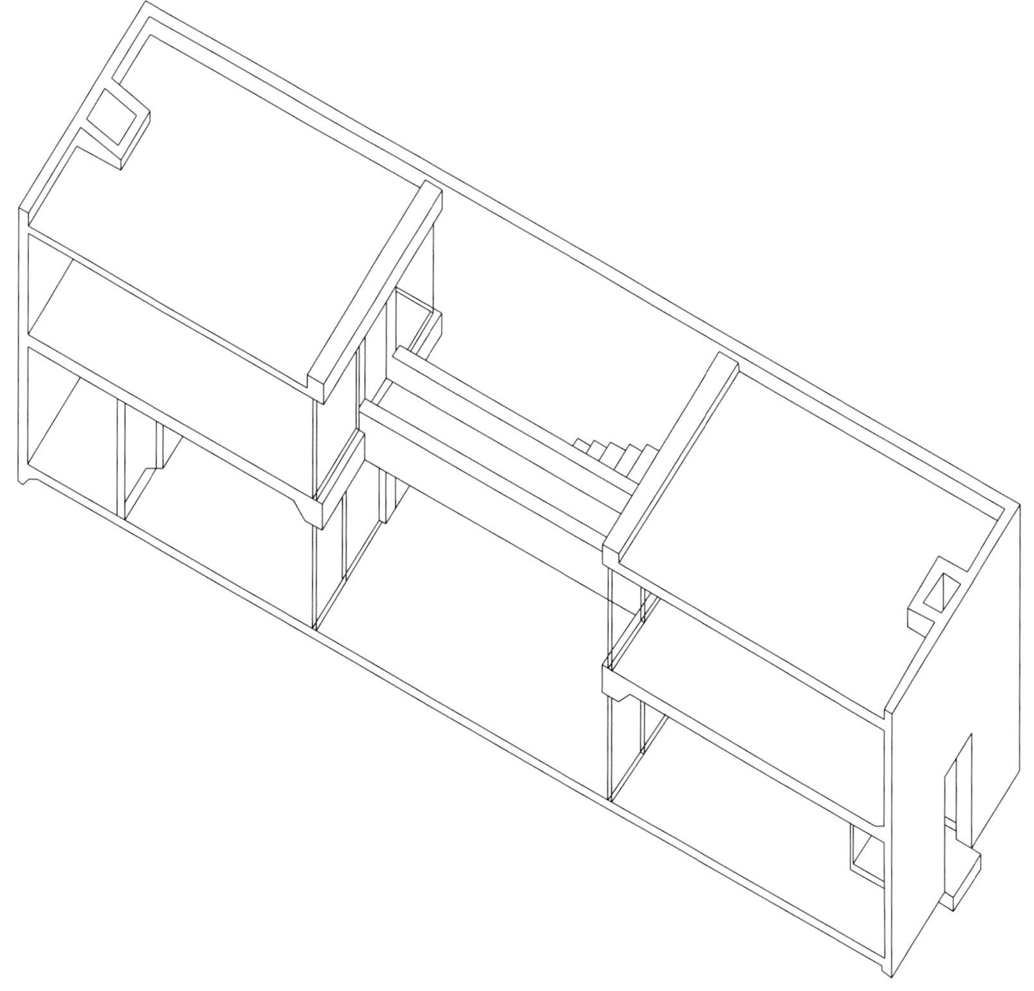

81A

26A

26B

Loadbearing brick structures are rarely used in buildings in the industrialised countries since they are more expensive than cavity walls, use more material and are slower to build. Over the past 100 years, brickwork has been designed to be ever stronger through the use of higher strength mortars even though this results in a brittle, less flexible structure. This has led to the need for movement joints in the material, usually at around 6.5 metres (20 ft) centres, with the effect of loadbearing walls becoming a series of linked panels. The gradual re-introduction of lime putty mortars, with their lower strength but greater allowance for movement, is set to change the nature of loadbearing brickwork, and its use is discussed in the Walls chapter.

The use of brick in these structures is restricted to the walls, since traditional vaulted brick floors have been superseded by timber or concrete floor construction, which can more easily achieve longer spans. Loadbearing walls are usually much thicker than their cavity wall equivalents, making the structure much heavier, in order to provide sufficient waterproofing, since the brick absorbs rainwater near the surface and later dries out. Traditionally, openings in walls have been made with arches of the same material as the surrounding wall and in flat, curved or pointed variations. Flat arches require considerable skill to construct, and are often reinforced with bed reinforcement in the form of steel rod or expanded mesh. In common with other types of masonry construction, brickwork cannot be loaded, and will fail in tension.

Loadbearing brickwork was used in the Indian Institute of Management at Ahmedabad, India, (completed 1974) and part of the National Assembly at Dhaka, Bangladesh (completed 1983) both designed by Louis Kahn. The architect had previously used brick slips (thin brick tiles that imitate brickwork) to conceal openings formed in either steel or concrete. In these two projects structural

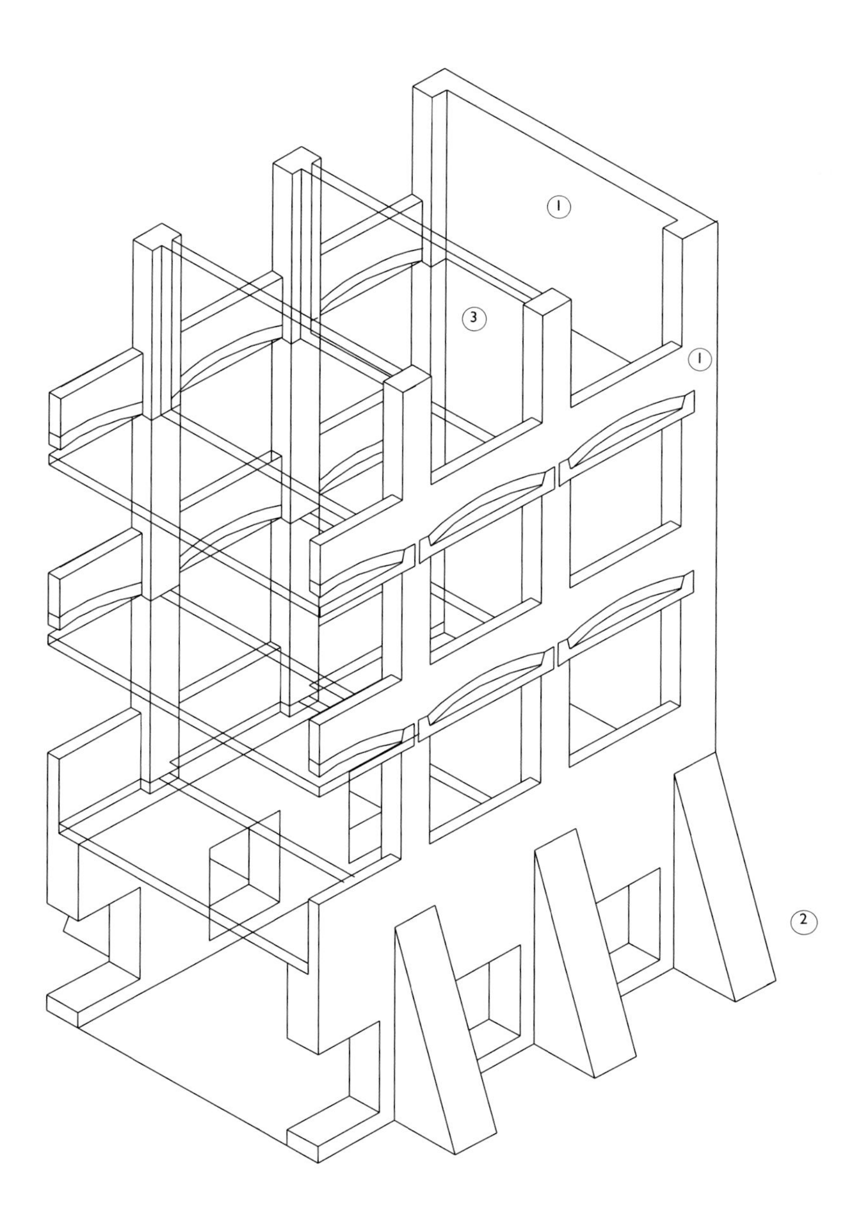


Details
1. Loadbearing brick wall
2. Buttress if required
3. Reinforced concrete floor slab

brick arches are used. Because of their large size, spanning up to approximately three metres (9ft 9in) they were combined with reinforced concrete lintels as tied arches. The ends of the ties were specially formed to bond into the adjacent brickwork, thus avoiding the need for additional steel connectors.

## A recent built example

Glyndebourne Opera House has loadbearing brick walls, which consist of two skins 220mm (8in) thick with a 50mm (2in) cavity providing acoustic separation. The inner leaf supports the back of the balconies. The brickwork forms a loadbearing drum, 33.7m (112ft 4in) in diameter, 17.7m (59ft) high, truncated on one face where it intersects with the fly-tower. From this wall radiates a series of pre-cast concrete panels that form the soffit to the balconies and partially support the seating above. Cast-in-place columns that are tied together with a cast-in-place topping and ring beam provide additional support. Where the structure runs alongside the side stages, the inner wall becomes an acoustic wall, 220mm (8¾in) thick, but elsewhere it is less dense, and where it is punctured by openings, providing access to the auditorium, it becomes a series of piers. The walls supporting the balconies are 334mm (1ft 1 3/8 in) thick continuous skin, supported by flat arches resting on gently tapering brick piers. The mix used for the lime putty mortar in the brickwork was 1:2:9 (cement:lime putty:sand) ratio by volume. The cement gave early strength to the wall and slightly improved durability and weathering. The compressive strength of the bricks was 27.5N/mm$^2$ ($5.7 \times 10^5$ lbf/ft$^2$) and that of the lime putty mortar was 1.5N/mm$^2$ ($3.1 \times 10^4$ lbf/ft$^2$). This provided an overall compressive strength of 6.2N/mm$^2$ ($1.29 \times 10^5$ lbf/ft$^2$), which is less than half that of conventional brickwork at around 15N/mm$^2$ ($3.1 \times 10^5$ lbf/ft$^2$).

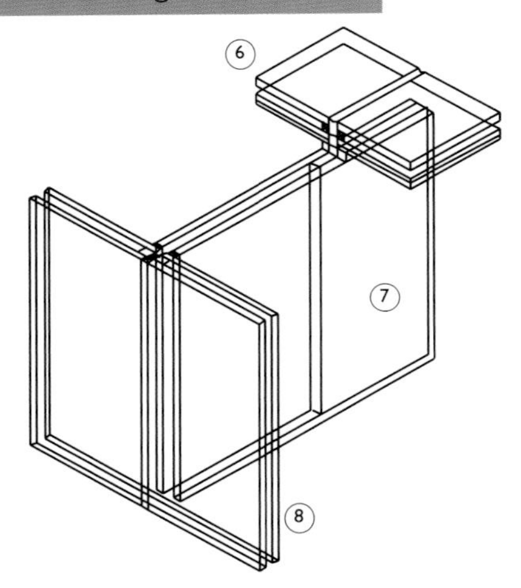


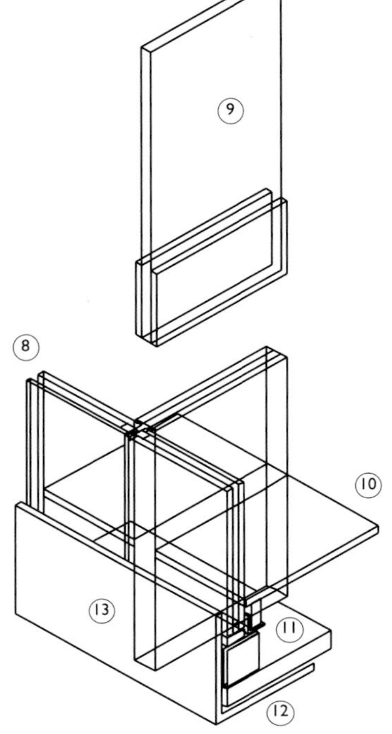

32B

All-glass loadbearing structures are a recent development. Toughened, laminated or toughened / laminated glass are bonded or bolted together to form beams, columns and panels in an all-glass structure. In common with bolt fixed glazing, an essential determinant of loadbearing glass structures is the avoidance of any stress concentrations that might lead to glass breakage while the structure is under normal load conditions. With the development of both resin-based laminates and of laminating glass techniques, this form of construction is set to expand dramatically in the next few years. Since the techniques used are very recent, no general principles for construction are set out here but two recent examples are shown.

## All-glass footbridge, Rotterdam, Holland

The Architect was Dirk Jan Postel of Kraaijvanger-Urbis working with Structural Engineer ABT Velpe. The bridge spans three metres (9ft 10in) with floor plates of 15mm (9/16in) laminated float glass supported on two laminated float glass beams. The side-walls consist of double glazed units made from 10mm (3/8 in) and 6mm (1/4 in) thick toughened glass sheet. In addition to being self-supporting, the walls support the glass roof, which has the same type of laminated construction as the walls. The all-glass components are fixed together with stainless steel brackets and plates.

## All-glass enclosure, Broadfield Glass Museum, West Midlands, England.

The Architect was Brent Richards of Design Antenna working with Structural Engineers Dewhurst MacFarlane. The enclosure is constructed without metal fixings or supports in any another material. It is 11m (36ft) long, 5.7m (18ft10in) wide and 3.5m (11ft 5in) high. The primary structure consists of glass beams 5.7m (18ft10in) long x 300mm (1ft) deep, at 1000mm (3ft 3in) centres. One end is supported by an existing masonry wall, whilst the other is supported on glass columns 3.5m (11ft 5in) high and 200mm (10in) deep.

The beams and columns are made from three sheets of glass laminated together, making them 32mm (1 1/4in) thick. At the rear, the beams

32A

Isometric view of assembly
glass loadbearing boxes

1. Glass roof panel
2. Glass wall panel
3. Steel frame at each end of bridge
4. Laminated glass beam
5. Glass sheet as floor deck

Details
glass loadbearing boxes

6. Triple glazed unit
7. Laminated glass beam with slot for column
8. Double glazed unit
9. Laminated glass column with projecting lamina that slots into beam
10. Glass at floor level to close gap between floor and wall.
11. Thermal insulation
12. Steel angle fixed to floor
13. Steel shoe at column base.

are secured by shoes in steel fixed to the wall, while on the glazed side they are connected to the columns in an interlocking of the glass layers. The junction was bonded on site with a catalyst-cured resin. The glass roof panels are bonded to the top of the glass beams. The double glazed roof panels have an outer layer of 10mm (3/8in) thick toughened glass, a 10mm (3/8in) air gap and an inner skin of two sheets of 6mm glass laminated together. A silver film on the inner face of the upper sheet reduces solar gain, as does the ceramic fritting on the inside face of the panel.

The roof panels are bonded to the beams with structural silicone bead with a rigid foam backing. The gap between the panels is sealed using a silicone-rubber weather seal with a foam backing. The roof slope is 1.5°. The roof can be walked upon for maintenance, and support expected snow loads. The front walls consist of 3.7 metre (12ft) x 1.1 metre (3ft 7in) high double glazed panels composed of two sheets of 8mm (5/16in) toughened glass sheet with a 10mm (3/8in) air gap.

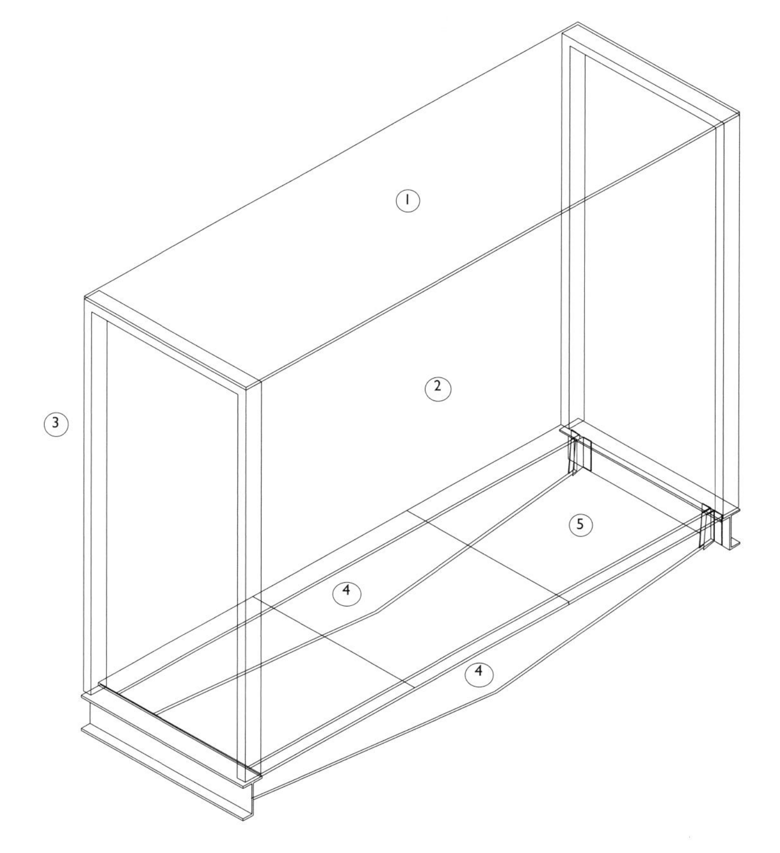

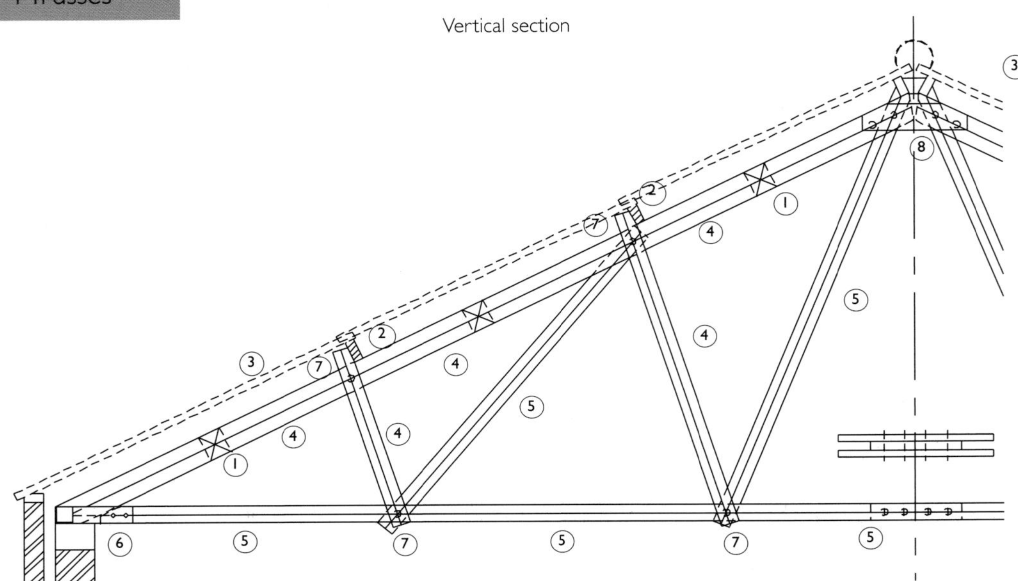


Details
1. Packing piece
2. Purlins spanning between trusses
3. Roof covering
4. Strut
5. Tie
6. Timber wall plate
7. Bolted steel connection
8. Steel connector plate

Truss types
9. Warren truss
10. Modified Warren truss
11. Pratt truss
12. Vierendeel truss

Trusses are used for large spans where a solid girder would be at a disadvantage due to its self-weight. Instead of using a deep solid section, such as a girder, a more economic solution for larger spans can be provided by transmitting forces across a series of diagonally set and connected members to take up the vertical, horizontal and diagonal shear stresses. With the exception of the Vierendeel type, trusses use pin-ended members that utilise tension and compression in a series of ties and struts. While trusses bend along their overall length, local bending within the truss can be largely avoided if loads onto the truss are applied at the node points. Loading trusses away from the nodes introduces local bending into horizontal members with a resulting increase in their design size, as members now have to resist bending as well as axial loads, resulting in trusses that are heavier, usually less elegant and more expensive. The voids between members are often used to accommodate services and mechanical ventilation ducts. These structures maximise the lever arm between the compression and tension flanges to generate a greater moment of resistance in order to provide a greater loadbearing capacity while minimising weight. Simply supported trusses exert no thrust at their supports allowing them to be easily supported on columns or supporting walls within a larger structure.

Truss types vary enormously; the most common types being the Warren trusses, where loads are carried mainly as axial loads in the members, the 'N' truss which has vertical and diagonal members in a rhythm of alternating tension and compression, and the Vierendeel truss, which has an orthogonal rather than a triangulated series of members. In the Vierendeel truss, forces are transferred between members by localised bending, called vierendeel action, with rigid connections between members. As such, it is not strictly a truss type. The bending makes members slightly larger than in an equivalent triangulated truss, resulting in their being heavier than a Warren truss spanning the same distance.

## Joints and Connections

Timber roof trusses are either large trusses set up to three metres (10ft) apart linked together with purlins or much lighter nailplate trusses set at around 450mm (16in or 24in) centres. Larger trusses often use double timber members forming the main rafters and joists which are spaced apart with secondary ties and struts which link the primary members. Timbers are bolted together using large steel washers that spread the load imposed by the bolt head and prevent it from crushing the timber locally as the connection is tightened. The exposed steel components are painted to avoid their corrosion. Timber trusses can also make use of ties made in steel rod where they are visible components. Smaller trusses are factory assembled with nailplates and split ring connectors as described in the Materials chapter.

Steel trusses can range from a modest set of angles bolted together with linking gusset plate connectors to welded tubular structures that can incorporate cast node connections to allow connections with complex geometries to be formed. These cast connectors allow the loads to meet at the intersection of the centrelines of members in order to minimise the diameter of the tubes. Modest steel trusses use angles set back to back and bolted together with gusset plates

set between them to allow several members to be joined together at a single node. They are in widespread use as economic long span supports to roof structures.

Reinforced concrete trusses are usually precast either as a series of prefabricated components which are stitched together or as a single completed component. They are much less commonly used than those in timber or steel due to their self-weight, which tends to reduce the lightness and economy associated with trusses.

61B

66C

## Interface with external envelope

Ideally, the external envelope is connected to trusses through the secondary members, or purlins, which connect trusses together at the node points. In addition to avoiding local bending in truss members, the fixing of cladding to secondary supports permits the use of simple bolted or screwed connections that avoids any complexity in fixing to a primary member. Cladding systems usually like to use a single connection type that can be used throughout, as is the case with a system of purlins.

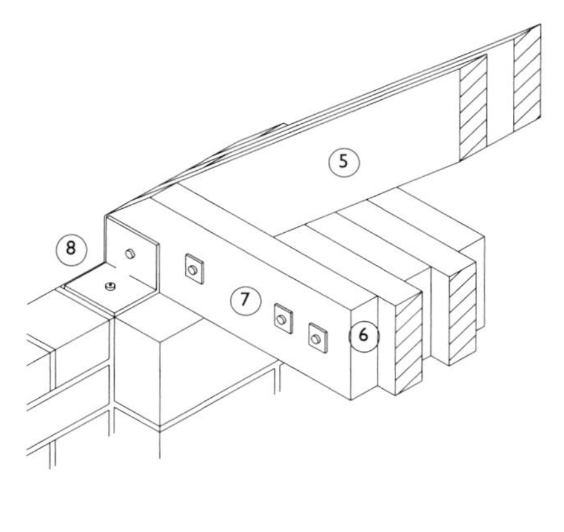


61C

2B

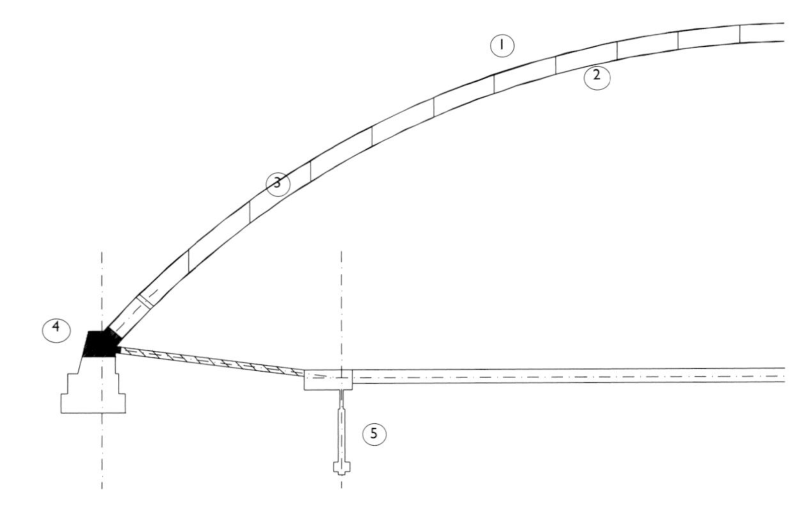


2C

The essential concept of the arch is to support loads primarily as compression forces in the plane of the structure, with abutments at the base to resist the outward thrusts generated by its shape. A variation is the tied arch in which a tension member links the base points across the arch to resist the outward thrusts and control spread of the arch. The three most common types used in modern construction are the rigid arch, the two-pin arch and the three-pin arch. These two-dimensional structures are linked together in bays with purlins to form a complete structure. Alternatively, the arch concept can be extended into three dimensions as a shell. Both arches and shells can be formed either from truss components or as monolithic structures.

## Arches

The rigid arch has no pin joints, but has rigid connections at both base points. It is more economical than the two other types, but has bending moments at its base that are transmitted into the foundations, which may need to be linked by ground beams to resist the thrusts, in the manner of a tied arch, unless the ground is sufficiently firm to withstand the thrusts without a beam. The two-pin arch has pin joints at the abutments only. It is more economical in material than the three-pin type and bending moments are more evenly distributed than in the three-pin type, minimising the amount of material used. The three-pin arch is hinged at three points-at the apex and at the two base points. Bending does not occur at the pin joints, which behave as pivots or hinges. Bending moments away from the pins are greater than in other types. As with the traditional masonry arch, there are horizontal thrusts at the supports. Thrusts at the base of an arch increase as the arch profile becomes shallower. Both the three-point and two-point arches suit larger spans which would present difficulties for transport and handling if made in a single piece. The most common materials used are steel, laminated timber and reinforced concrete.

6

7

8

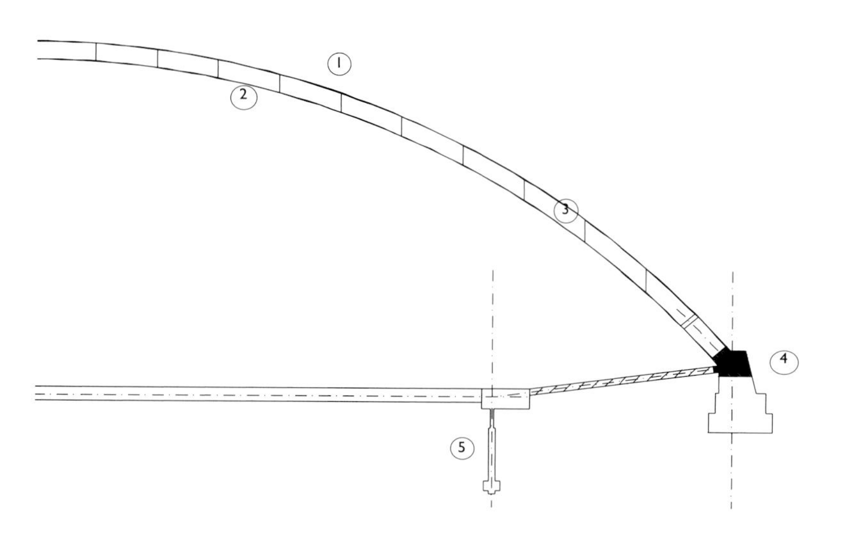


60B

Section through shell
1. Outer shell
2. Inner shell
3. Concrete ribs
4. Ground level
5. Anchorage piles

Arch types
6. Three pin arch
7. Two pin arch
8. Rigid arch

## Shells

Shells are of two generic types; those made from framing members to which cladding is applied and monolithic shells which are made either by casting concrete in place over complex formwork or as pre-cast sections which are stitched together. In both cases, the loads are carried in the plane of the structure with the outward thrusts being taken out at the base. Shells constructed from framing members often have a ring beam at the edge to take out the outwards thrusts. Monolithic concrete types may either follow the same principle or have a continuous abutment at their base to deal with the higher loads associated with concrete. Shells in which the structure and enclosure are combined in a single form, as found in an egg shell, are referred to as monocoques. Shell forms vary enormously, from the relatively simple shell vault resembling an extruded arch, to the complex ribbed structures resembling seashells.

## Joints and Connections

Arches and shells made from framing members are formed using the pinned and moment connections associated with each material and are illustrated here.

## Interface with external envelope

Monolithic arches and shells are covered externally in continuous membranes where rainwater is collected at the base of the structure. Membranes used vary from polymer-based types that are bonded to the structural substrate, to standing seam metal sheeting.

Trussed arches and framed shells provide the opportunity for transparency, where cladding panels can be used. The type adopted is usually expected to be suited to all conditions of the curved geometry and is designed to suit the most onerous conditions for weatherproofing. A single cladding panel type is normally used to provide visual continuity across the surface of the structure.

29A

60A

30

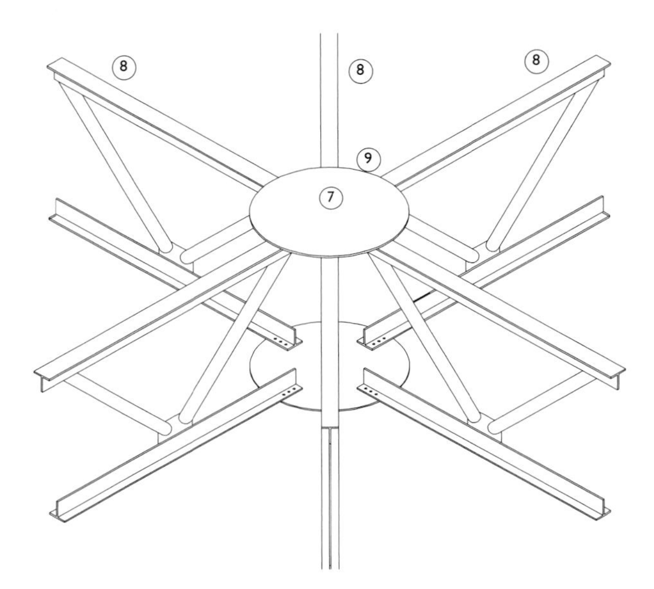


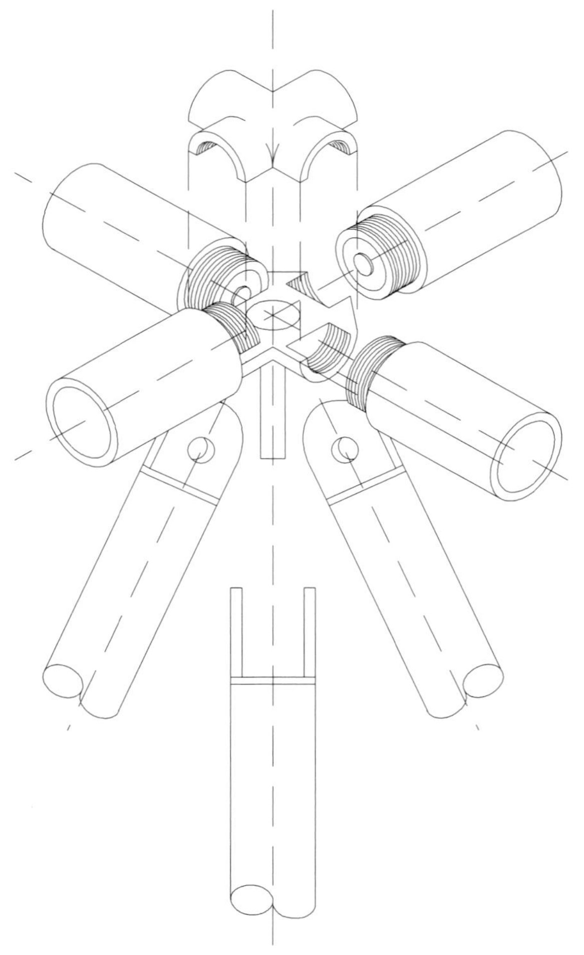

Space frames and geodesic domes are the most well known types of space grid. They are effectively three-dimensional trusses. One of the earliest space frames was the Mero system designed in the 1940's by Max Mengeringhausen, and is still used today. A development by Konrad Wachsmann in the 1950's was based on the tetrahedron. The space frame comprises a series of triangulated members linked by node connectors. The general principle is to use a single node type throughout, though some geometries may require a different connector at the edges. Geodesic domes were developed by Richard Buckminster Fuller in the 1940's. Fuller believed that structures in nature behave in a way that internal forces usually act in the direction of minimum effort with a maximum gain achieved with a minimum of energy input. Rather than apply this idea to the form of the plant leaf or the animal shell, Buckminster Fuller interpreted triangular crystalline forms found in nature, using triangulated frames. Like the space frame the geodesic dome is essentially a continuous triangulated structure but their geometry often allows them to be used as single-layer structures without the need for introducing triangulation into the depth of the structure. Space grids provide long-span structures for roofs and walls. Space grids can take on a huge variety of forms from vaults to supporting structures for large glazed walls. Deflections in space frames are small due to the overall stiffness of the frame. Steel is most commonly used because of the strength and stiffness of the material combined with its comparatively low cost which would make aluminium a less viable option.

## Joints and Connections

The simplest structural element is the pyramid, made as a frame of connected rods or tubes. More complex versions use square- and polygon-based geometries. Rods are connected together using a node joint which is repeated throughout the space frame.

The essential component of the space grid is the node connector. Very simple types with simple geometries use a flat plate connector to which steel angles are bolted in the manner of a truss. Complex geometry types use a cast or machined node that has threaded holes to receive the ends of the threaded connectors of the ties and struts. These form part of proprietary systems designed by manufacturers.

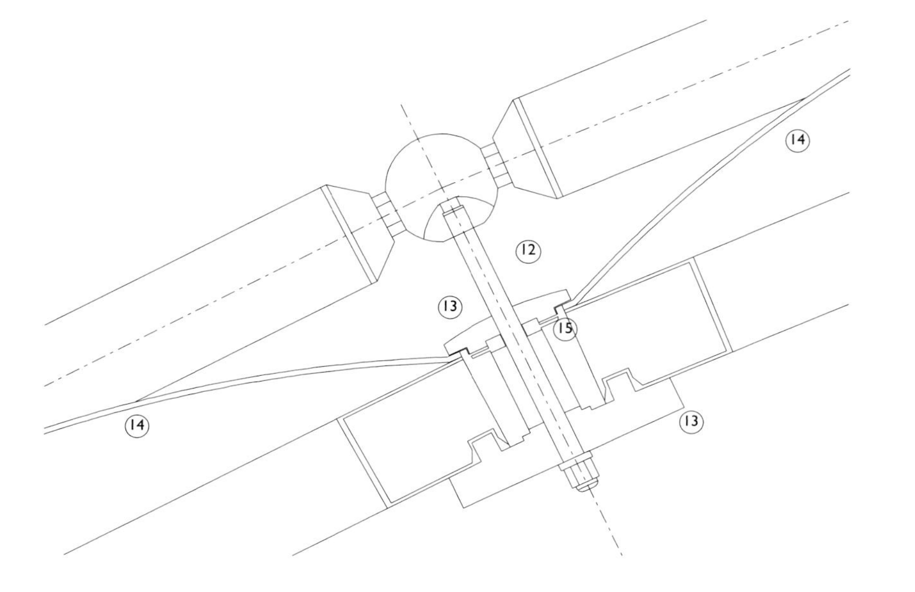


Plan and section
1. Brackets to support space frame
2. Node connectors
3. Space frame supported on loadbearing wall
4. Space frame supported on columns
5. Line of top chords of frame
6. Line of bottom chords of frame

Details
7. Cleat to connect roof covering
8. Prefabricated beam
9. Metal plate node connector
10. Externally fixed structure
11. Spherical cast connector
12. Fixng bolt
13. Pressure plate
14. Drained and ventilated cladding panels (acrylic rooflights shown)
15. Rubber-based seal

## Interface with external envelope

An essential advantage of the space grid is that cladding panels can be fixed directly to the structure without the need for additional rails or purlins. Cladding is fixed either with lugs or brackets welded to tubular members, or is fixed directly to the node connectors. Where rectangular hollow section is used, cladding can be fixed directly to structural members without the need for welded lugs, provided that the potential corrosion caused by penetrating the tube is taken into account. Bolt-fixed glazing is particularly suitable for space grids since the glazing connectors or spring plates can be fixed directly to the node connector.

Cladding conditions can vary from flat, to inclined, to curved. Cladding panels follow the geometry of the supporting structure, to which they are fixed. In flat conditions, the cladding drains to gutters which run between panels and follow the lines of structural members beneath. In inclined conditions, the cladding is drained like a glazed roof, with water running quickly down gutters with the possibility of junctions in the outer skin becoming submerged beneath water. Great attention to detail is required in this application.

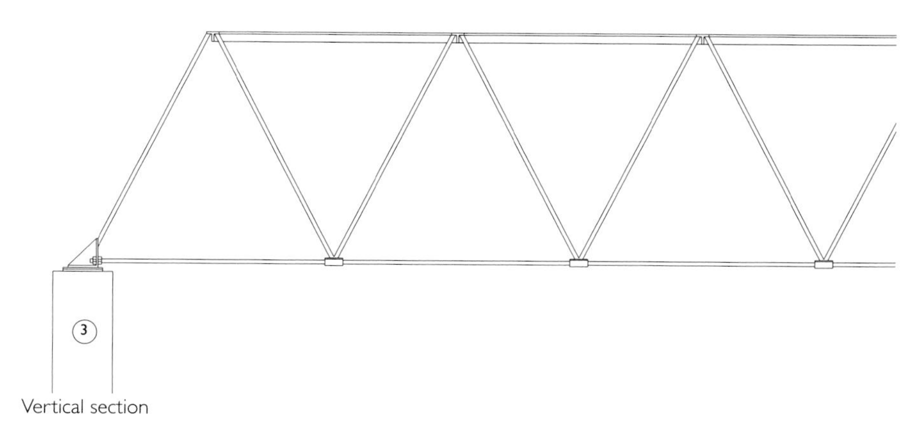


Vertical section

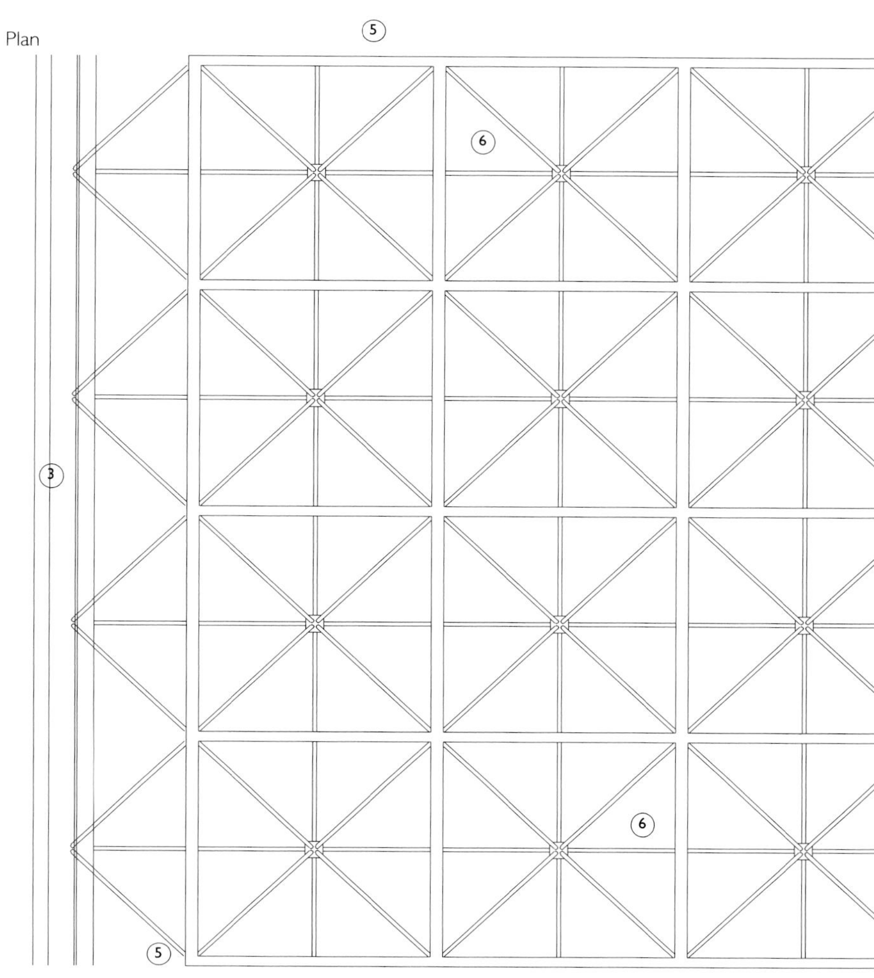


Plan

104D

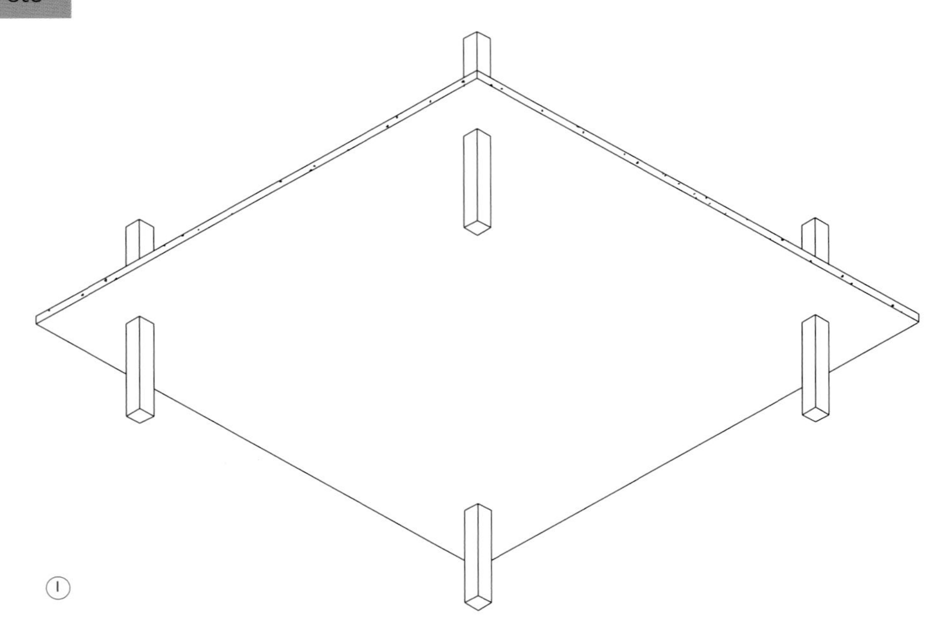


Sections through floor slabs

4. Screed
5. Concrete
6. Damp proof membrane (DPM)
7. Sand blinding
8. Hardcore
9. Earth

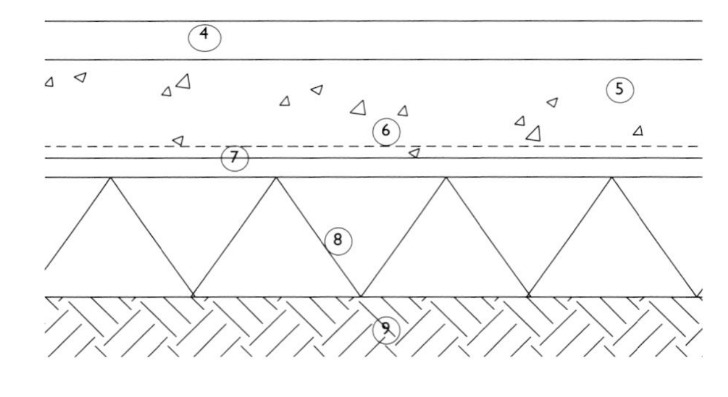


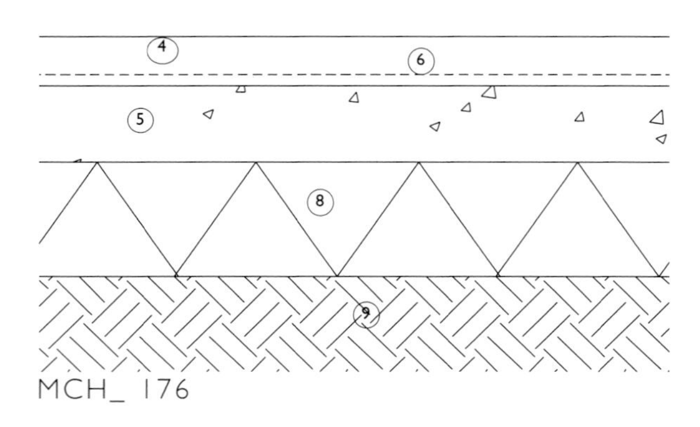


The flat slab is the simplest type of cast-in-place floor and is suitable for spans up to around nine metres (30ft). It can be used in a one-way or two-way span, approximately 300mm (1ft) deep, depending upon span and loading. In a one-way span, steel reinforcement is introduced into the bottom part of the slab and the floor is designed to span in one direction only, like a beam. In a two-way slab, the reinforcement is laid in a perpendicular grid so that the total load of the floor is more evenly distributed at its perimeter as the floor spans in two directions. Reinforcement is concentrated on the lines between columns to create beams connecting columns within the depth of the slab. The soffit is flat, providing a smooth ceiling and allowing a straightforward type of formwork to be used. Providing the tops of columns with protruding caps where the transfer of forces is concentrated can reduce the depth of flat slabs. These caps provide greater rigidity for the structure and reduce the span of the slab between columns.

Floor spans can be increased economically from six metres (20ft) to 15 metres (49ft 2in) by forming a series of downstands in the floor soffit to create a ribbed floor. Steel reinforcement in the bottom of the ribs makes this type of floor lighter than the flat slab because of the structurally efficient ribbing but the formwork is more complex.

The two-way coffered floor slab is a ribbed floor for large spans up to around 17 metres (56ft 8in) in two directions. The hollow coffers are often used to house lighting and service outlets. The floor structure gives better resistance to shear when the beams are set diagonally to the structural grid of columns but as with ribbed floors, the formwork must be specially made.

Ground-bearing slabs may be either reinforced or poured as mass concrete, depending on the degree of strength and rigidity required. Where ground-bearing conditions permit, they are supported directly by the ground beneath. Where the ground is too soft, a suspended floor is used. Ground-bearing slabs can be designed either as a raft or as a slab supported by separate foundations beneath. Ground slabs are laid on a compacted base of hardcore (gravel),

Generic types

1. Flat slab
2. Slab with downstand beams
3. One way spanning ribbed slab
4. Two way spanning ribbed slab

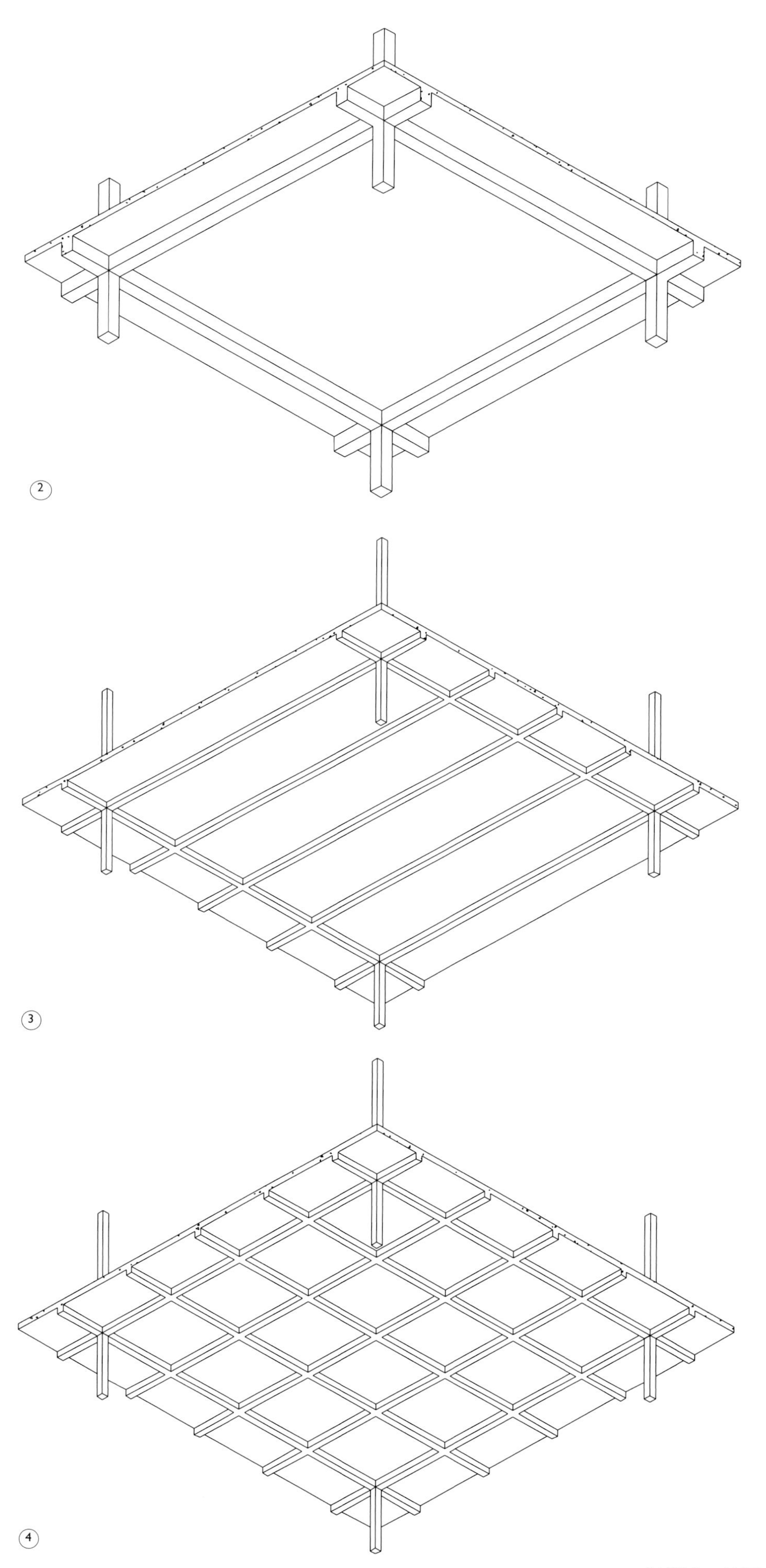


which provides a level and well-drained base. The hardcore is blinded with a smooth layer of sand onto which is laid a damp-proof membrane (DPM: vapour barrier). The sand prevents the DPM, usually a thick polythene sheet, from being punctured by the hardcore. The DPM prevents moisture rising up through the slab. The concrete is poured directly over the DPM. In addition, a DPM can be laid between the concrete slab and the screed. A liquid material is used to provide a bond to the surface of the slab. This results in a loss of bond between the screed and the slab, and a thicker screed (about 75mm) is used. The DPM is joined to the wall DPC (dampproof course: vapour barrier) to provide continuity of dampproofing. A DPM laid on top of the concrete, usually in the form of a liquid-applied layer, is used as an alternative where sealing the underside and edges of the slab is not practical. Distribution bars (reinforcing mesh) are set near the reinforcing bars in order to avoid cracking in the underside of the slab due to its being in tension under load. Reinforcement used in concrete floors can be either conventional reinforcement bars or welded mesh fabric.

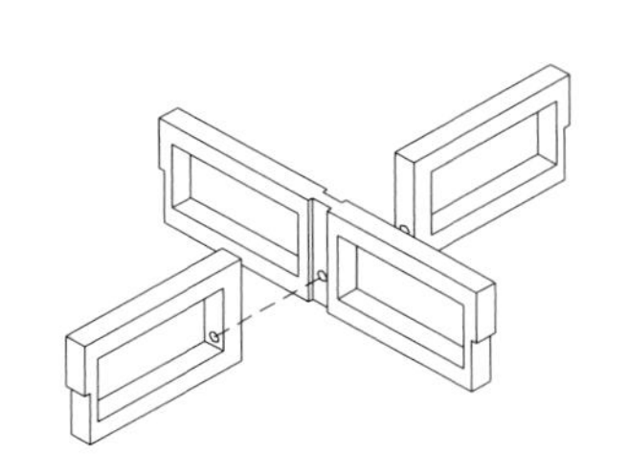

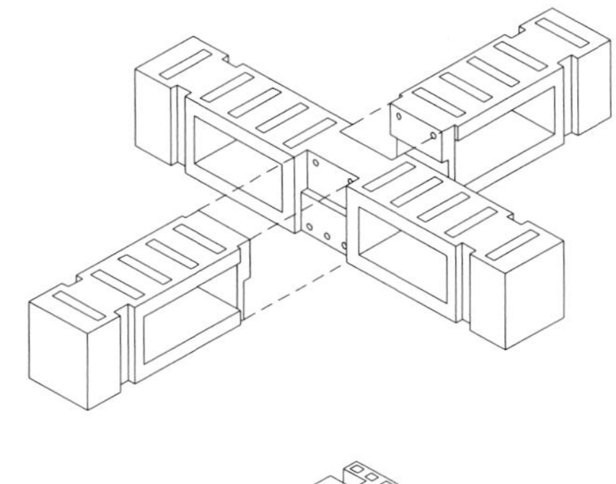

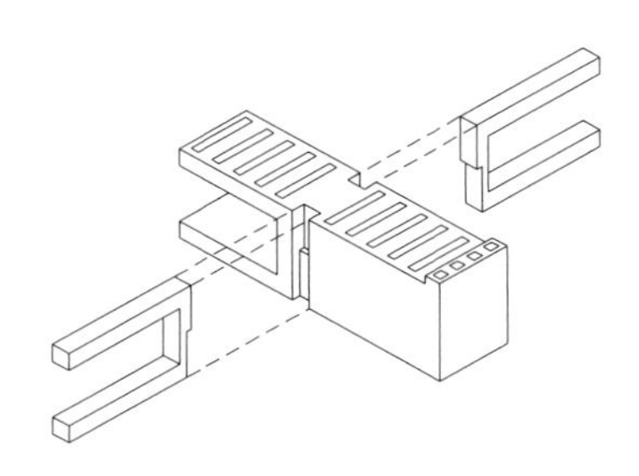

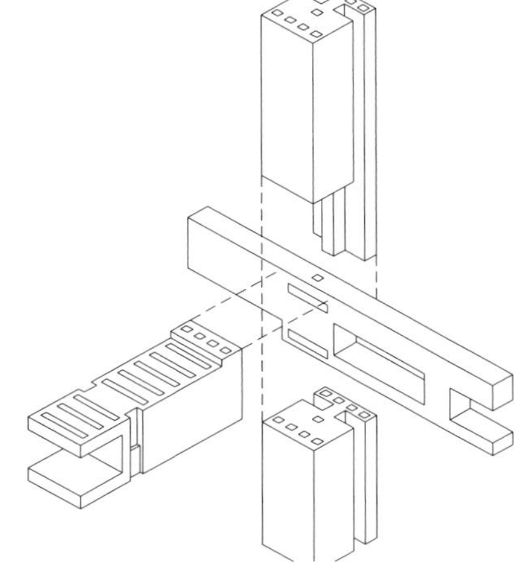

Richards Medical Research Laboratories

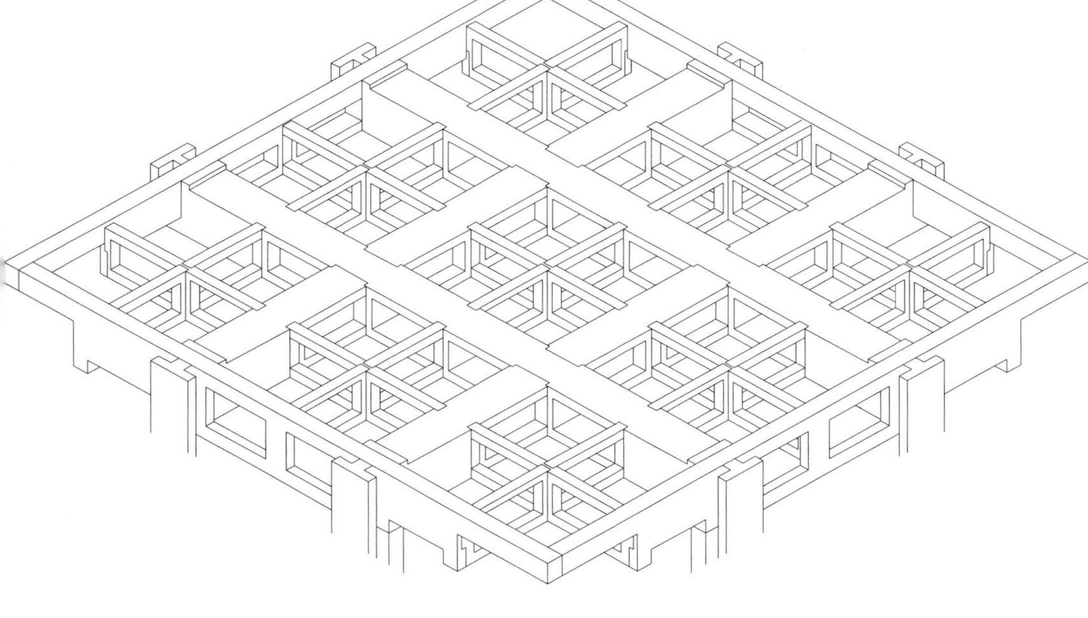

Precast concrete floors consist of prefabricated planks or beams connected together to form a deck. Proprietary systems provide longer spans than cast-in-place concrete floors but they span in one direction only. Precast floors that span two ways have been developed for specific projects by the respective design teams on the Lloyds building and the Richards building and go beyond the constraints imposed by standard systems. Two examples are shown here.

## Proprietary systems

The four generic types produced by manufacturers are as follows:

1. Solid flat slab, spanning to around 7.5 metres (25 ft).
2. Hollow core slab, spanning to around 12 metres (40 ft).
3. Double tee, spanning to around 15 metres (50 ft).
4. Single tee, spanning to around 20 metres (65 ft).

These are all essentially prestressed beams, which are stitched together to form a complete deck. Ends are supported either on beams or by a loadbearing wall, typically forming part of

Proprietary systems

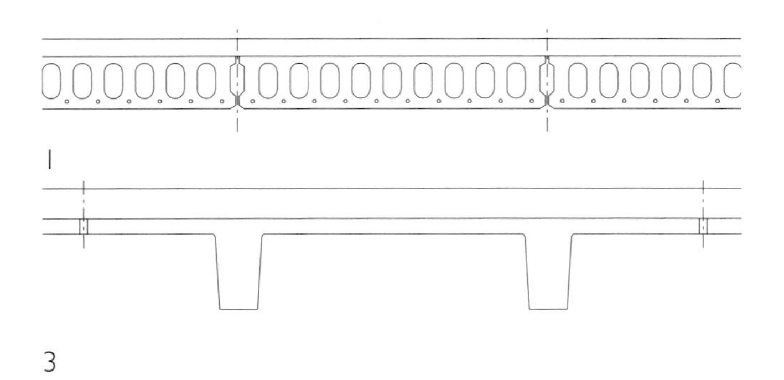


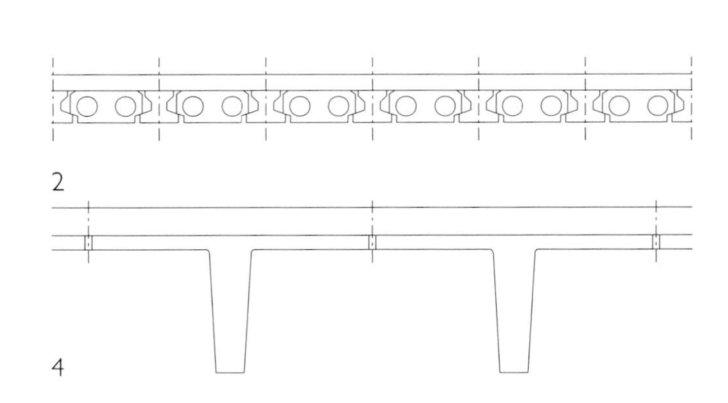

an overall precast concrete construction, which has the advantage of a rapid rate of construction. All types can be left either exposed, as is sometimes the case with parking garage construction, or be topped with a layer of concrete to form a composite structure. Openings for service runs need to be fixed at an early stage, since the size and position of openings in precast floor systems is much more constrained than is the case for cast-in-place floors.

## Richards Medical Research Laboratories

Richards Medical Research Laboratories, University of Pennsylvania, Philadelphia, USA. Louis Kahn. A series of interlocking beams provides a floor structure whose exposed soffit could be clearly articulated as bays formed by primary and secondary beams. This approach contrasts with that of the flat slab.

## Lloyds Building

Lloyds Building, London, England. Architect: Richard Rogers Partnership. The high degree of prefabrication of precast components was used in the floor bays. Each structural bay is divided into a 1.8 x 1.8 metre (5ft10in x 5ft10in) sub-bay into which lighting and ventilation are installed. Like the Richards Medical Building, the soffit is fully exposed. Two-way spanning cast-in-place beams are connected to columns and pre-cast support brackets. The structural frame is based on a 10.8 x 18.0 metre (35ft6in x 59ft) grid. The floor structure consists of 550mm x 300mm (1ft10in x 1ft) beams supported on post-tensioned inverted U-shaped beams, which span between brackets. The connection between the floor and brackets is made using a bearing assembly, the yoke, designed so that vertical loads are carried on elastomeric bearings and horizontal loads by steel dowels. Overall, stability against lateral forces is achieved by six sets of diagonal bracing. Pre-cast concrete stub columns use a steel permanent formwork to support the floor deck, providing a zone for services distribution. The permanent formwork also provides support for services and acoustic attenuation.

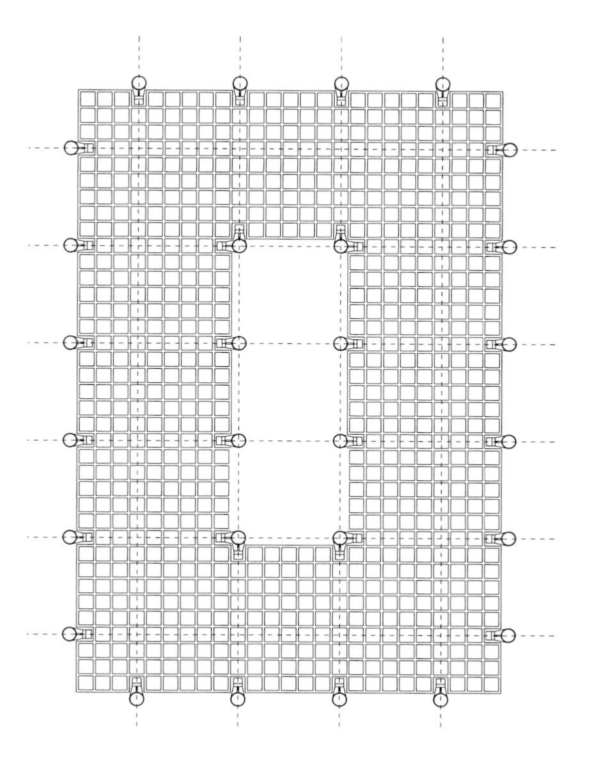

Lloyds Building

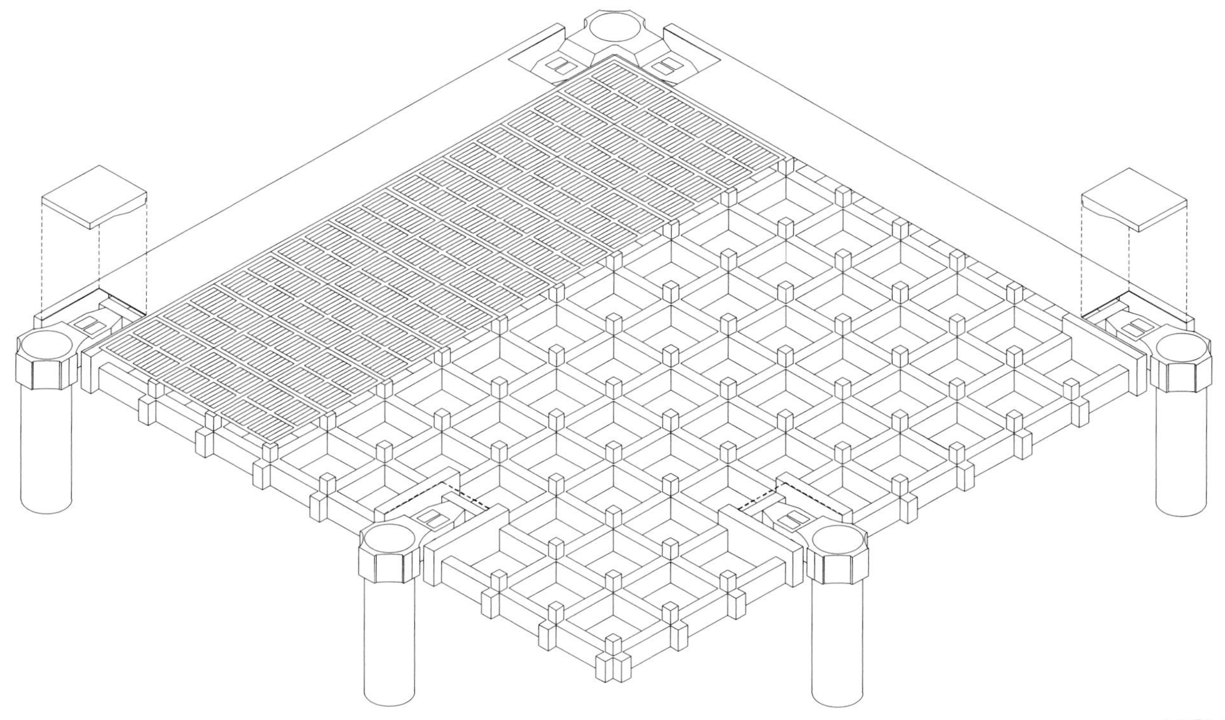

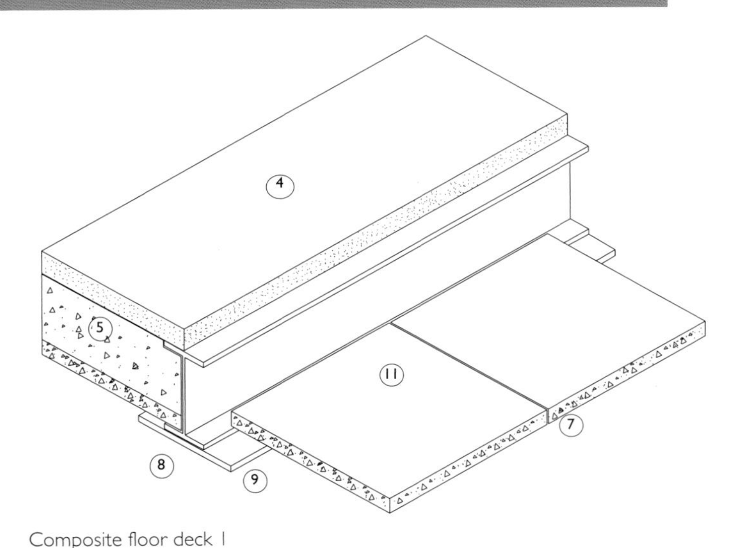


Composite floor deck 1

Composite floor deck 2

## Steel mesh details

1. Steel mesh
2. Supporting steel structure, typically I-sections or channels
3. Fixing bolts

## Composite floor details

4. Concrete topping
5. Concrete
6. Light reinforcing bars
7. Profiled steel decking
8. I-section beam
9. Steel plate bolted or welded to I-section
10. End filler piece to form edge of poured concrete
11. Precast concrete planks
12. Precast concrete beams

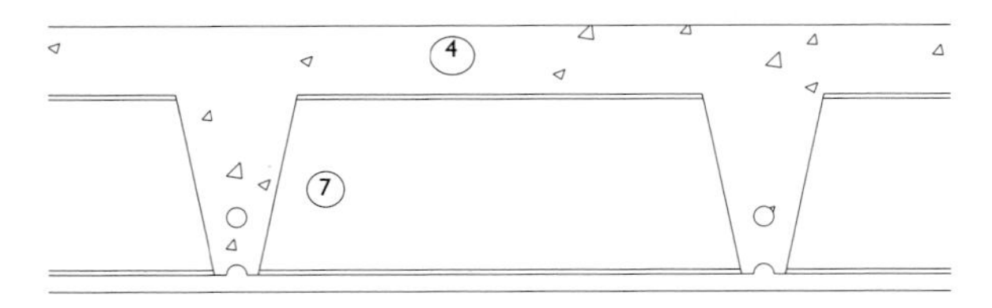


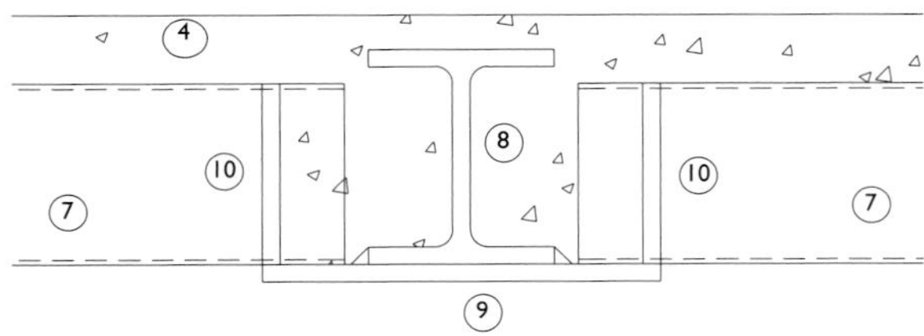


Vertical sections
Composite floor deck 3

Although it is possible to make an all-steel floor, in practice it is structurally inefficient and difficult to provide the necessary fire resistance, impact resistance and sound insulation. In addition, the amount of steelwork fabrication required would make it very expensive. The use of all-steel floors is restricted to steel grating in areas that are not required to provide fire resistance. All-steel floors in sheet and plate are most commonly used in industrial buildings.

## Steel composite floors

An alternative to precast concrete floors is to use a composite deck made from deep profiled steel sheet with concrete poured on top. During construction, the profiled decking provides permanent formwork to the concrete. Since the steel deck requires little or no temporary propping when the concrete is cast, construction time is reduced in comparison to other cast-in-place techniques. The steel deck and concrete perform structurally in a composite action where, in a simply supported situation, the concrete is in compression and the steel is in tension. The economic use of this type of floor is limited by its span. Over large distances, the supporting beams become very deep, resulting in an increased floor-to-floor height.

The profiled decking is fixed to the supporting steel beams with shear studs which provide the composite action between slab and beam. The concrete grips the shear studs, transmitting the shear forces through the metal deck to the supporting structure. The bond between the profile and the concrete is improved by the additional ribbing on its surface. The profiled sheet can be set either onto the top flange of the beam or onto a plate projecting from the bottom flange. The second method reduces the overall structural depth of the floor, with a consequent reduction in floor-to-floor heights; it also stiffens the web of the beam, increasing its performance.

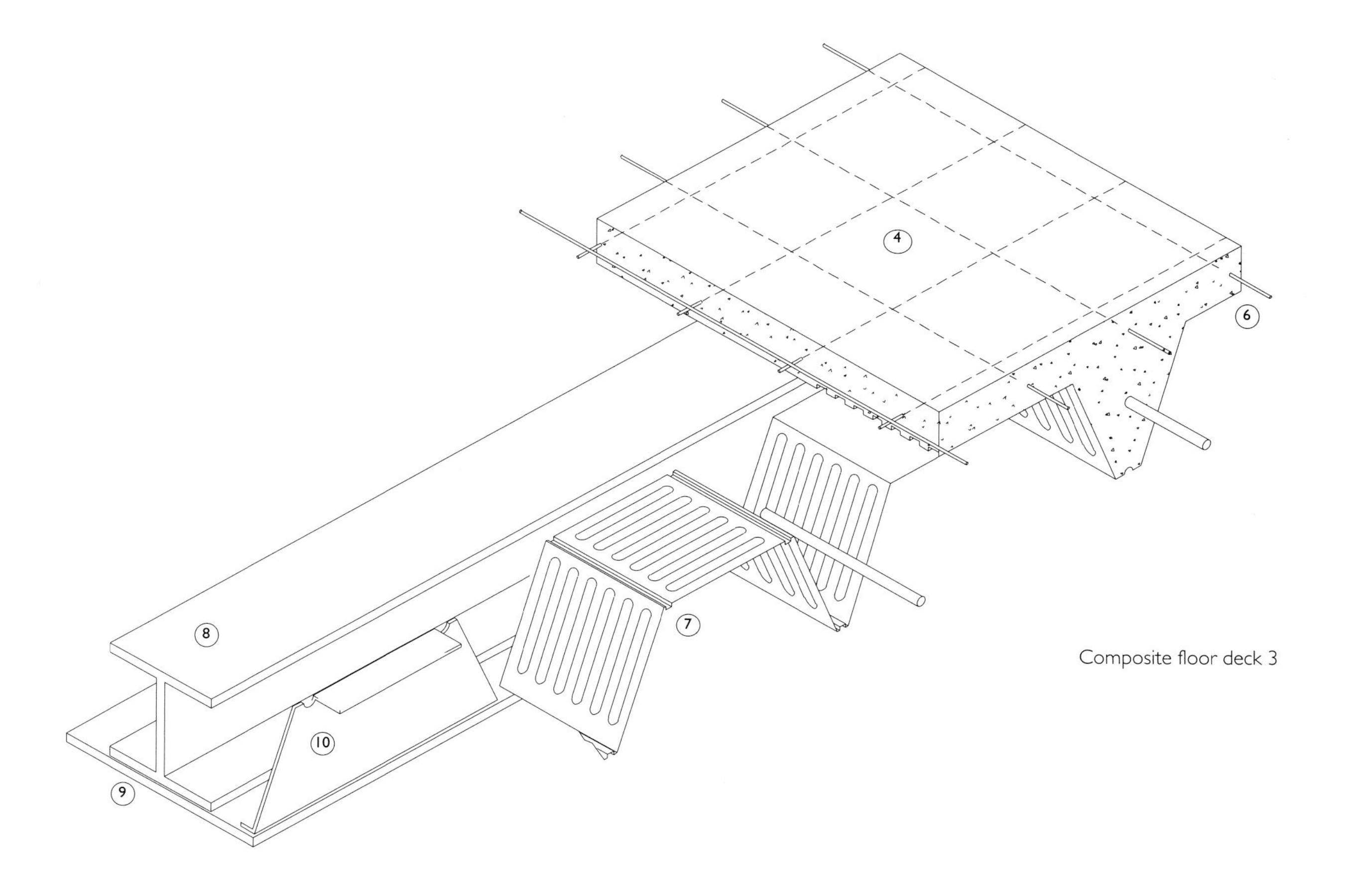


Composite floor deck 3

## Steel mesh floors

Steel grating is used primarily in industrial buildings and on maintenance access decks in other building types such as offices. It provides a lightweight, economic deck material that allows rainwater to drain off it immediately, making it less susceptible to corrosion when painted or galvanized. Steel grating is used to make structural decks by making a span between supports up to about two metres (6ft), depending on the depth of the grating. The choice of grating depends upon the spacing between the bars and their depth. Steel grating is made by one of two processes: welding or pressing. Lightly loaded small panels are manufactured by welding flats and rods together. Larger panels are made by a process that involves pressing together rows of notched flat bars positioned at right angles to one another to form a grid.

Steel mesh deck

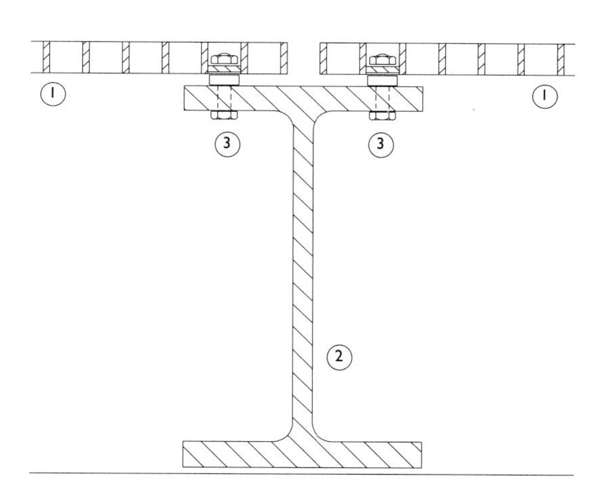


64

Timber floors are used with load-bearing masonry walls or as part of a timber-framed building. Timber floors have inherently poor fire resistance and sound insulation when compared to reinforced concrete floors. A typical floor for residential use might consist of softwood joists supported at each end by galvanized metal shoes or timber plates. Timber struts provide lateral stability and rigidity. Softwood boards are typically used as decking which is either butt-jointed or tongue-and-grooved. Alternatively, thicker boards are used, which have the advantage of bracing the floor structure horizontally. Plywood sheet provides superior diaphragm action but its use makes access to the void beneath more difficult, particularly if the floor void is used for the passage of services.

Fire resistance can be improved by introducing plasterboard sheets on the underside that typically forms the ceiling beneath. The traditional method of improving sound insulation by adding sand pugging between the joists is used rarely today since it increases considerably the weight of the floor. Instead, various systems of floating floors incorporating layers of sound absorptive quilts and resilient channels are commonly used.

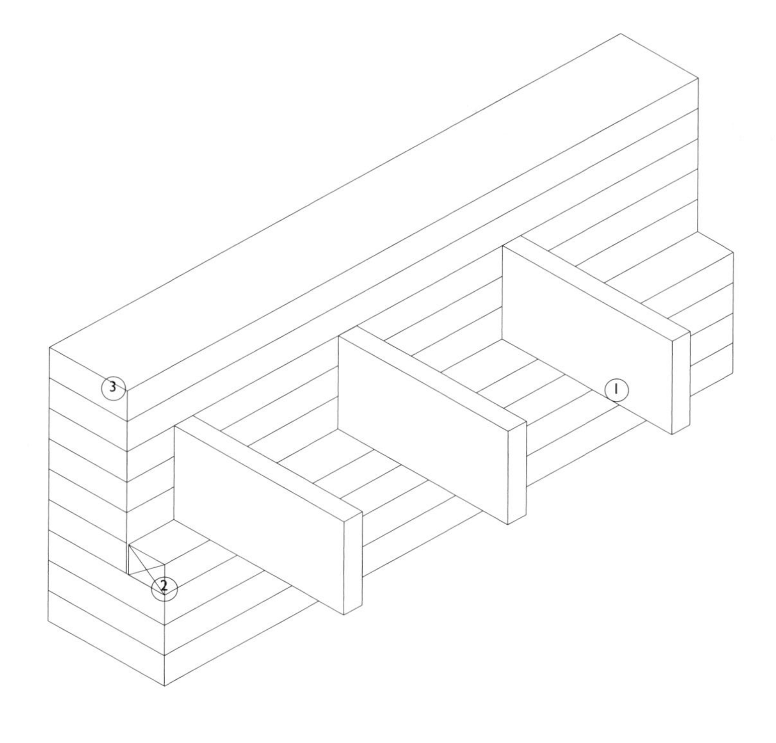

Details

1. Softwood floor joists
2. Softwood floor plate
3. Loadbearing wall (brick shown)
4. Floor boards or plywood/chipboard
5. Strutting to stiffen floor construction

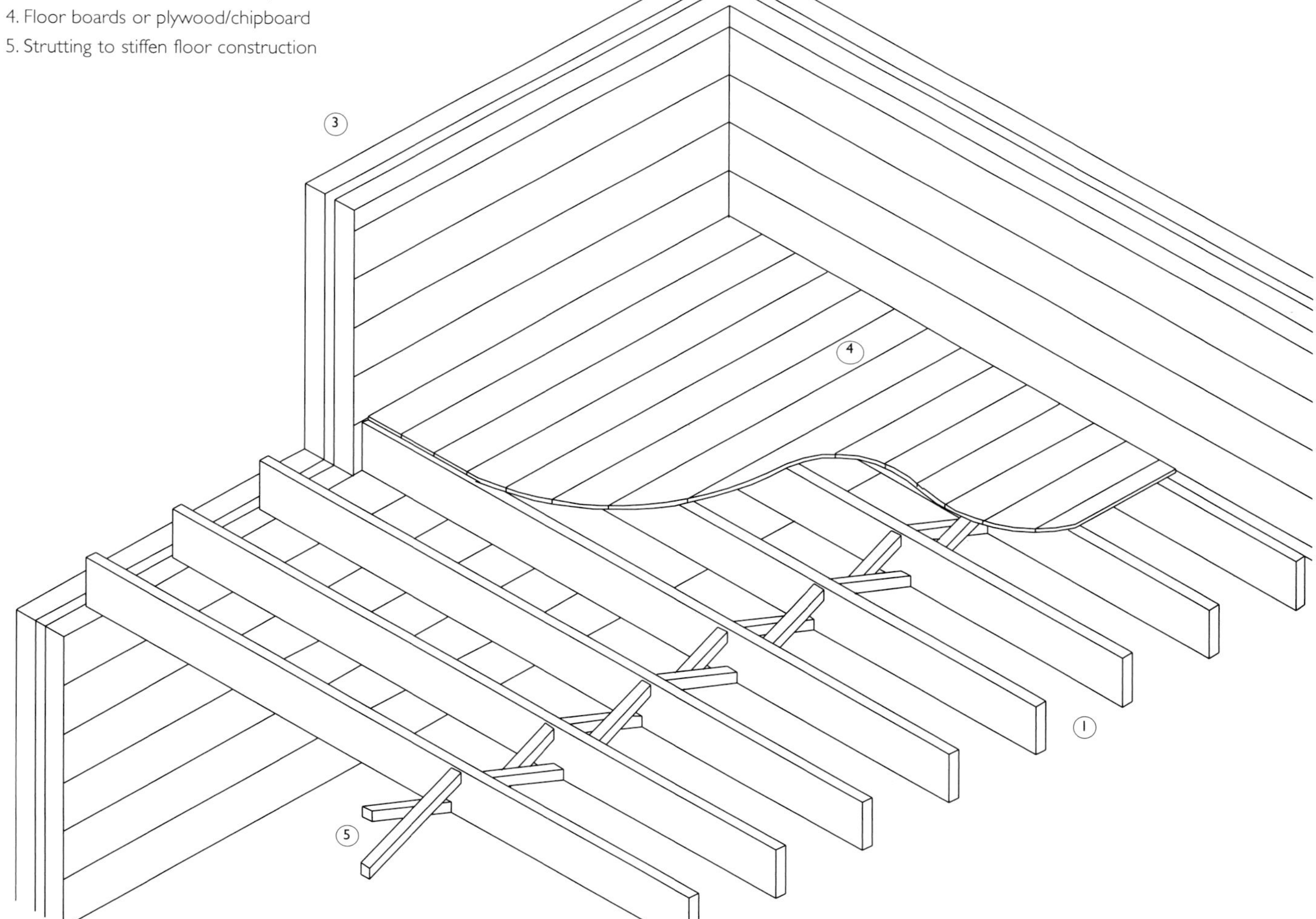

16B

Details to glass sheet floors
1. Laminated glass sheet, single glazed or outer layer of double glazed unit.
2. Silicone seal.
3. Stainless steel angle.
4. Spacer.
5. Silicone bond.
6.. Supporting structure.

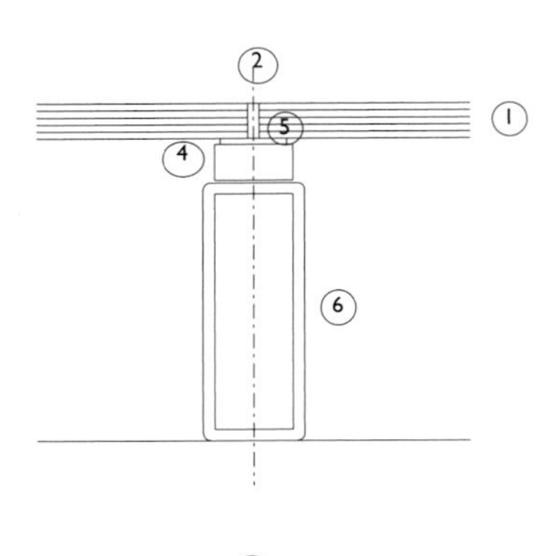


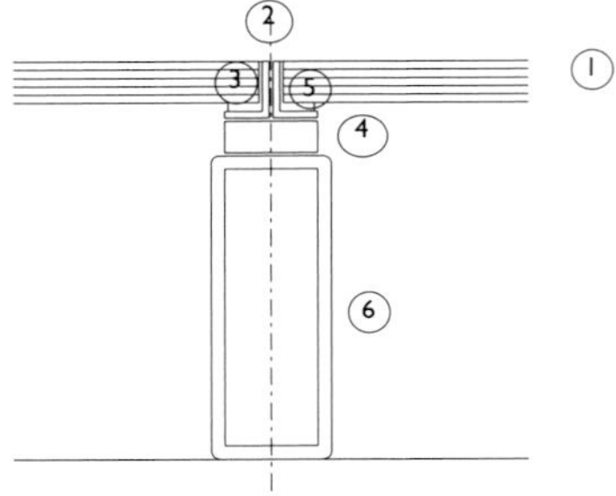


Vertical sections
Alternative fixing methods

16A

Glass floors are used to enhance naturally lit spaces by allowing light to pass through the deck. This can be done using either glass blocks or glass sheet. Where glass blocks are used, they are set into a supporting frame of steel or concrete with reinforcing bars. Each block is individually supported, giving it a limited effect of lightness and limited transparency. Where glass sheet is used, laminated float glass is set into a steel frame, giving it much greater visual lightness and transparency.

## Glass block floors

Glass blocks with textured or ribbed surfaces are often used as they have good slip resistance. This type of construction has good resistance to shock, and is capable of very high loadings. Because each glass block requires support, the number of junctions in the frame becomes very high. Since joints in materials are usually expensive to form in relation to the cost of components, junctions are kept as simple to make as possible. For this reason, frames are often made as castings, where the material is poured in place to form a frame. Cast iron, cast steel and reinforced concrete are commonly used. An advantage of concrete is that it can be cast directly against glass blocks which become a permanent formwork. Reinforcing bars are laid in a grid in the joints between the blocks.

## Glass sheet floors

Economic spans for glass sheet are currently in widths of one metre, but panel sizes of 1.2 metres by 2.6 metres (4 ft x 8ft 6in) can be achieved but at considerably higher cost. The glass is usually bedded on a flexible, rubber-based material (such as neoprene or EPDM) within the frame, or bonded to it with silicone. This allows the supporting structure to move independently of the glass, as well as all components to move with thermal expansion. Junctions with the frame are closed with a silicone sealant. Sandblasting grooves into the glass can provide transparency and slip resistance. In addition, a translucent laminated interlayer can be used to control views through the glass. Sheet glass floor panels are currently restricted to relatively low loadings. Supporting structures can be in either steel or reinforced concrete. If the supporting structure is constructed to the correct height and fall, if used externally, then the laminated glass sheets can be bonded directly to it rather than setting them into a steel sub-frame.

40C

92E

Details to glass block floor

1. Glass blocks
2. Mortar joints
3. Junction between adjacent concrete floor and frame to glass block floor sealed with rubber-based strip
4. Steel frame

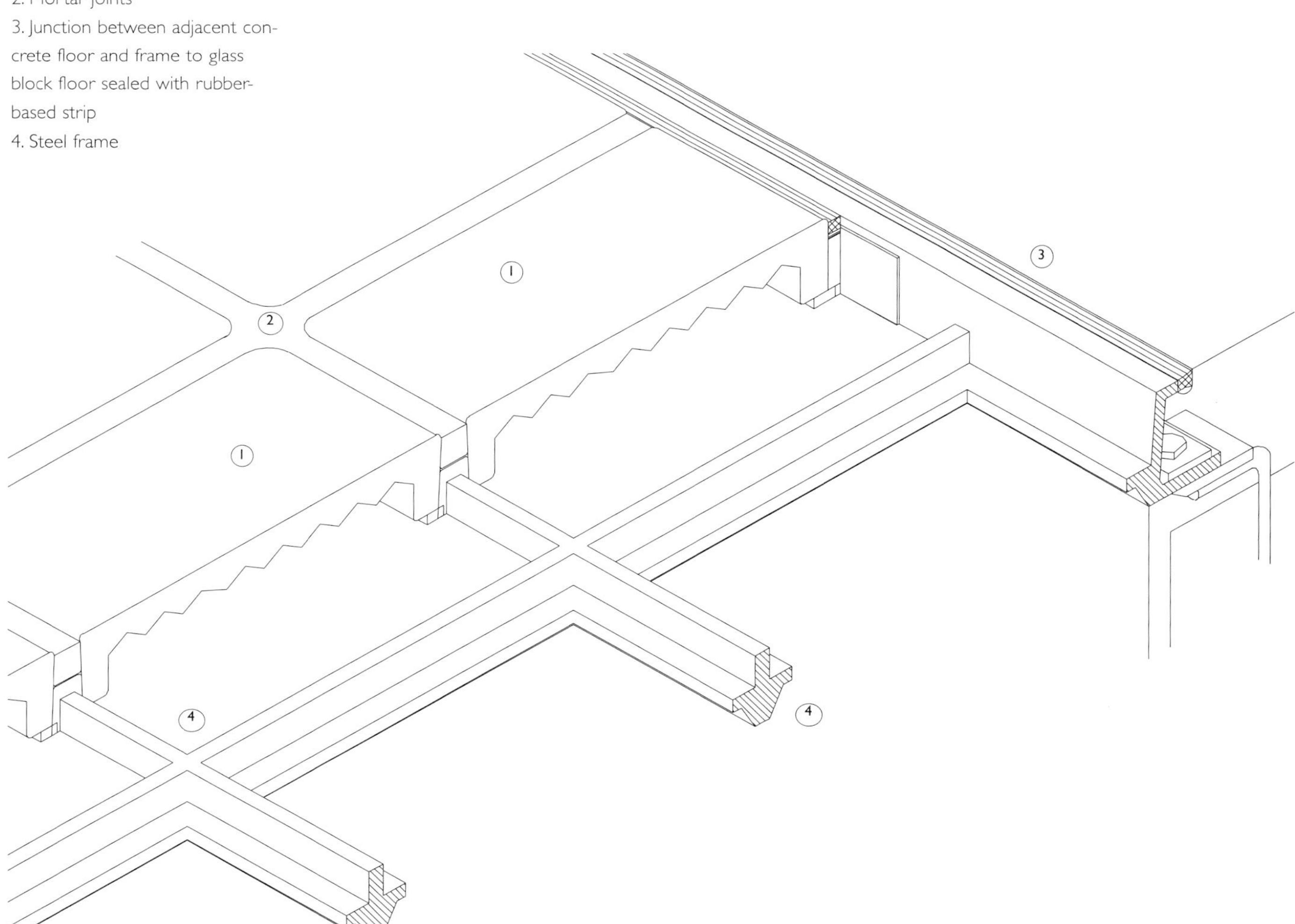

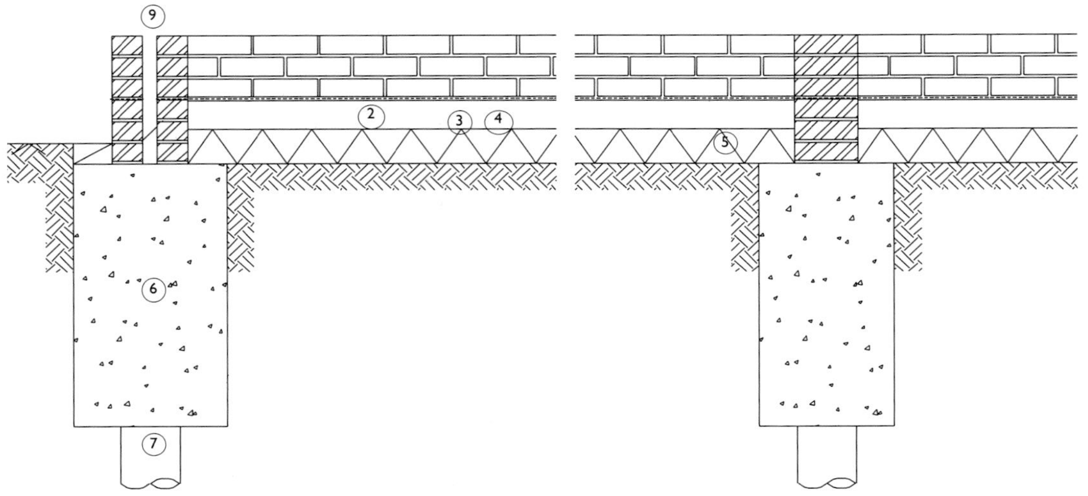


Horizontal section

Vertical section

Details
options for strip foundations

1. Screed
2. Concrete slab
3. Damp proof membrane (DPM)
4. Sand blinding
5. Hardcore
6. Concrete strip foundation
7. Short bore piles if required
8. Earth
9. Loadbearing wall (brick cavity wall shown)
10. Thermal insulation

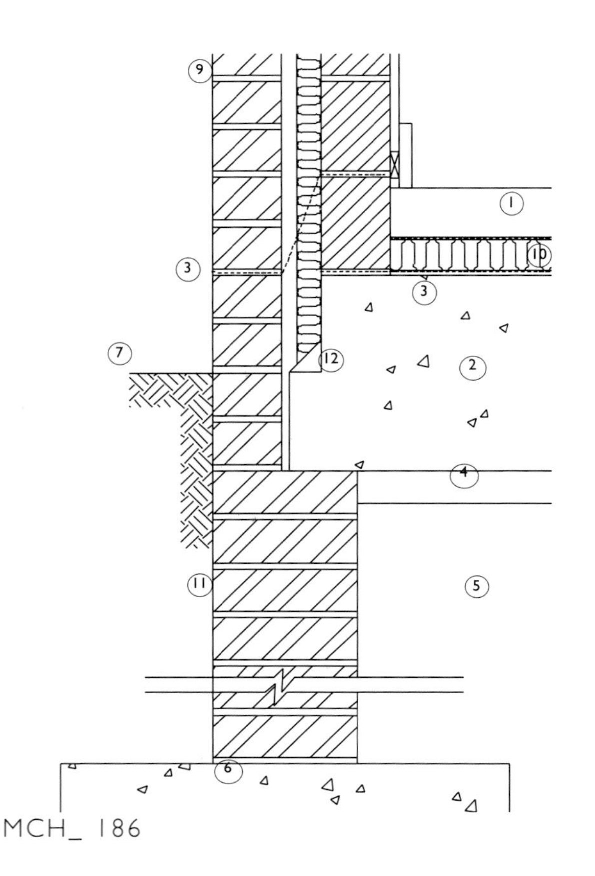


Foundations are cast in place, though prefabricated foundation components have been used in pioneering projects. Since the digging of foundations and laying of drainage and utility services is itself a lengthy process, the use of precast concrete components for foundations currently brings little reduction in construction time. Concrete is the material used for almost all foundations. The varying types reflect the concentration of loads from the building at ground level as well as the way these loads are distributed to the ground. Different foundation types are rarely mixed in a single building due to the differential settlement caused by the different loading conditions.

## Strip

Strip foundations are used to support loadbearing walls. The foot of the wall requires continuous support at its base so that loads may be evenly transmitted to the ground. Loads at ground level are distributed along the full length of the loadbearing walls.

Groundbearing conditions may vary along the length of the wall, and the strip may require additional support from beneath. Short bore piles are used, consisting of short 'columns' of concrete. These are made by pouring concrete into a hole bored into the ground. The short piles help to increase friction with the soil. The strip foundation itself may be reinforced with steel bars to resist bending along its length. Strip foundations are usually designed to be wide and shallow, or narrow and deep, depending on ground conditions. The design of the strip foundation needs to be co-ordinated with the passage of piped services that may cross through it.

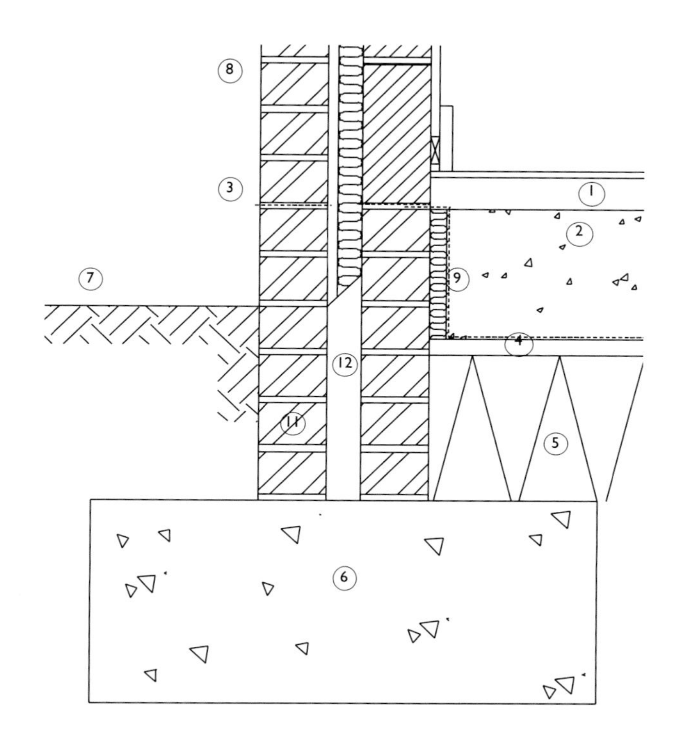


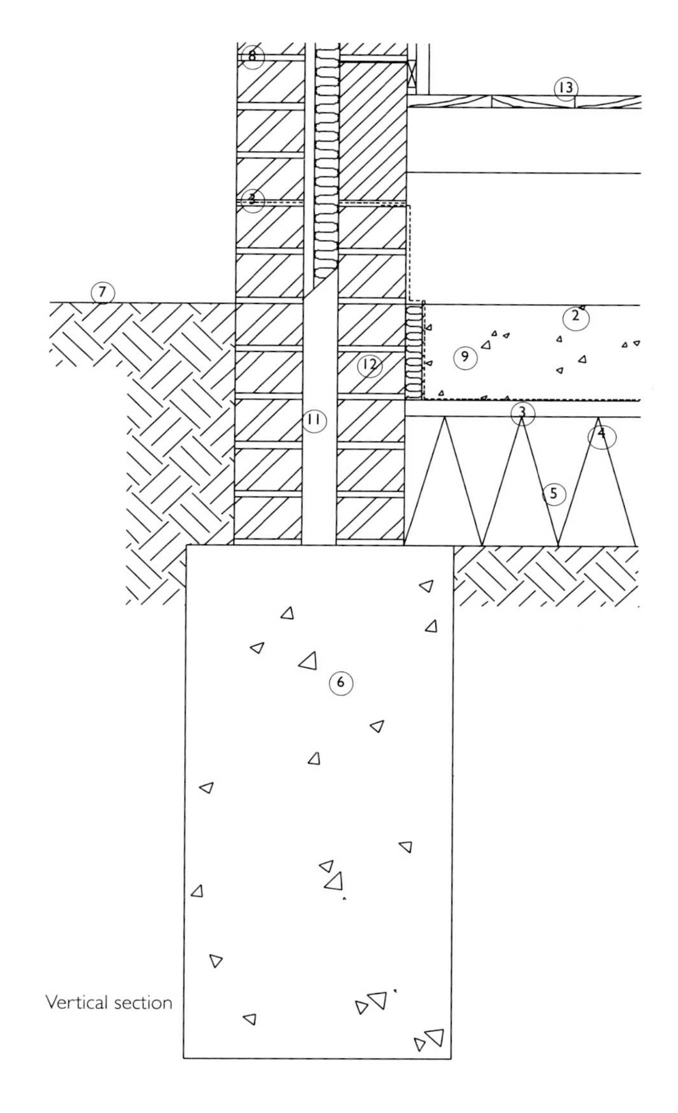


Vertical section

## Raft

Raft foundations are used where the bearing capacity of the soil is either very poor, or where there is a tendency for the ground to move seasonally in a way that would cause cracking in strip foundations. A frame or loadbearing wall structure may be supported on a reinforced concrete raft foundation. In effect, this pad foundation extends across the site and also provides the slab floor of the lowest level in the building. Rafts can be built as a slab on flat ground or may require additional ground beams. Of all the foundation types, this one has the greatest impact on piped services entering the building below ground, since they must penetrate the slab and then traverse any beams extending below the slab. Rafts are often used with lightweight construction, particularly in timber construction.

### Details

1. Screed
2. Concrete slab
3. Damp proof membrane (DPM)
4. Sand blinding
5. Hardcore
6. Concrete strip foundation
7. External ground level
8. Loadbearing wall (brick cavity wall shown)
9. Thermal insulation
10. Vapour barrier
11. Engineering bricks
12. Mortar cavity fill
13. Raised timber floor construction shown.

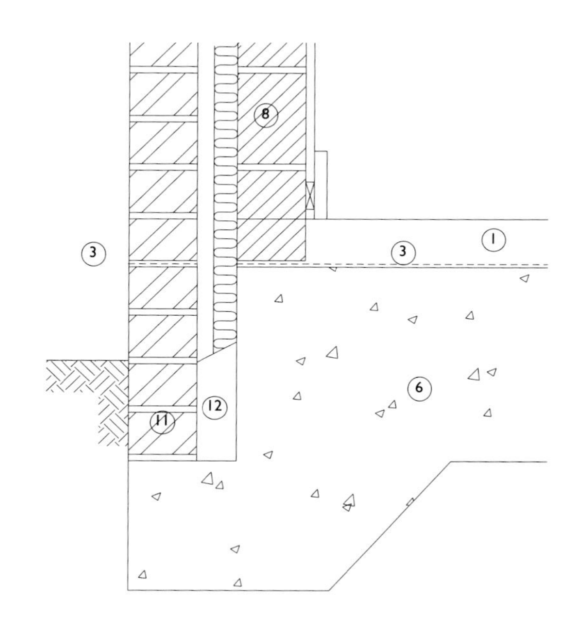


Vertical section

Details
options for pad foundations
1. Screed
2. Concrete slab
3. Damp proof membrane (DPM)
4. Sand blinding
5. Hardcore
6. Concrete ground beam
7. Concrete foundation pad
8. External ground level
9. Loadbearing wall (brick cavity wall shown)
10. Thermal insulation
11. Mortar cavity fill

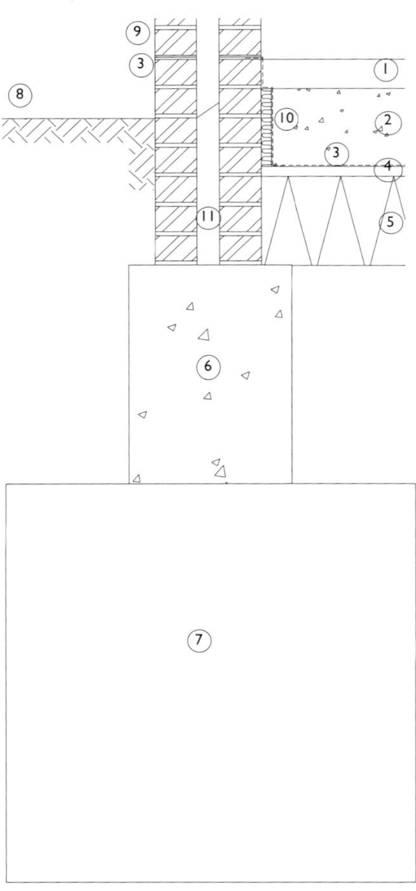


## Pad Foundations

Pad foundations are used to support light framed structures on reasonably strong soils. Loads from a frame are concentrated at points and are distributed to the subsoil by concrete pads. These may be connected together by ground beams, which consist of reinforced concrete strips cast in place and act to hold the pads in place. The beams span between the pads and are not supported on the ground beneath. Pads consist of either mass concrete or reinforced concrete. The use of ground beams also allows some columns to be supported in cantilever from the pads. The column is supported by a ground beam which is in turn supported by a pad. Cantilevered foundations are sometimes used where support to a frame or loadbearing wall cannot be provided from directly beneath, such as at a party wall.

## Piles

Piles are columns cast into the ground to support a building where either ground conditions are extremely poor or the loads from the building structure are very high, as in tall buildings. Traditionally, piles were made of wood, but today they are more likely to be formed of concrete or steel. The frame rests on pile caps which may in turn be supported by several piles. The cap transmits the load evenly to the piles beneath which transfer the loads down to a loadbearing strata. The piles support the building by transmitting its loads to the ground by a mixture of friction with the surrounding earth and by end bearing, where loads are transmitted through the ends of the piles like columns. Concrete piles are usually reinforced with steel bars for this purpose. Piles are either bored (drilled) into the ground or, in the case of steel piles, they can be driven into place.

8U

8T

8E

# 5

## ENVIRONMENT

55C

33C

In this chapter, environmental design is discussed in terms of the passive and active methods that can be used to modify environmental conditions within a building. Active controls are provided mostly by mechanical ventilation systems that are used to heat, cool and ventilate spaces. Passive controls can be provided by natural ventilation, by the use of the building fabric as a thermal mass to slow down the rate of heating and cooling, and by solar shading or passive heating. A combination of these methods helps to reduce the energy consumed by buildings in use.

Fifteen years ago, combinations of structure and cladding were generally limited to a use of materials that were constructed either from the same base material (either steel, concrete or timber) or from materials having compatible amounts of thermal movement. Today, the use of the thermal mass of the structure, and materials with low embodied energy have led to a much wider range of structure/cladding combinations. Where thermal and structural movement was the main issue in detailing fifteen years ago, the idea of 'energy' is considered important today and detailing becomes more complex in response to these changing priorities.

Until recently, the expression of structure and construction (particularly of joints) was considered one of the primary architectural intentions in cladding design. This situation is slowly changing. More buildings are clad externally with insulation, where the rainscreen panels that conceal it from view have both a waterproofing and a decorative function. Rainscreens are a recent development that allows a simplified approach to be taken to movements in the primary supporting structure of the building. A continuous, uninterrupted appearance can be achieved with a consistent gap between panels that extend across movement joints in the building structure. Both movement joints and breaks in the construction need not be 'registered' on the facade. Joints can be registered on the facade but they serve only as a decorative device. The real movement joint concealed behind the panel can be

93A

93B

achieved using very simple techniques that are not required to be visible.

The increased use of rainscreens and a variety of solar/daylight shading have led to a shift in the aesthetic expression of facades. The expression of 'energy' is taking over from the expression of 'structure', the latter having been a major preoccupation in building design since the adoption of modernism in architecture at the end of the Second World War. In addition to the widespread use of rainscreens and shading devices that has occurred over the past 15 years, two new issues seem set to transform cladding design. These are embodied energy and photovoltaics.

The embodied energy of a building's external envelope can be considered the sum of the energy used in the manufacture of its components, their transportation, their final assembly on site and recycling capability at demolition. Timber has low levels of embodied energy when compared to steel and reinforced concrete, and is becoming a popular cladding material once again. Notable examples are the Cultural Centre at Noumea, designed by the Renzo Piano Building Workshop, the Henley Rowing Museum, designed by a team led by David Chipperfield and Nottingham University, designed by Michael Hopkins. Reliable information about levels of embodied energy in typical forms of building construction has yet to become available. However, various guides are available which help to steer designers towards suitable material selection.

Photovoltaic cells are used to generate electricity from solar radiation (sunlight). The power generated is then used inside the building. Electricity is generated in panels of 'semi-conductor' devices containing a glass substrate coated with tin oxide, forming a transparent electrode, which is covered with layers of silicon together with a coat of aluminium film which forms the other electrode. Particles of ultraviolet light called photons interact with electrons in the semi-conductor to convert sunlight into direct current electricity. Panels are orientated towards the path of the sun. The photovoltaic process has the advantage of requiring no moving parts, requires no fuel and needs relatively little maintenance. Photovoltaic panels can be used as semi-transparent panels in glazed wall systems, though they perform more efficiently in generating electricity when inclined towards the sun. The first large photovoltaic panel installation is at the University of Northumbria, UK. The output of electrical power fluctuates with changing weather conditions. For this reason the building is supplemented by a conventional electricity supply. Operating periods range from fifteen hours per day in May to eight hours in January. The installation is expected to pay back the energy used in the manufacture of the cells within three years, then produce electricity for a further twenty years. The electricity generated is expected to meet 50% of the building's needs in summer and 10% in winter, averaging around 30% over a year. The cost of generating electricity through photovoltaic means is about four times the current commercial rate.

41

20B

MCH_ 196

The return to natural ventilation in buildings, and away from dependency on mechanical ventilation (including air conditioning), has gathered momentum in recent years. This has come about from the desire to reduce both the energy consumption in buildings and the wider issue of carbon dioxide ($CO_2$) emissions from the primary energy generation through to building use. The link between global warming and the build-up of greenhouse gases has yet to be conclusively proven, but the circumstantial evidence seems convincing. Nearly half of $CO_2$ emissions generated in the UK alone arise from energy consumed within buildings. The reduction of energy consumed in buildings has a huge benefit for future generations. The decline in the supremacy of air conditioning can also be linked to 'sick building syndrome' due mainly to systems dependant on the re-circulation of used air in buildings.

This has resulted in the increased use of opening panels and window systems. Previously, building types other than housing were often designed as hermetically sealed boxes which aimed to provide a constant temperature and steady rate of ventilation. In addition, air conditioning systems would control humidity as well as temperature. This high degree of environmental control led to very high energy consumption from running the equipment. In urban areas, the use of operable windows to provide natural ventilation is often considered inappropriate due to high levels of noise and atmospheric pollution. This has led to the development of 'layered' systems in which functions of sound insulation, weatherproofing and air infiltration are separated.

Natural ventilation in buildings is generated either by wind or by a 'stack effect'. In most buildings a combination of both occurs. As air flows across a building, a difference in air pressure will occur as a result of varying wind pressure. This difference in pressure causes the air in the area of higher pressure to flow towards the area of lower pressure. Provided openings are appropriately located in these pressure regions, the movement of air creates a natural ventilation flow through the building. 'Stack effect' ventilation is caused by a difference in density of air due to internal heat input from casual and solar gains to the ventilating outdoor air. The temperature, and consequently density difference between indoor and low level supplied outdoor air results in an upward flow (buoyancy). Air moves from low level to high level in a building with a higher temperature than that outside whilst the reverse will occur if the air in the building is cooler than that outside.

Natural ventilation has two main benefits; reducing the amount of mechanical ventilation needed during normal daytime use, and providing free nighttime cooling of the building structure. Heat generated during the day will be absorbed by the structure if exposed, which leads to a slower rise in resultant temperatures within

110A

*Southern Zone Offices – Cross Section*

INTERNAL BLINDS TO CONTROL GLARE
WINTER SUN CAN ENTER AND WARM INTERNAL SPACE
EFFECTIVE SHADING FROM OVERHANG
WHITE SHADES ACT AS LIGHT SHELVES INCREASING DAYLIGHT LEVELS
ELECTRIC PERIMETER HEATER
SUPPLY AIR DUCT
LOW VELOCITY OUTLET IN FLOOR
RADIANT COOLING FROM SLAB
CONVECTIVE GAINS FROM PEOPLE AND EQUIPMENT
RADIANT COOLING FROM SLAB
POWER DATA OUTLET BOXES IN FLOOR
PROPRIATRY FLOORING SYSTEM WITH AIR SEAL
EXTRACT DRAWN BY STACK EFFECT THROUGH ATRIUM
OUTLET FROM SLAB INTO FLOOR VOID
DATA/POWER SUPPLY RUN IN FLOOR VOID
150mm CONNECTER THROUGH SCREED
AIR PASSES THROUGH HOLLOW CORE SLAB

*Mid-Season Operation*

PREVAILING WINDS ASSIST NATURAL VENTILATION
SOLAR ENERGY ENHANCES STACK EFFECT VENTILATION
WEATHER STATION CONSTANTLY MONITORS CLIMATE & ALLOWS CONTROL OF VENTILATION & TEMPERATURE
SHADING SYSTEM PROVIDES PROTECTION FROM SUNLIGHT ENSURING GOOD DAYLIGHT PENETRATION
COOL SLAB MAINTAINS THERMAL MASS
21-24°C
21-24°C
CELLULAR OFFICES IN NORTH ZONE HAVE SEPARATE SUPPLY & EXTRACT
SOUTH SIDE NATURALLY VENTILATED
OPEN PLAN AREAS ON NORTH SIDE CAN RELY ON NATURAL VENTILATION OR A COMBINATION OF BOTH NATURAL AND MECHANICAL
Not to scale

the building. In order for nighttime cooling to benefit daytime internal comfort, the building structure requires a medium to high thermal mass which should be exposed to the occupied zone. A structure with high thermal mass, such as concrete soffits, cooled at night will absorb much greater amounts of heat generated within the building during the day than will a lightweight structure with low thermal capacity such as steel. A considerable amount of research into the effect on thermal mass performance due to different slab construction methods was recently published. Where the contribution of thermal mass is to play a part in the internal environmental strategy, the preference for exposed concrete soffits is particularly important given that external walls tend to have a high percentage of fully glazed walls.

In addition, during winter months, natural ventilation can be used to redistribute heat stored in the building structure to help keep the building warm at night. There are a number of proven methods whereby thermal mass is used to exchange heat with ventilating air streams. These are generally mechanically driven due to the airflow resistance normally encountered. This approach can be via hollow core pre-cast slabs as used on the Ionica building in Cambridge, England, which are linked to form concrete looped passages to optimise heat transfer before supplying the air to the occupied zone.

110B

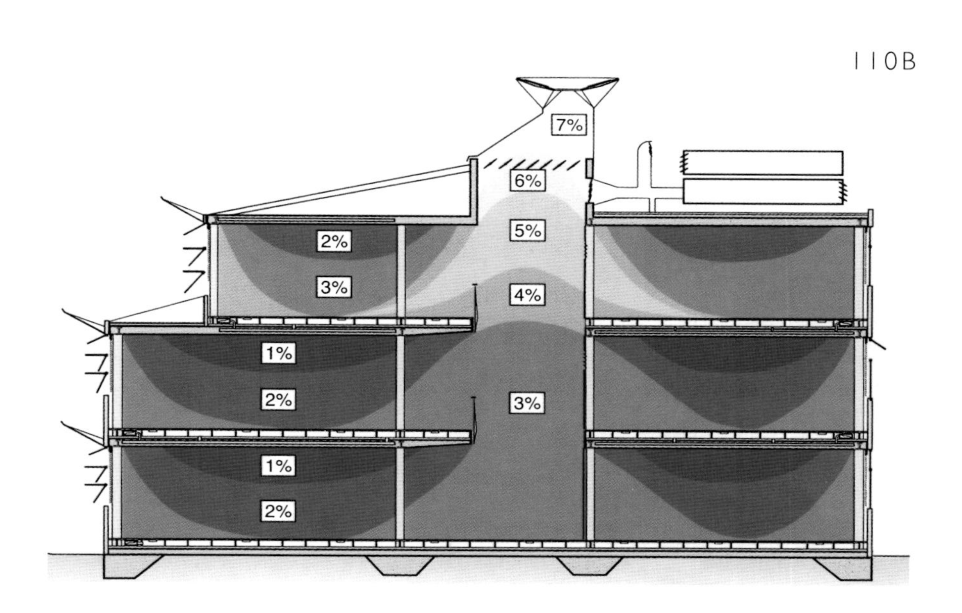

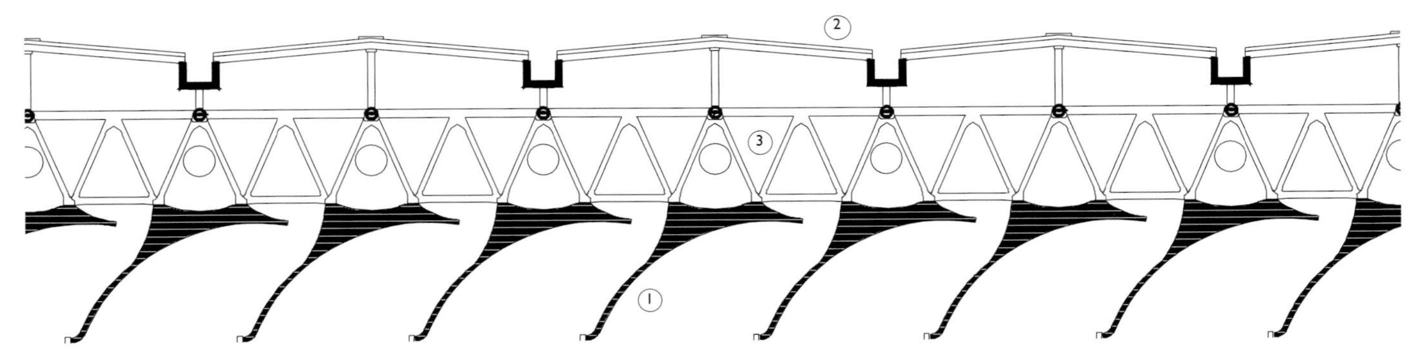


Details

1. Fixed ferro-cement louvres
2. Glazed rooflights
3. Support trusses for louvres

90A

A space within a building lit by natural daylight produces about one tenth the amount of heat created by an equivalent incandescent lighting installation and about one half of that of a discharge lamp installation. This, along with the desire to offer greater naturally lit space, has led to the increased use of daylight as part of the drive towards reducing the in-use energy costs associated with mechanical cooling. In addition, there has been a move away from reflective glass and body-tinted glass to increased transparency and admission of daylight. Increased daylighting can increase the problem of glare. Viewing two highly contrasting levels of light at the same time causes glare. The huge increase in the use of computers in buildings and the need to see monitors clearly has made the problem all the more important. Conventional solar shading, which projects horizontally from the wall plane reduces incident solar radiation and therefore reduces heat build-up in glazing, but it only reduces glare to the extent that it obscures part of the bright sky. It cannot obscure the sky at the horizon, nor can it deal with bright reflections from objects at eye level. These two sources of glare are particularly characteristic of very open environments. Glass treatments such as fritting, silk-screening and sandblasting merely diffuse brightness and can even make the problem worse. The solution requires the introduction of a mechanism to control overall light transmission and therefore light intensity, thus enabling a reduction in perceived contrast.

With increased daylighting, there is often an increased amount of solar gain particularly on those facades facing the sun. This has brought the issue of solar shading into sharper focus. Even facades with low percentages of glazing (10 - 40%) are increasingly using solar control devices instead of relying heavily on mechanical cooling. A popular method of solar shading has been the use of externally mounted canopies and screens. A variety of horizontal continuous canopies, vertical blinds, fixed louvres and movable screens is used. External shading has the advantage of both controlling the passage of sunlight and reflecting heat away from the sun outside the building envelope. The positioning of solar shading inside the building envelope results in some of the heat being reflected back out of the building, but more than 50% is radiated into the building, providing little advantage in controlling solar gain. Although external shading provides better reduction of solar gain than an internal fixed type, the components will require cleaning and maintenance. This has led to the

31A

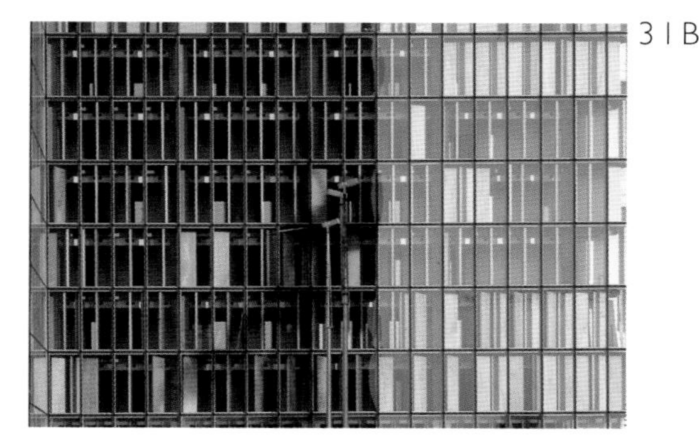

31B

95B

increased use of access decks or walkways, which also provide a degree of horizontal shading. However, this method imposes a high maintenance budget and commitment on the building occupiers. An alternative solution is to set the shading behind an outer glazed wall such as a rainscreen, which draws radiated heat directly away from shading components prior to its reaching the inner glazing. This heat gain can be drawn away by natural ventilation, when external air temperatures permit, or by mechanical means.

## Solar shading and daylight controls

Shading devices used to control solar gain and the admission of daylight fall into two generic types: 'integrated' shading and 'applied' shading.

'Integrated' shading uses components set within the depth of a cladding panel. One of the most technically sophisticated examples is the Institut du Monde Arabe in Paris, designed by a team led by the architect Jean Nouvel. The principle is based on photographic iris diaphragm aperture control, an instrument that precisely controls the intensity of light, and this was used in glazed cladding panels on the south facade. A set of computers responds to direct solar radiation by operating shutters in each panel. These shutters control the amount of sunlight entering through the south facade of the building. . This allows both the admission of daylight and solar gain to be closely controlled with a single mechanism within the cladding panel itself. More typical examples use motorised blinds set within a sealed glazed cladding panel.

'Applied' shading sets components as a separate layer to the (usually glazed) weatherproof layer. A typical example is the Menil Museum in Houston, Texas, designed by the Renzo Piano Building Workshop. Daylight baffles, which also provide solar shading, are made in a lightweight concrete construction, mounted beneath the glazed roof. Besides controlling the amount of daylight and diffusing it, the white-finished louvre blades are shaped to reduce internal temperature variations from solar radiation. Their horizontal tops reflect heat but allow light to reflect off the surface and into the spaces beneath. A more modest approach to applied shading is the 'light shelf' principle, a horizontal reflective shading device that projects through the facade both to shade sunlight and reflect light off its upper surface and across the adjacent ceiling. This helps to even out daylight levels within deep spaces with a shallow floor-to-floor depth that require good natural light, such as in office buildings.

This 'layered' approach is used in the Bibliotheque Nationale in Paris, which features all-glass facades. The project team, led by Dominique Perrault, tackled the problem by introducing a fine metal mesh on an independent frame offset from the external face of the glass. Views are maintained through the light veil of stainless steel which creates a gentle hazy quality internally. The mesh material provides a textured and sparkling surface externally, which complements the smooth, reflective appearance of the glass skin. Other recent examples include the Berlin Velodrome also designed by Dominique Perrault and the Euralille Commercial Centre at Lille by Jean Nouvel. In order to facilitate glass cleaning, mesh screens can be placed at a distance from, or tilted away from the plane of glazing. Alternatively, the screens can be designed to unclip for easy removal, or can be motor-controlled in an overhead or roller-shutter format.

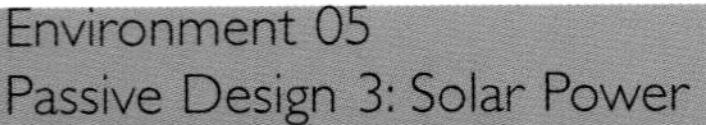

49C

49D

49F

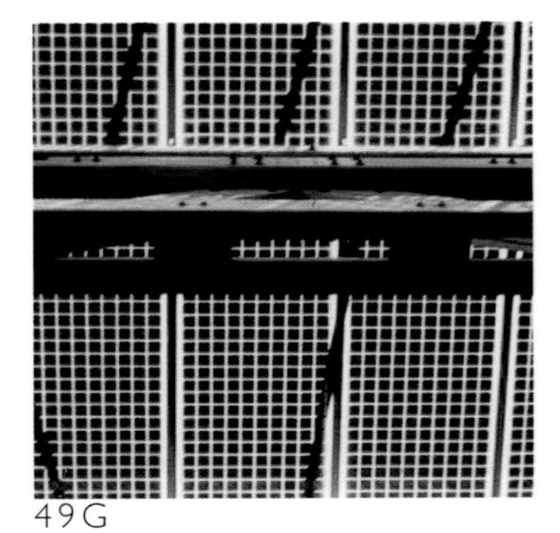

49G

49E

## Photovoltaic panels

Photovoltaic (PV) cells are used to generate electricity from sunlight. The power generated is then used inside the building to contribute to the electrical demands of the building and, in some cases, the power generated can be sold back to the electricity supplier. Electricity is generated in arrays of cells set in panels on roofs and facades. The orientation of panels is important. Those inclined close to the horizontal produce more electricity annually than those inclined vertically. Panels are set as close as possible to the angle at which the most amount of electricity is generated over a one year period, taking into account the varying path of the sun during that period and the effects of diffused solar radiation. Panels comprise a glass substrate coated with tin oxide, forming a transparent electrode, which is covered with layers of silicon together with a coat of aluminium film, which forms the other electrode. Particles of ultraviolet light called photons interact with electrons in the semi-conductor to convert sunlight into direct current electricity. PV cells are of three types: multi-crystalline, which produces the most amount of power per unit area, mono-crystalline, which produces less but is cheaper, and the thin film type which is currently made only in one fixed panel size but which is yet cheaper and can be fixed into a glazed walling system.

1
2
2
3
4
5
1

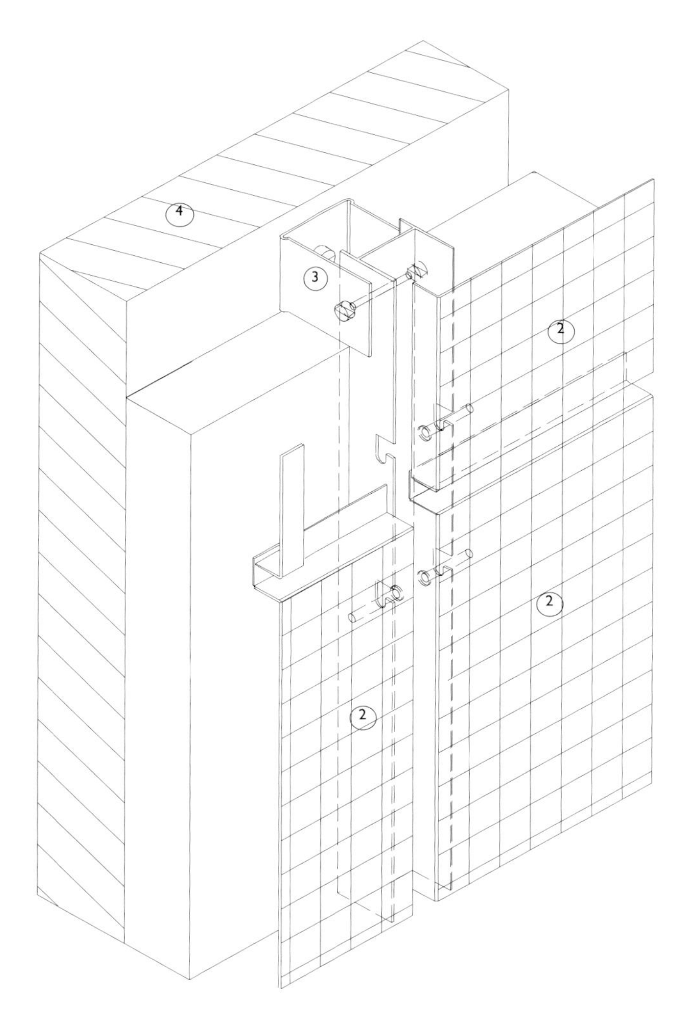

The photovoltaic process has the advantage of requiring no moving parts. It requires no fuel and needs relatively little maintenance. The amount of energy required to manufacture the PV cells and deliver them to site is estimated by manufacturers to be equivalent to the amount of energy delivered in the first three years of use; the PV installation will continue to produce power for the next twenty years or more.

## A sampled photovoltaic panel installation

An early large-scale PV installation, from which a relatively long performance has been recorded, is at the University of Northumbria in England. The output of electrical power fluctuates with weather conditions and it is necessary to supplement the supply with electricity from conventional sources. Daily operating periods range from fifteen hours in May to eight hours in January. It is estimated that the installation paid back the energy used in the manufacture of the cells within 3 years, and is still expected to produce electricity for a further 20 years. However, the cost of generating electricity is about four times the current commercial rate. The PV cells are integrated into the facade with a rainscreen cladding system. The cladding is inclined at 26° to the vertical, also providing passive solar shading. Ventilating the void behind the panel disperses heat, a by-product of the photovoltaic process.

Here, the array of PV cells is divided into a series of units 3.0 x 1.36 metres (9ft 10in x 4ft 5in). Each panel is demountable for ease of maintenance or replacement. The cells are bonded to an extruded aluminium frame with structural silicone. Excess electricity can be exported to the electricity supply company. The electricity generated meets 50% of the building's electricity needs in summer and 10% in winter, which averages out at 30% over a one-year period.

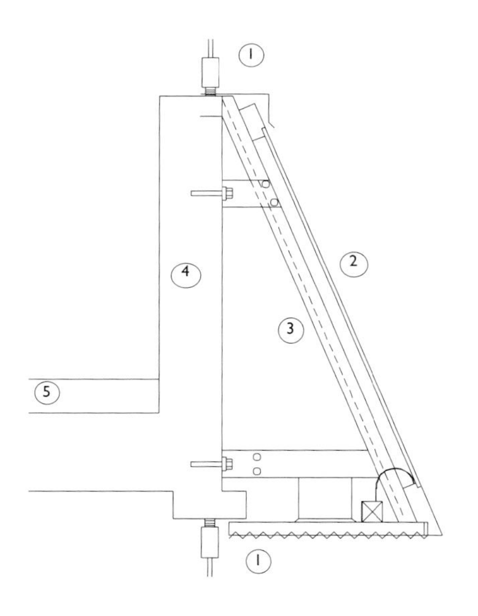


Details

1. Window
2. Photovoltaic panel
3. Supporting frame
4. Structural wall
5. Floor construction

116A

116C

116B



Solar heating systems consist of collector panels that absorb heat from solar energy to warm circulating water. Heat absorbed by the water is transferred to a storage tank by pumping or when positioned above the collector the two components form a thermal circulating system. In the latter case, the thermo-siphon effect of water movement is created by the density difference between hot and cold water. The heated water is normally used to exchange heat with a building's system water, which is then distributed around the building as required for either heating water or as domestic hot water. The amount of heat generated is significant and whilst it is currently suited to climates that have abundant sunshine for much of the year, the performance efficiency has developed to the extent that Northern European states can also benefit.

Collector panels should ideally be orientated to face south and in the UK, they should be angled at 45° from the horizontal. Panels traditionally consisted of sheets of clear high light transmission glass that allows maximum solar radiation penetration and trapping this radiation to be absorbed by the receiver surface. This surface is a copper base plate that transmits the heat to a continuous pipe fixed to its rear face. The pipes are enclosed in thermal insulation to reduce heat loss to the air. The most popular types of panels currently in use are; a) flat plate collectors, b) vacuum flat plate collectors and c) evacuated tube, (the latter being the most efficient over a typical annual temperature/solar radiation yearly profile, whilst the addition of a vacuum void to flat plate collectors significantly improves their efficiency).

108

Solar absorbers have been used for many years to provide warm water of up to 50°C for swimming pools. They consist of special solar absorbing high-grade rubber mats with integral water flow channels for circulating the pool water. They can also be used for domestic warm water supply.

A recent example is the Solarchis Housing System in Japan, set up by the Solar Architecture Studio and Maeta Concrete Industry Ltd. Demonstration houses built in the Yamagata Prefecture, Japan, use roof-mounted solar water heaters in conjunction with a photovoltaic array on the roof and a wall construction with high thermal mass internally and thermal insulation fixed externally. The Iyama House (illustration 116c) has a solar water heater installed in a transparent roof that produces domestic hot water at 50°C in a 300 litre storage tank. Food is cooked mainly by a solar steam cooker with the support of an electromagnetic induction heater which uses high-temperature solar domestic hot water from a 20 litre storage tank. The House of the Sun (illustration 116a) has a system of evacuated tubular solar collectors heating water stored in a 300 litre tank. The cooking system is similar to that used in the Iyama House.

Devices for obtaining hot water directly from solar energy have been in development since the 1850's. An early example was a boiler and power plant for water pumping built near Cairo in 1913. It consisted of five long parabolic mirrors, each 4.2 metres wide (14ft) by 62 metres (203ft) long. Steam was produced at slightly above atmospheric pressure and used to drive a 20hp water pump. It was abandoned during the First World War. Modern solar furnaces still use a similar system of mirrors. Recent developments, prompted by the space programme in thermo-electricity and photo-electricity, are used both to run water-heating systems and to generate electricity. Only a few large-scale solar furnaces have been built. The 10.5 metre (35ft) diameter mirror and heliostat at Montlouis in the French Pyrenees are used for research into materials at high temperatures rather than as an economic source of heat.

A recent installation is that at Daggett Solar Farm, California, USA where electricity is generated from an array of parabolic mirrors, arranged in long rows, that heat liquid in pipes set into them.

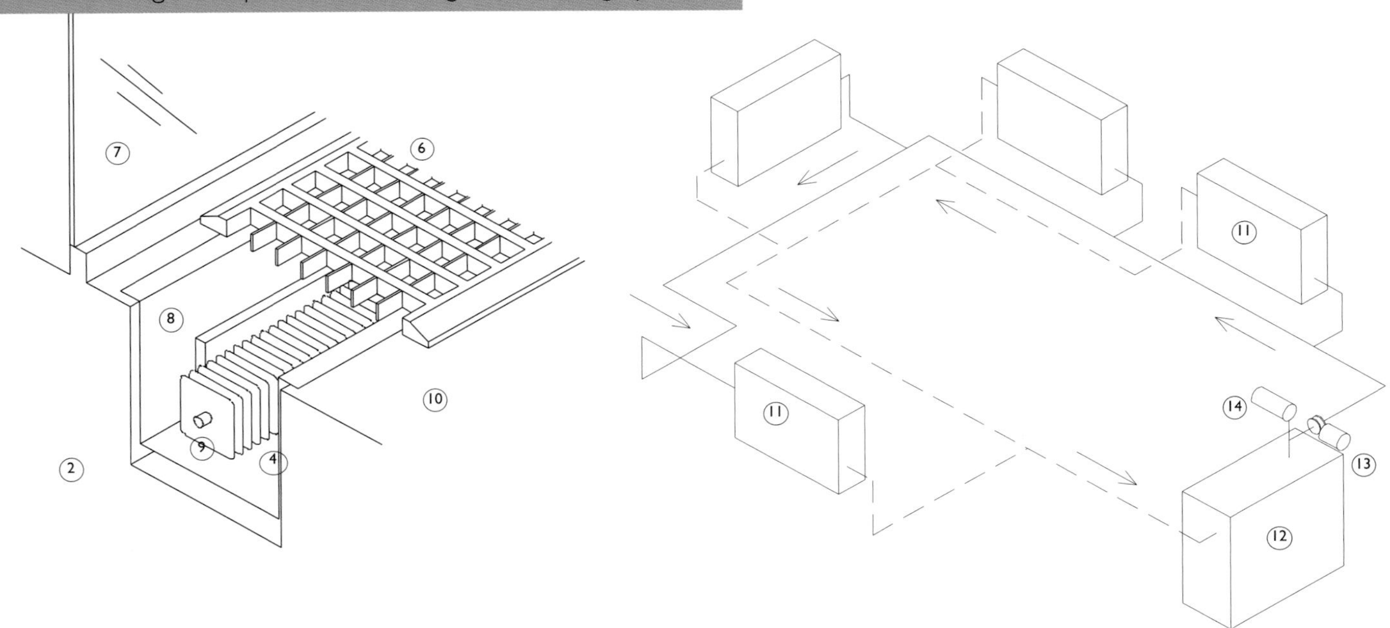


Details
1. Floor mounted fin radiator
2. Convector in floor recess
3. Wall mounted fin radiator
4. Fins
5. Water flows through panels
6. Floor grille
7. Glazed external wall
8. Metal tray to reflect heat
9. Water flows through pipe
10. Floor
11. Radiator
12. Boiler
13. Pump
14. Thermostat

Liquid based heating systems are provided mainly by radiators, convectors or underfloor heating. Commonly used liquid cooling systems are chilled ceilings, chilled beams and related items which are mainly used in office buildings where heat loads from building occupants and equipment impose cooling requirements for much of the working year even when solar gains are minimal.

## Radiators

Radiator systems pump low-pressure hot water through a pipe circuit, and distribute the heat through radiators which have a large surface area to radiate and convect heat. Heat is generated by boilers, usually fired by gas or oil. The pressed steel radiator which has generally replaced the traditional cast iron type, is made from pressed metal sheet welded together and has become the most commonly used type. They are generally ribbed to encourage convection currents, many have fins on their rear surface to enhance performance and panels can be supplied as single double or even triple to match duties required. Most of the heat is given off as convection, the remainder as radiation. Radiators are positioned to maximise convection currents and minimise pattern staining. They can be individually temperature-regulated with either a manual or thermostatic type valve on a constant temperature circuit or linked on a compensating distribution system, which automatically varies heat supplied to different facade orientations and adjusts in response to ambient variations.

## Fan and passive Convectors

Fan convectors are used where air needs to be heated very quickly, such as in countering downdraughts adjacent to large areas of glazing or above building entrances. Passive convectors are more static and are generally used where the loads are more stable. Convectors utilise a constant temperature low-pressure hot water pipe circuit. They are also used in conjunction with mechanical ventilation systems. Convectors consist of thin metal fins radiating from a central pipe supplied with circulating hot water. The fins are enclosed in a metal casing to increase the effect of convection, caused by the movement of air around the fins. Convectors can be floor or wall mounted and, like radiators, they are positioned to maximise air currents. When used in con-

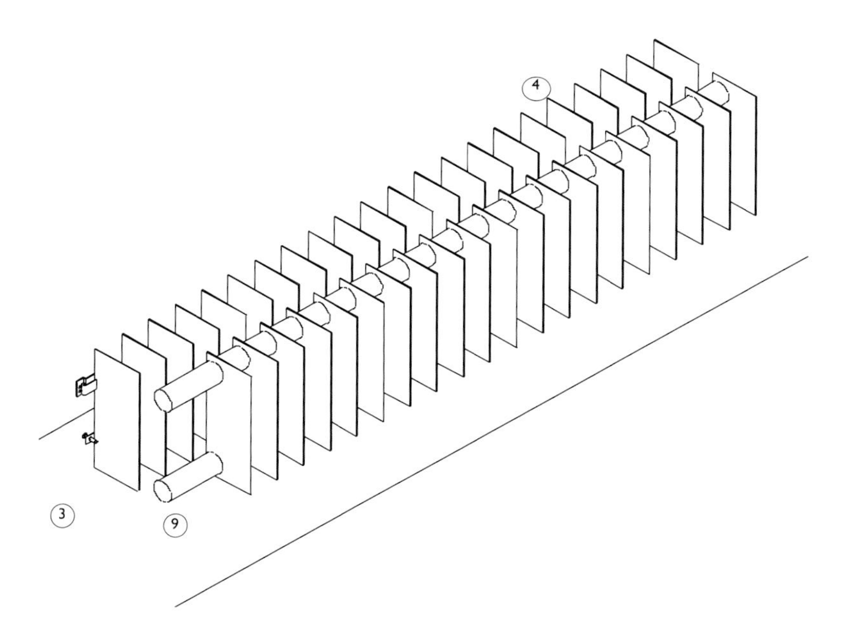


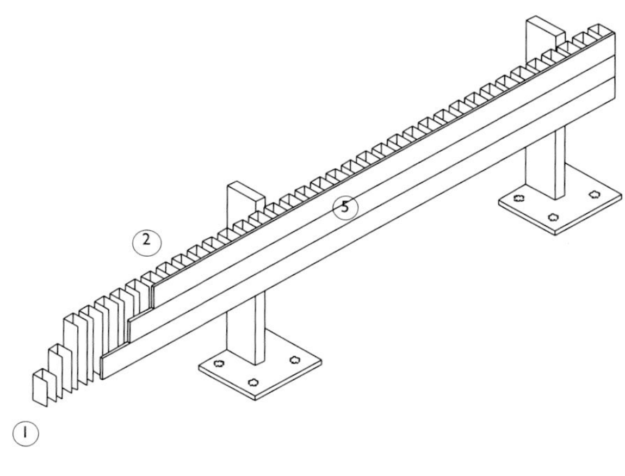


junction with glazed walling they are often recessed into the raised floor. Fan-assisted units distribute the air more quickly, but can create noise. Convectors are generally controlled by valves operating in response to temperature sensors and, because they require less water than radiators, react more quickly to changes in temperature if thermostatically monitored. A convector type used in conjunction with mechanical ventilation systems is the fan-coil unit, used primarily for cooling.

## Underfloor heating

Underfloor heating is another alternative to radiators. Heat is radiated from PVC pipes set into the screed of a concrete slab or located within the void in a timber floor. Insulation is laid under the pipes to minimise heat loss and, in the case of multi-storey buildings, avoid heating of the floor below. Water in the pipes is kept at a constant temperature of about 50°C, which is only warm to the touch. This system is often used as low-level background heat or where large floor surfaces in materials such as stone would be cold to the touch.

## Chilled ceilings

Passing air through a building during the night at outside temperature can help to cool the building structure. When a building is designed for this strategy the structural soffit is exposed to the occupied zone providing a passive cooled ceiling to the space thereby slowing the rate of rise in internal resultant temperatures. However, this approach is hard to control and is dependant on reliable diurnal temperature differences during warm summer periods.

Chilled ceilings have been developed to provide a controllable means of achieving overhead static cooling. These systems consist of chilled radiant panels, cooled by a chilled liquid, which are fixed to ceilings providing cooling by both convection and radiation. There are a number of types but generally chilled water is circulated in very small diameter pipes connected to panels which act as the heat exchanger to the internal space. As cooling is provided, the temperature of the water rises as it flows through the panels. The water is pumped back to a refrigeration plant or heat exchanger. Chilled ceilings are used as an alternative to full air-cooling and are used in conjunction with ducted mechanical fresh air ventilation systems.

## Chilled beams - active and passive

These devices are often grouped with chilled ceilings but they are quite different in operation. Active chilled beams are supplied directly with treated fresh air to induce added convection across the heat exchanger surfaces and generally provide greater cooling capability than a chilled ceiling system. Passive chilled beams rely on warm room air from the surrounding volume creating a downward flow as it is cooled when in contact with the beam heat exchanger surface; they provide considerably less cooling at the same density when compared to active beams. With passive beams, the ventilation air is often provided by an underfloor or displacement system. A chilled liquid, normally water, cools both types of chilled beam.

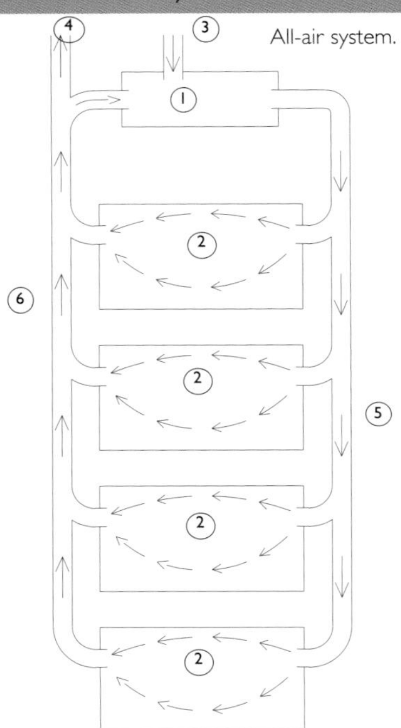


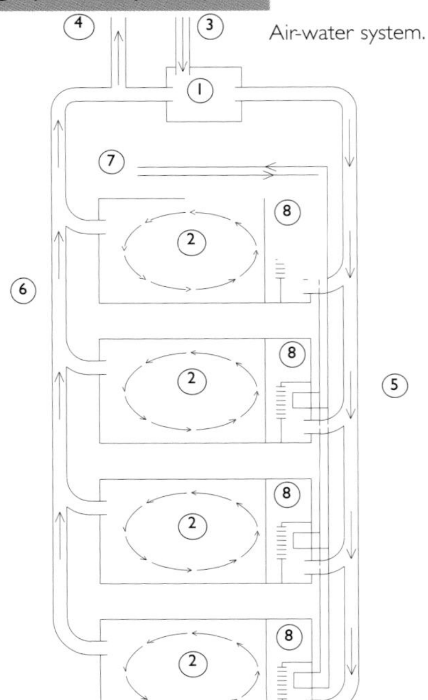


Details
1. Air handling unit
2. Space being ventilated
3. Fresh air
4. Exhaust air
5. Supply air
6. Return air
7. Supply and return warmed and/or chilled water
8. Induction unit or fan coil unit

Mechanical ventilation systems use ducted air that is heated or cooled either by a central plant or at the point of use or by a combination of these two methods. Mechanical ventilation is used primarily for its ability to provide fresh air to large areas within a building where natural ventilation cannot be used successfully. This is often due to the need for closer temperature control, or because of a deep floor plate where natural ventilation would be less effective, or where windows cannot be opened due to external noise and/or air pollution. Air from a central plant is ducted in prescribed volumes, based on either the number of people or the room thermal load, to where it is needed around the building. Air conditioning modifies the state of the ventilation air, which can include re-circulated air, primarily changing the temperature and humidity of the ducted air. As is the case with radiator systems, the air temperature can be modified at the point of delivery. Simpler systems are available whereby local air supply units are used which deliver the air directly into the space.

There are two basic approaches to mechanical ventilation, which are as follows:

· Constant air volume (CAV) system, in which the temperature of the air is altered.
· Variable air volume (VAV) system, in which the volume of air delivered is altered.

## Constant air volume (CAV) system

The constant air volume (CAV) system delivers air to a space with no variation in volume regardless of load change. As various parts of a building have different heating and cooling requirements at any given time so in theory separate CAV systems are required for each zone. This causes problems in large developments with multiple zones so a central supply system is used with local temperature adjustment. Traditionally these systems relied on central plant supplying a premixed volume of return air and fresh air treated to a level that coped with the highest sensible heat gain in summer or lowest net heat loss in winter. Re-heat coils in the terminal ducts cater to local variations in load in response to room temperature sensors. This type of system fell out of favour during the energy conscious nineties due to the waste of energy inherent in cooling the air centrally only to re-heat it locally.

The dual duct system allows air at varying temperatures to be supplied to different zones. Separate ducts supply heated and cooled air, which is mixed at the point of delivery in response to room temperature sensors. This enables a much greater degree of control, but it is much more expensive than the single duct system and requires much more space in the building services zones.

The constant air volume system utilises air-handling units that supply either warm or cool air. Some units can supply both hot and cold air simultaneously by dividing their supply. One half is warmed and the other cooled, after which they are remixed to provide the required temperature. Ducts supplying different zones can be connected separately to the air-handling units, with the required temperature controlled by dampers.

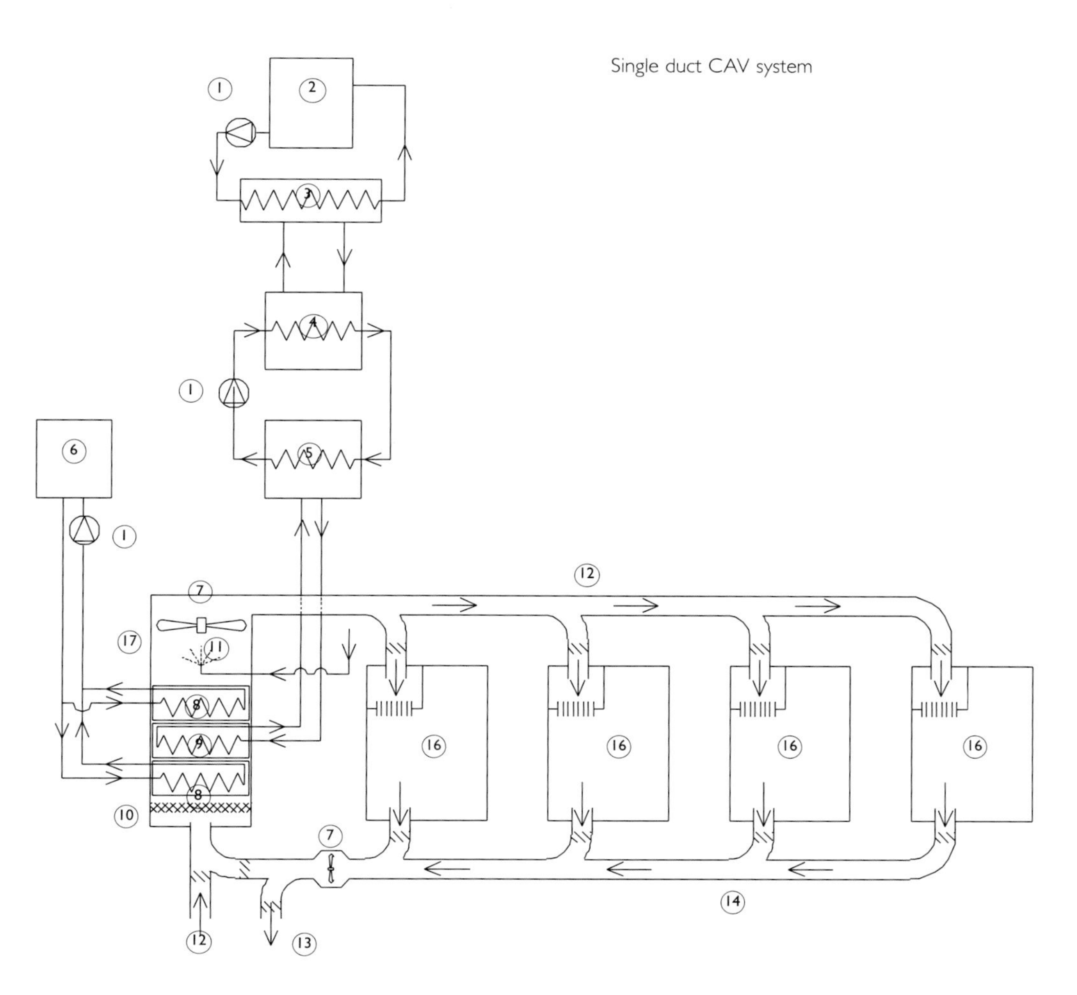


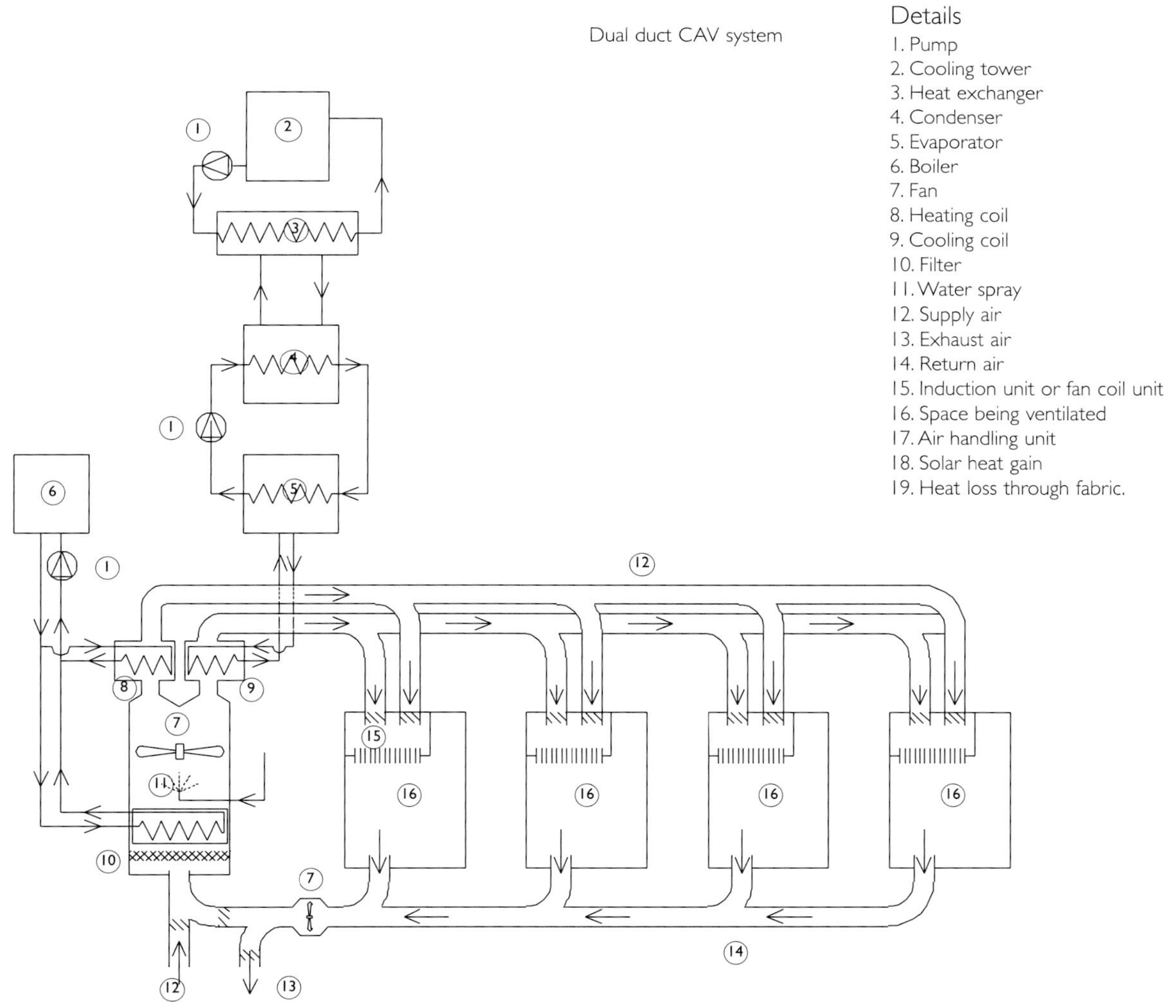


## Details

1. Pump
2. Cooling tower
3. Heat exchanger
4. Condenser
5. Evaporator
6. Boiler
7. Fan
8. Heating coil
9. Cooling coil
10. Filter
11. Water spray
12. Supply air
13. Exhaust air
14. Return air
15. Induction unit or fan coil unit
16. Space being ventilated
17. Air handling unit
18. Solar heat gain
19. Heat loss through fabric.

107

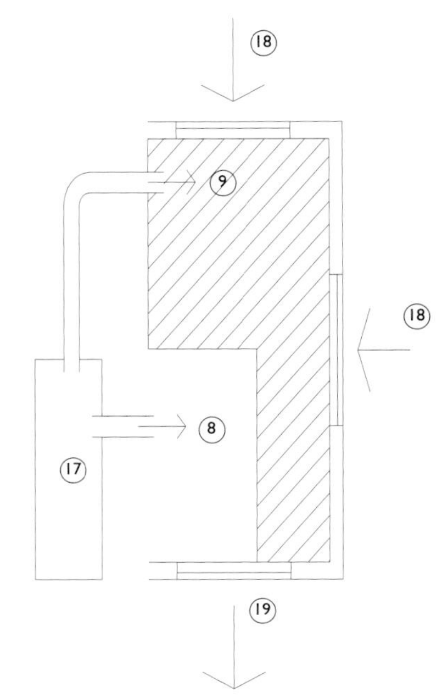


Principle of VAV

## Variable air volume (VAV) system

The variable air volume system (VAV) is used where the heating and cooling requirements differ within the same building, or even within the same treated volume. VAV systems use a single supply duct distributing air at a constant temperature. It responds to different temperature requirements in different spaces by altering the volume of air supplied. For example, the air volume supplied to a space will be increased if the temperature begins to rise, whilst the reverse occurs for dropping temperatures. The system provides local temperature control without the need either for separate air-handling units, or for water to be supplied to terminal units to warm or cool the air. The heating of perimeter spaces can be provided for by a liquid perimeter heating system. It is important when designing these systems to ensure that the blend of fresh air to return air is sufficient to maintain adequate fresh air at minimum volume turn down for ventilation purposes.

Air supply to each space is controlled by variable volume terminal units which adjust the volume supplied whilst maintaining adequate velocities for mixing. Variable speed electric fans are used in the central plant to maintain constant duct pressures thus ensuring that terminal units can maintain volume control.

## Mechanical ventilation equipment

Mechanical ventilation equipment consists of air handling plant and a network of supply and extract ductwork. Air is blown through supply ducts from the air-handling unit to a series of supply grilles or diffusers and returned via extract grilles or from a ceiling void. Air that is recovered in extract ducts which return it back to the air handling unit can be filtered and partially re-circulated depending on the system and time of day or year. The remainder is discharged to outside. Air is usually re-circulated because the quantity of air required for cooling or heating is often several times greater than that required for ventilation.

## Air handling units

An air-handling unit draws outside air in through a filter using an electric fan, where dust particles are removed. The air is then warmed or cooled to the required temperature for distribution. The temperature and relative humidity of the air can be controlled by a combination of warming, cooling and humidifying devices. The air is warmed by drawing it through a heating coil, which is heated by heating fluid, or through a heat recovery device, heated by the return air stream. To cool the air it is drawn across a cooling coil cooled

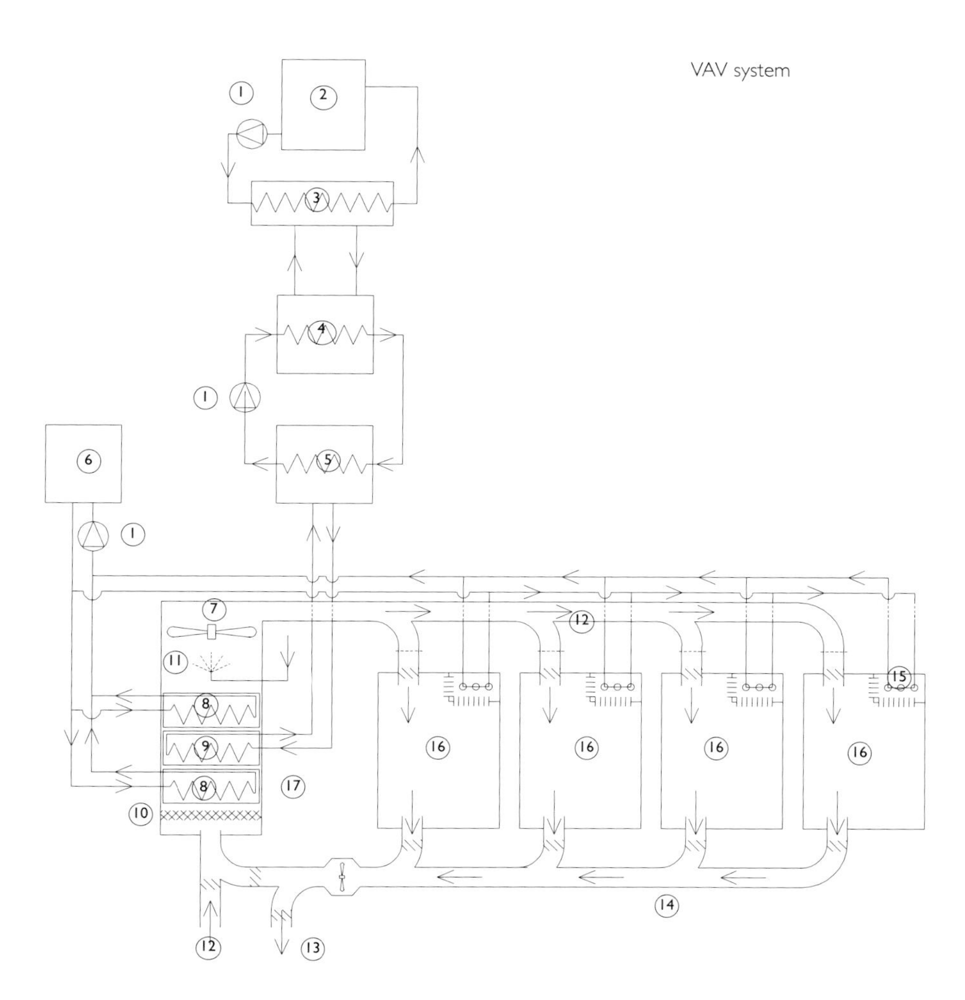


VAV system

Details
1. Pump
2. Cooling tower
3. Heat exchanger
4. Condenser
5. Evaporator
6. Boiler
7. Fan
8. Heating
9. Cooling
10. Filter
11. Water spray
12. Supply air
13. Exhaust air
14. Return air
15. Induction unit or fan coil unit
16. Space being ventilated
17. Air handling unit
18. Solar heat gain
19. Heat loss through fabric.

either by a cooling fluid, water or directly by a refrigerant. The humidity of the air can be controlled by spraying water into the air stream to increase its humidity, or by de-humidifying it as it is cooled. Air handling units either are installed in a plant room or are self-contained, waterproof package units that are located externally.

## Heat pumps

Heat pumps are used in a wide range of applications. The heat pump is a device that makes use of both the evaporation and the condensing stages of a vapour compression cycle. The refrigerant in the cycle absorbs the heat when vaporised (evaporator) and releases it when the gas condenses (condenser). These heat pumps can be reversed to deliver either heating or cooling to the same air stream or directly into the space.

## Heat recovery

Heat recovery is an important aspect of low energy design for ventilation systems. There are a number of methods of heat recovery used which generally transfer heat from the return air to the supply stream, which are as follows:

- Run around coil, capable of up to 50% efficiency.
- Recuperators or cross-over plate exchangers, capable of up to 65% efficiency.
- Thermal wheel, capable of up to 80% efficiency for dry heat using the non-hygroscopic type and 65-90% total heat recovery (latent and sensible) for the hygroscopic type.
- Heat pipes capable of 50-65% efficiency.

77E

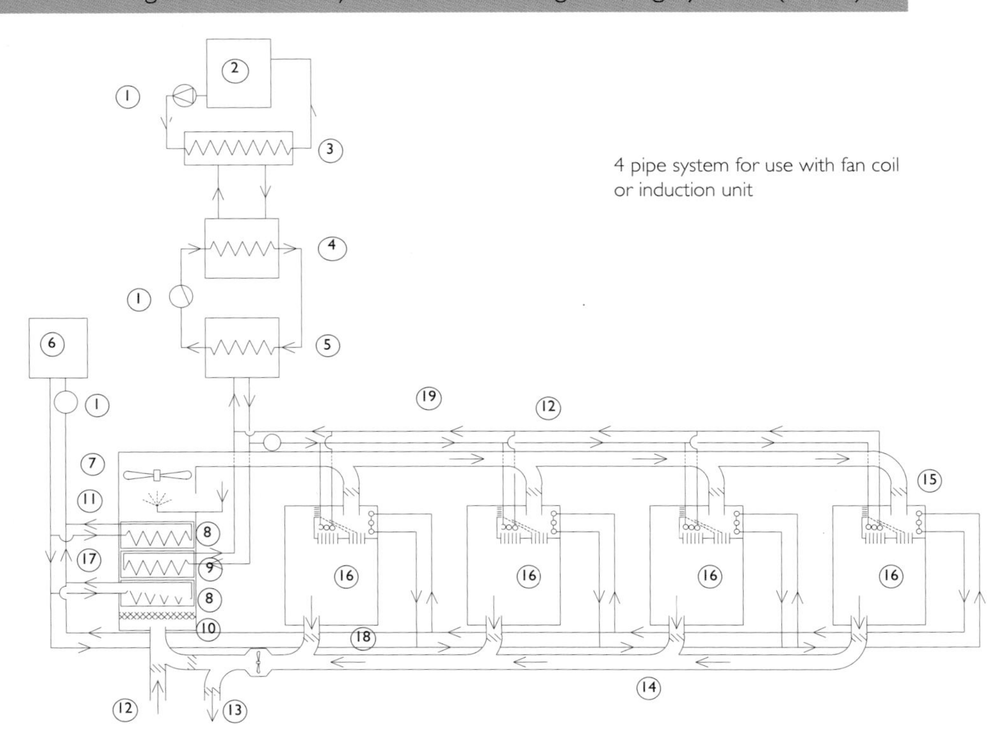

4 pipe system for use with fan coil or induction unit

Details
1. Pump
2. Cooling tower
3. Heat exchanger
4. Condenser
5. Evaporator
6. Boiler
7. Fan
8. Heating
9. Cooling
10. Filter
11. Water spray
12. Supply air
13. Exhaust air
14. Return air
15. Induction unit or fan coil unit
16. Space being ventilated
17. Air handling unit
18. Warmed water supply/return
19. Chilled water supply/return

98H

Run-around recovery consists of a series of coiled pipes with water circulating in them, set into the extract duct, usually near the exhaust air outlet. The water absorbs heat from the extract air and releases it into the supply duct. The thermal wheel consists of a wheel made from metal blades that spins in an enclosed jacket between the supply and extract ducts. The wheel absorbs heat from the extract duct and releases it to the supply duct. The same principle applies in the cooling mode although with less benefit, where heat from the warm incoming air stream can be slightly pre-cooled by exchanging heat with the exhaust air from the extract duct. Heat pumps can also be used to recover heat from exhaust air, helping to reduce energy costs. They are often used in swimming pool ventilation applications.

## Ducting

Air is circulated in ducts of galvanized sheet steel, used because of its ease of fabrication, light weight, smooth internal finish to assist air flow and is economic. Ducts are usually insulated externally in order to reduce heat loss, or gain, and to prevent condensation forming on the outside of the duct when cold air is circulating. Circular ducts have the

80H

77C

least surface area and the lowest frictional losses, however rectangular section ducts are more commonly used in commercial developments due to the fact that they require less depth when fitted above suspended ceilings and are economic. Square sections perform better than elongated sections; the maximum normally accepted depth/width ratio being 1:3. Flat oval ducts are also used in certain applications as they offer an advantage over rectangular ducts in being able to handle the same volumes with smaller sections due to lower frictional losses. Flat oval ducts are however more expensive to manufacture and require more specialised fabricators than circular or rectangular. The cross-sectional size of a duct is a balance between volume of air required, pressure drop and air velocity. Small volumes of air can be delivered at high speed or, alternatively, larger amounts of air can be delivered at lower speeds all with the same pressure drop using the same duct size.

## Supply and extract plenums

The need for extract ducting can be reduced where the void provided by a suspended ceiling or raised floor is used as a supply or extract plenum. When used as a plenum, the ceiling or floor void can house air-handling equipment that recirculates air at the required temperature. Alternatively, extract air can be drawn back to the central plant for treatment. Sometimes light fittings are used as air extracts to draw away the heat they produce. Extracted heat from these fittings can be recirculated, or exhausted from the building as required, which helps to reduce the direct impact on room heat gains due to lighting.

## Package Units

Package units do not require a network of ducts. They are essentially air handling units which draw air from outside and either cool or heat it, releasing the treated air directly to the space served. Due to its relatively small capacity the package unit is unsuitable for large spaces, since too many units would be required.

## Fan coil units

Fan coil units (FCU) work in conjunction with central fresh air plant. Generally, the fresh air required is treated at a central air handling plant and supplied at a constant temperature and volume either directly to the FCU or into the room/space served. The FCU is normally provided with cooling and heating coils and an inlet filter for the return air. The heating coil can sometimes be an electric coil if low usage is anticipated or no heating fluid is readily available. The fans inside these units can be speed regulated, normally by step changes, they simply draw room air in, mix this with fresh air where directly supplied, and distribute the air at the required temperature and volume to the space served.

FCU's can be either horizontally mounted above a false ceiling or vertically mounted concealed or exposed against a wall. They are sometimes concealed below a raised floor working as part of a floor supply system.

## Induction units

Induction units come in a variety of forms although they are less popular nowadays. The most commonly used units were perimeter mounted, normally part of a perimeter office cabinet section. They are supplied with high velocity centrally treated fresh air which discharges via supply nozzles within the units, inducting room air from low level to be drawn across cooling and/or heating coils, then mixing with the fresh air and discharging vertically into the room.

76H

72C

## Electrical Lighting.

The two main types of electric light source are tungsten filament lamps or incandescent lamps, and discharge lamps.

## Incandescent filament lamps

These lamps produce light by passing a current through a thin tungsten filament in a glass bulb containing either a partial vacuum or an inert gas. The filament glows and produces a mixture of light and heat. When an inert gas such as halogen is used, the lamp runs more efficiently as more of the electrical power is converted into light rather than heat.

The most common lamp is the general lighting service (GLS) light bulb. It has clear, coloured or pearl glass and its shape can vary from the common bulb form to spherical or linear. Some bulbs have their end silvered to reflect light upwards, reducing glare. The bulb used for spotlights or floodlights is the internally silvered reflector, or IS lamp. Its characteristic shape is designed to reflect light in a particular direction. A bulb with better directional focus, due to a reflector and a lens at the front, is the parabolic aluminised reflector, or PAR lamp, which is particularly suitable for spotlights and floodlights. Some lamps have dichroic filters in front of the bulb to reduce the amount of heat projected forward of the fitting. PAR lamps are also produced in a tungsten halogen type.

All these three lamps run on mains power (house current). The GLS and IS types last approximately 1,000 hours, while the PAR lamp lasts up to 2,000 hours.

Another type called low voltage tungsten halogen lamps, run off 12 volts and require a transformer. They produce sharply directional beams of light, making them useful for illuminating displays. They can last up to 4000 hours. They are also used for car headlights.

## Discharge lamps

These lamps work by discharging an electric current through a gas or vapour in a sealed glass tube or bulb, which causes it to glow. These produce more light and less heat than a tungsten lamp for the same amount of power. The most common type is the fluorescent tube, which is filled with low-pressure argon and proportionally small amount of mercury. The tube is coated with a fluorescent powder, and lengths vary between 320mm to 2400mm (13in to 96in). There is no standardisation in the production of these tubes, and each manufacturer designs lamps to their own specification. Tubes can last up to 10,000 hours, but their efficiency is slightly diminished with use. The low-pressure sodium discharge lamp is another common type used in many countries for road lighting, which has a characteristic yellow colour. It is extremely efficient and lasts approximately 12,000 hours.

High-intensity discharge lamps have a wider range of colour rendering, use much less electricity and last longer than an equivalent tungsten filament lamp. Mercury vapour lamps are used for lighting large spaces, such as factories, where colour rendering is not a determining factor as they have a yellow/blue colour. Metal halide lamps have excellent colour rendering and are commonly used in all types of building from commercial offices to public buildings. Both types of lamp last approximately 10,000 hours. Compact fluorescent lamps, which are used as a substitute for GLS and are compatible with tungsten fittings, offer significant energy savings as a direct substitution for incandescent fittings.

## Luminaires

Luminaires are used both to hold the lamp and direct light to the point of use. They are used for both incandescent filament and discharge lighting. A huge range of luminaires is available, each suited to its intended

85A

7E

use, location and method of directing light by diffusion, reflection or absorption. Manufacturers' catalogues use polar curves to plot the intensity of light radially around a cross-section or long-section through a light fitting. This helps the design team to understand how light from a single fitting will be directed.

## Ceiling mounted types

Luminaires on ceilings are either pendant (hanging), surface mounted or recessed. The pendant is the most flexible, allowing light to be thrown up, down or a mixture of both. Surface mounted luminaires can direct light to the side or downward. Recessed fittings throw light downward by concealing the lamp and luminaire behind the ceiling plane. Pendant luminaires are often used where there is no suspended ceiling. They can be fixed directly from the ceiling or housed in a track that is attached to the soffit. Huge varieties of ceiling tracks are available, and some incorporate both fluorescent strips and tungsten filament spotlights in a single length.

Ceiling fittings are often co-ordinated with low energy mechanical features such as chilled ceilings and beams. Fittings suspended from a thermally exposed soffit can be designed to incorporate acoustic absorbent material to help address reverberation. Ceiling light fittings can also form part of integrated servicing fingers, often used to service exposed soffit areas, incorporating sprinklers, air supply, chilled beams and high level electrical services.

## Wall mounted types

Luminaires on walls can also be pendant, surface mounted or recessed. Pendant fittings are mostly used in conjunction with a track hung from the wall on a cantilevered bracket. Most wall-mounted fittings are surface mounted which can direct light in all directions away from the wall, either directly or by reflection. Recessed wall lights have either a diffuser in front of the lamp or a reflector to avoid glare.

## Floor mounted types

Freestanding floor mounted uplighters are often used in offices to provide or supplement office lighting. They are generally better for glare control than ceiling mounted lights. Energy efficiency is lower with this type of fitting as the light has to be reflected from the soffit to illuminate the space.

A recent development in luminaire design is the recessed floor type. These have a cover that is strong enough to walk on and which is designed to minimise glare when looking down at the floor. This type has to be capable of dissipating its own heat generated in use, and as a result the casings are often made from metal alloys with high thermal conductivity such as brass.

## Exterior luminaires

Exterior lighting utilises all of the above fixings. Luminaires must be either waterproofed or sheltered from the weather and power supplies must be planned well in advance so as to incorporate cable ways and access boxes into the external works.

## Emergency lighting

Emergency lighting allows escape routes and stairs to be sufficiently illuminated in the event of a fire or other emergency. The lighting can be permanently illuminated or switched on automatically by a fail-safe system in the event of a power failure. Both tungsten filament and discharge lamps are used. Power to emergency lighting is normally supplied by batteries charged continuously by the mains using separate circuits. Emergency generators should be used in buildings where mains failure can affect life safety and other essential equipment. Batteries are usually housed within plant rooms.

## FUEL SUPPLY

### Natural gas

Natural gas is a non-renewable source formed in pockets beneath the earth's surface. It is brought to its point of use by a network of pipelines. Natural gas has no smell, so a tiny quantity of odour is added in order that dangerous leaks can be detected. Gas is supplied to buildings under pressure normally via a single service pipe where the supply terminates in a stopcock and meter. Beyond this point, the gas supply forms part of the building services, and is usually connected directly to the gas-fired boilers. There are design limits for internal pressure drop between the meter and the point of use. In some cases gas pressure is insufficient for the plant needs and special gas boosters are used to raise the pressure to a satisfactory level. Geographical areas that are not served by natural gas require large gas tanks that are refilled by deliveries.

### Electricity

Electricity is generated in a variety of ways, including coal burning, nuclear and hydro-generation. Electricity is usually produced as alternating current (AC), which is cheaper to produce and transmit than direct current (DC). Electricity is transmitted regionally at high voltage and low current except in large countries such as the USA where low voltages are used to reduce wastage across long distances. The voltage is stepped down at substations for use in buildings.

Power is distributed from substations as a three-phase supply. Smaller buildings are connected to one of these phases (single-phase), whilst larger buildings are supplied with all three phases plus an earth (three-phase). Smaller installations in buildings have a consumers unit (circuit breaker panel) distributing power to a primary ring circuit. Socket outlets (for portable appliances) and spurs (for fixed appliances) are connected to the ring main. Spurs can also be used to connect other sockets in order to minimise wiring.

In larger installations, a distribution board allocates circuits for lighting, sub-mains and individual appliances such as lifts. Sometimes individual distribution boards are provided to serve separate areas within a building. Some areas require single-phase, while others will require three-phase supply.

The sub-main must be capable of being modified. This is facilitated by a cable-tap system. The four cables of the three-phase supply are supported in a metal trunking and connections are made with connectors called tap-off units. For heavier loads, or where it is necessary to make connections without interrupting the supply, a busbar is used. This consists of solid conductor rods, without insulation, supported in a metal trunking. Connecting units enable leads to be run off from the conductors to tap-off boxes without switching off the current.

## WATER SUPPLY

Fresh water (from reservoirs, lakes, rivers or ground wells) is treated, filtered and distributed by a water supply network, which terminates at the boundary of each property, usually as a single pipe. Beyond that point, the supply forms part of the services installation. A stopcock separates the supply grid from the individual distribution system.

## Cold water distribution

Water distribution within a building is either gravity fed from storage tanks at or close to the roof of the building, from a pressure booster pump set at a lower level or by direct mains supply. Although water storage can require a large space, tanks provide a limited reserve in the event of an interruption to the mains supply. Traditionally this is based on a capacity relating to 24 hours of predicted consumption. Cold water tanks are generally made from galvanized mild steel or glass reinforced plastic (GRP); they are normally insulated to prevent condensation forming on the outside face. Supply to the tanks is controlled by a float-operated ball valve with overflow pipe to protect from overfilling the tanks. In tall buildings where the water pressure in the supply pipe is too low to achieve adequate pressure at the draw off points, water can be pumped. This can fill a series of smaller tanks, each serving separate parts of the building, or pressurised throughout with pressure reducing stations at intermittent levels to avoid over-pressure at draw off points. Various regulations govern the method of maintaining uncontaminated supplies in buildings. These generally stipulate the method of connection to appliances and draw off points with an emphasis on protecting against reverse flow.

## Hot water storage and distribution

The choice of hot water system is based on the location and density of draw off points, size and configuration of the building and the location of boiler plant. Large office buildings with fairly big distances between closest and farthest draw off points are generally suited to localised water heaters, whereas buildings such as hotels are more suited to centralised boiler powered hot water calorifiers.

The method of distribution is determined by the length of run-off from the point of heating the water to the point of use. Hot water can be fed by gravity or pressurised, this will normally depend on the strategy determined for the cold water system. If the storage cylinder is unable to provide adequate pressure to the highest appliance from a gravity supply then pressure boosting will have to be considered. If hot water is stored in calorifiers, the distance from draw off points is minimised to reduce the cold water run-off from the pipe. Circulating pumps are used to maintain high temperatures at all points on the network via flow and return pipes. Trace heating is used to maintain temperature on remote draw-off points fed by a single hot water pipe and is sometimes used in place of running a flow and return network.

Hot water can be generated by either primary hot water from a boiler, by direct gas fired storage heaters, by an electric immersion heater or locally via electric or gas hot water heaters. In some cases, it is preferable to provide hot water from a storage tank that is fed directly from the mains water supply rather than from a cold water tank. This system is unvented and requires a series of valves to ensure that the water pressure and temperature are closely controlled.

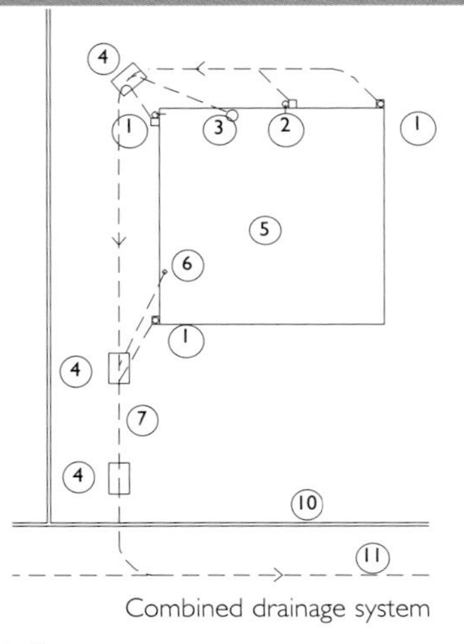

Combined drainage system

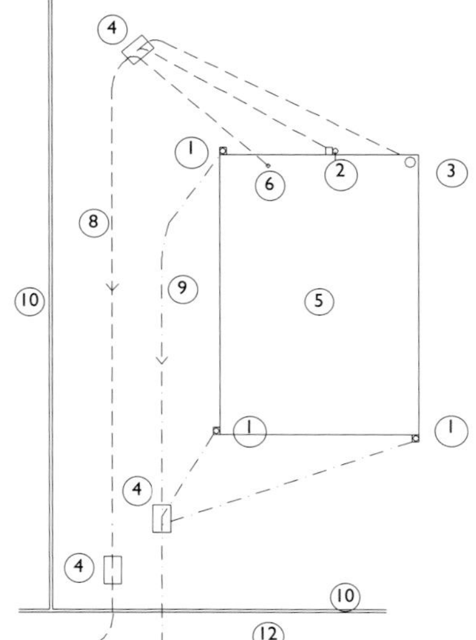

Separate drainage system

Details
1. Gully to rainwater pipe
2. Gully from kitchen
3. Soil and vent pipe
4. Inspection chamber
5. Building
6. WC
7. Combined drain
8. Foul water
9. Surface water
10. Property boundary
11. Sewer
12. Foul sewer
13. Surface water sewer

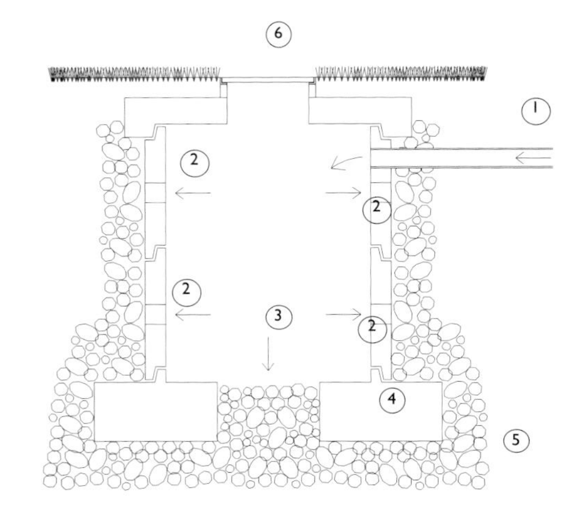


Details
1. Supply to soakaway
2. Drainage opening
3. Drainage out
4. Foundation
5. Hardcore
6. Access cover

## Drainage below ground

Drainage is the second half of the supply cycle. Drainage can be provided by a separate surface water (storm water) and foul (sanitary) drainage system, or combined so that rainwater is drained with the foul water. In some areas, particularly towns, the separate arrangement is preferred so that during storms rainwater does not overload the drainage system, preventing the discharge of foul water from buildings.

Underground drainage is laid to falls in straight lines to allow for ease of cleaning by rodding. Access points are provided either where there is a change of direction or where pipes join, as blockage tends to occur at these points. Rodding eyes or inspection chambers are provided for this purpose. Inspection chambers are also provided along the length of the pipework. They consist of a brick, plastic or concrete lined box. The top is sealed with a removable cover. The upper halves of the pipes in the chamber are cut away to provide access for cleaning by rodding. The connecting pipes in the chamber do not need to be at the same level. This allows pipework to follow levels and gradients which suit other parts of the drainage layout and meet at different levels. The lowest level at which the water runs is known as the invert level of the manhole. Ladder steps are provided if the chamber is deeper than approximately 600mm (2ft). Where a pipe runs below a building and an inspection chamber is required, special measures, in particular double sealed chamber covers, are taken to prevent foul air escaping. Not all pipe intersections require inspection chambers. An alternative is to provide sufficient rodding points along the length of the drainage run.

Pipework below ground consists of a series of pipe lengths with a spigot at one end and a socket at the other. The pipes slot into one another, and are sealed with rubber rings. Sometimes separate sleeves are used to connect pipes instead of sockets. The rubber rings also provide a degree of flexibility for movement. Pipes are made from vitreous clay, concrete, plastic (UPVC) and cast iron. Pipes are laid in trenches on a well-compacted surface. Concrete is sometimes used depending on the local ground loading conditions. They are enclosed in gravel wherever excessive movement is expected and rocker pipes, short lengths of pipe with connections at both ends, are used to allow movement to occur.

## Septic tanks

Septic tanks are used where there is no local sewer system. These can be set either at or below ground level. They are manufactured as complete units prior to installation on site and consist of a series of chambers which separate solid from liquid waste. The liquid waste is allowed to leach back into the soil through a series of perforated pipes called a 'leaching field'. The solid waste is periodically emptied into tankers and removed for off-site treatment.

## Soakaways (drywells)

These are a method of providing surface water drainage directly to permeable ground adjacent to the building. Soakaways are used in rural areas where there is no benefit in draining away rainwater into the sewer system. They consist of a large pit into which the surface water drain discharges, permeating the rainwater into the surrounding soil. The most common type is a concrete tank made from a set of stacked rings. Simpler versions consist of a pit filled with hardcore (gravel) that prevents it from collapsing, whilst allowing the water to drain away.

## Sanitation and drainage above ground

Water in a building is drained as either soil or waste. Both are drained either separately in a two-pipe system or together in a single-pipe system. The drainage is usually installed

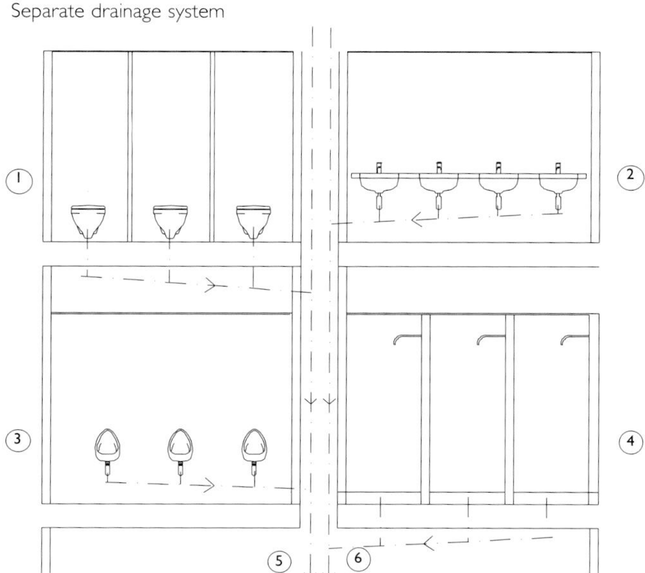


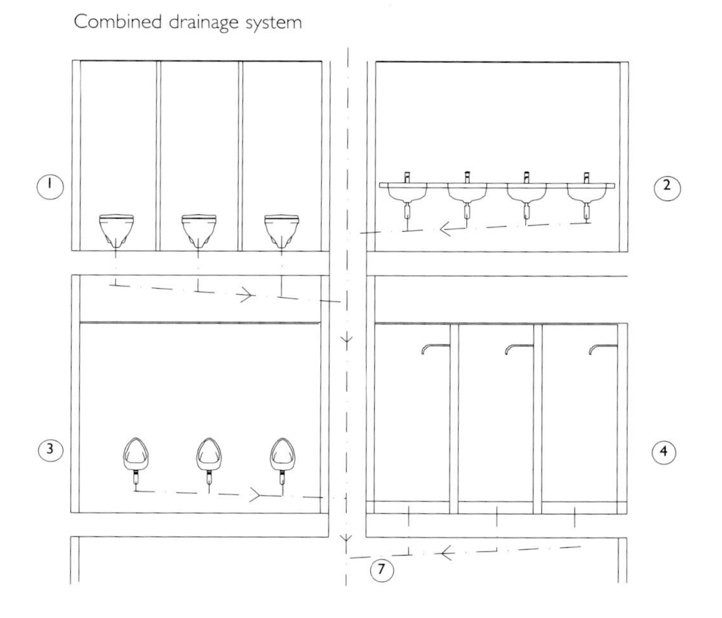

Details

1. WC
2. Wash basins
3. Urinals
4. Showers
5. Soil pipe (may require additional vent pipe)
6. Waste pipe (may require additional vent pipe)
7. Soil and vent pipe

and connected to the below ground drainage from inside the building to avoid the pipes freezing and to avoid multiple street sewer connections.

Water closets (WC's) and urinals are drained to a soil pipe. Sinks, basins, baths and showers are drained to a waste pipe. Rainwater is drained separately. Below ground, the soil and waste pipes combine as foul water drains. Rainwater either drains into separate surface water drains or it joins with the foul water via a special trap forming a combined system. These pipes run below ground to the main sewer. In areas not served by a sewer network, a septic system can be installed for soil and wastewater treatment.

Air from these pipes is prevented from entering the building using traps fitted to each sanitary and waste appliance. Pipes slope downward to the sewer, since they work entirely by gravity. Pumps are only occasionally used where adequate falls cannot be provided. For this reason, pipes with smooth internal bores are used to reduce friction. Falls in soil pipes below ground are normally around 1:60, depending upon location, design flow rate and size.

## Balancing the system

Water flows down the inside surface of vertical pipes, leaving an air gap in the middle. In horizontally laid pipes, the water runs down the bottom or invert of the pipe due to a shallow fall. The air inside the pipework is designed to remain at atmospheric pressure to equalise air pressures on either side of traps. If the pipe bore is filled with water then air cannot balance the system, causing traps to pull due to the siphon effect, or blow through their water seal due to over-pressure. Consequently, pipes are designed to be large enough to accommodate a percentage of air and are ventilated to atmosphere close to the sewer outfall and at the top end of the drain stacks.

## Single-pipe system

This system, also called the single stack system, combines the soil and waste pipes. Each sanitary and waste appliance is connected to a single pipe, one above the other in a stack. The pipe used is generally larger than in the two-pipe system, helping to avoid pressure fluctuations that might cause the traps to lose their seal. The use of a straight stack, that is, without bends above ground level, ensures a controlled flow of soil and waste, allowing air to move freely in the pipe. Lengths of waste pipes from appliances to the stack are limited to approximately two metres (6ft). Beyond this distance, the trap to the sanitary appliance is fitted with a separate vent pipe known as an anti-siphon pipe.

## Rainwater

There are a number of rainwater systems in use today, the choice of which is based on the type of areas being drained. Large areas where space for high level collection rainwater pipework is limited can be drained using siphonic drainage systems. These systems are designed to create a suction pressure within small-bore horizontal rainwater pipework and use specially designed outlets. Large glass roofs are often drained by flat channel systems that have special tapering outlets to cope with the large flow rates needed to drain the roof efficiently. Insulation is applied to rainwater pipework where it passes within internal areas to avoid condensation.

Rainwater is generally collected below ground into either a combined or a separate drain, which is then connected to the combined foul and surface water sewer or a surface water only sewer. In both cases, the connection is made via a vented intercepting trap. Rainwater is sometimes collected into storage tanks for reuse in irrigation applications or as a supplement to grey water systems. Bypass pipes are needed to avoid overfilling storage tanks.

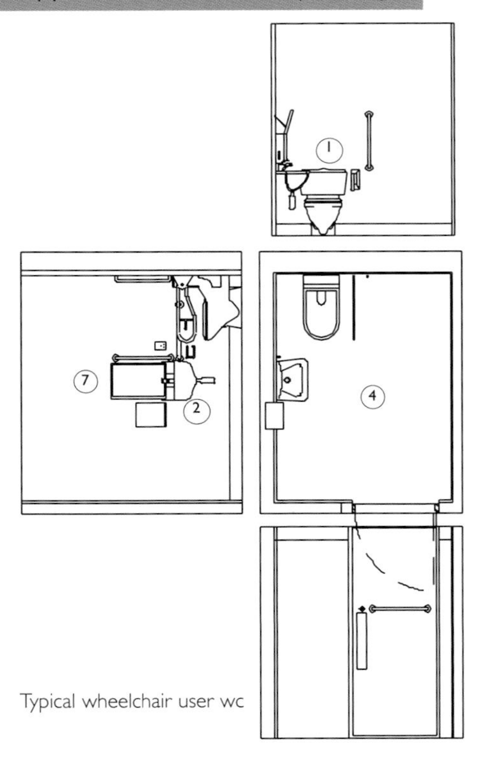


Typical wheelchair user wc

39A

14C

14H

Sanitary fittings are designed to avoid the possibility of contaminating the water supply by positioning the supply above that of the outlet. For this reason the taps on a wash basin are set above the top of the rim in order to avoid the risk of back siphonage into a supply under low pressure. There is a variety of standard fittings on the market from a large number of manufacturers.

## WC's

Washdown and syphonic WC's are the most commonly used types. The water in the pan forms a trap, sealing contaminated air inside the pipes. Unlike waste pipes, there is no gully or trap to separate the sewer from the soil pipes. WC pans are made from ceramic glazed earthenware, vitrified china or pressed stainless steel. In many countries, a flushing valve is used instead of a cistern, which may be fitted with a non-return valve to prevent contamination of the water supply. The water is supplied either from the mains or from a cold water storage tank.

## Urinals

The bowl type is the most common urinal. It functions in a similar way to a WC. A separate trap is used to connect them to the soil system. Materials used are the same as those specified for WCs.

## Wash basins

These are set on brackets, pedestals or into a flat surface, such as a vanity unit. Types for extensive washing have the taps set well back to allow maximum use of the bowl. Taps used for washing hands extend towards the centre of the bowl to make them easier to utilise running water. Basins can be plugged using a traditional bung (stop) or with a 'pop up waste' which comprises a lever mechanism fitted into the waste pipe. Materials used are the same as those specified for WCs.

## Vanity units

A vanity unit is a generic term for a set of panels that enclose wash hand basins in order to conceal the plumbing beneath. Units may comprise a vanity top into which basins are set, or may be part of a complete cupboard unit. An important consideration in the design of the top is

Details
1. WC
2. Wash basin
3. Urinal
4. WC for wheelchair user
5. Removable duct panel
6. Fixed duct panel
7. Mirror
8. Towel roll holder

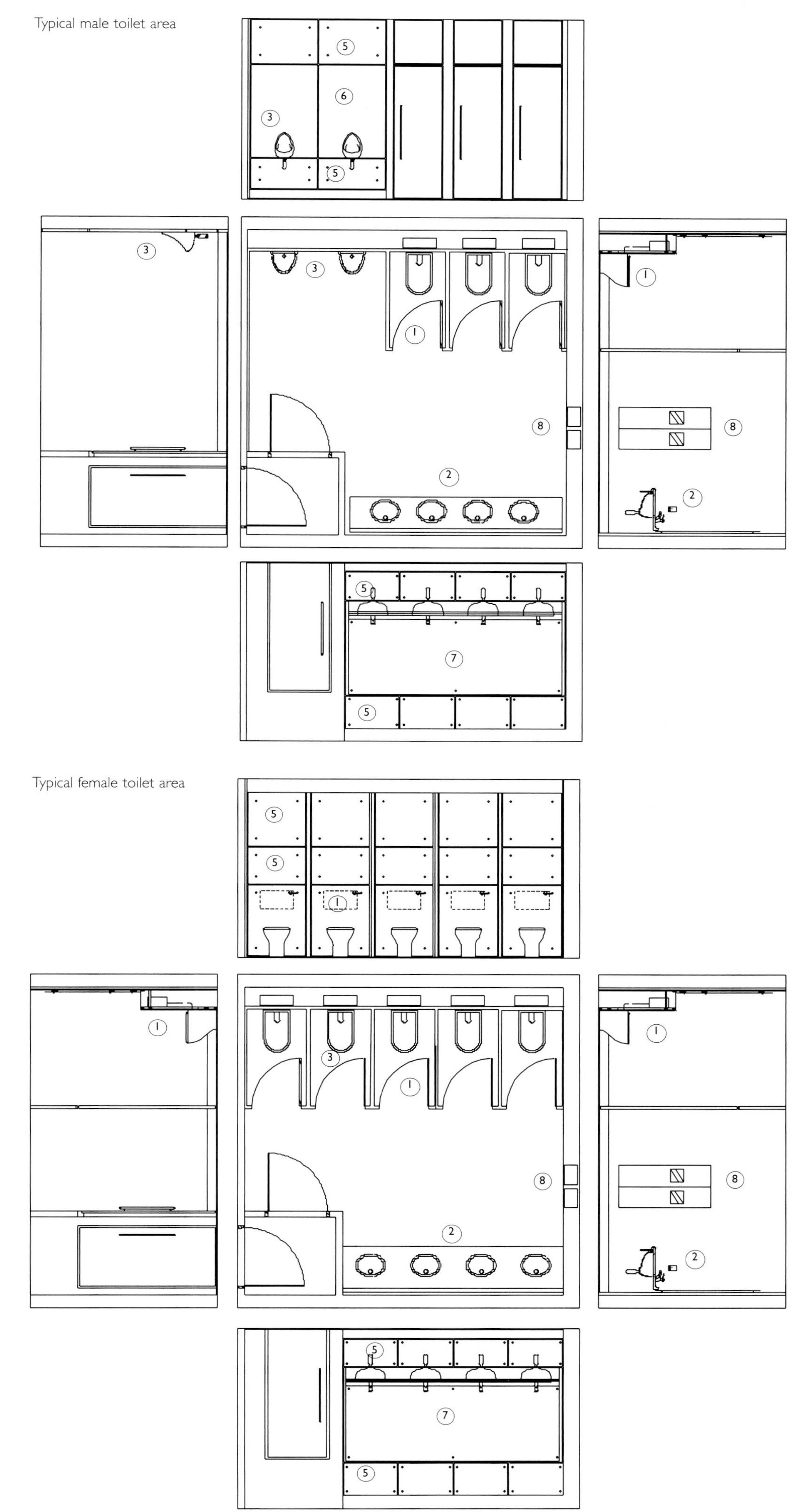


dealing with water that splashes from the basins. Some have grooves around their edges to prevent water from running onto the floor beneath. The plumbing either is completely exposed, typically as chrome-plated traps with chrome plated or stainless steel pipework, or is concealed with a panel using more economic pipework materials. If the plumbing is left exposed then its design requires a close attention to detail.

## Concealed fittings

Ancillary components to sanitary appliances are often concealed in voids behind accessible panels. Pipework, WC cisterns and towel dispensers are often partially concealed. Panels are supported on a frame made from pressed steel or wood studs. Panels that require regular access are usually the lift-off type, though more expensive hinged types are also used. Panels requiring only occasional access are usually secured with screw fixings which have a cap applied to the screw head in order to conceal it. Domed caps are often used to conceal screws used to fix large wall mirrors.

## Fire prevention

Most fires are caused by electrical problems. Since electrical cabling and installations are usually hidden from view, an early detection system is essential. Medium and larger sized buildings are required to install an electrically operated fire information panel that will indicate the position of any fire within the building. Once a fire has started it must not be allowed to spread unchecked. For this reason the choice of materials inside a building is extremely important both in terms of their combustibility, spread of flame across their surface and the production of poisonous fumes.

## Fire protection

Structural walls, columns, floors and staircase enclosures must be furnished with a fire resistant outer layer in order to remain stable during a fire, for between one and four hours, depending on the component. Concrete naturally has a high fire resistance and consequently requires no further protection. Steel structures are encased in concrete, covered in a fire resistant board or sprayed with a fire resistant material such as vermiculite.

Buildings are usually divided into fire compartments, which are used to limit the volume in which a fire can spread. Although separate floors are isolated, shafts that pass between them, such as service risers, require fire breaks either at each floor level to ensure the integrity of the compartments or the shafts are enclosed by suitable fire rated walls and all penetrations are fire stopped. Fire breaks are constructed from a fire retardant material such as cement mortar or mineral wool quilt. To stop fire spreading between buildings, external walls are protected with fire resistant applications or utilise specifically designed dividing walls.

Recently, some of the prescriptive rules for fire protection are slowly being tempered by fire engineering studies undertaken at the design stage. These involve an overall fire protection strategy that considers both the building structure and the means of escape together and sometimes involves some computer modelling or testing. Each study allows alternative combinations of fire protection and prevention to be considered before the design is finalised and thus forms an integral part of the building when complete.

## Fire escapes

The most important function of fire escapes is to ensure that people can leave the building quickly and safely. Individual compartments have protected areas that allow people to escape without coming into contact with either smoke or flames once they have reached these areas. Protected areas usually discharge to the outside of the building at ground level. The usual minimum requirement for a protected area is 30 minutes. This is judged to be the maximum time required to evacuate everyone from a building. Legislation in most countries for stair widths, door widths, and corridor widths is based on occupant flow rates that will clear everyone past a given point within two minutes. Emergency lighting and signs are provided to ensure that this evacuation can occur even during a power failure.

## Fire detection

The provision of fire detection equipment varies with size and type of building. Break-glass units are the most common types of manual alarm where occupants can activate the fire alarm system. In addition, automatic smoke detectors or heat detectors are installed to initiate alarms. Tall spaces have beam detectors where the presence of smoke or fire might otherwise go unnoticed. Buildings with large fire compartments tend to utilise special heat detectors, which close fire shutters when activated. The strategy for evacuation forms an important part of the planning for detection and alarm signalling in the event of a fire. Phased evacuation buildings require addressable voice alarm systems and most large capacity buildings need stages in alarm signalling to confirm whether it is necessary to evacuate the building.

## Fire fighting

Sufficient access and facilities must be designed into all buildings so that safety for fire fighters can be provided when tackling a fire. There are extensive prescriptive rules, which must be adhered to, that are based on fire-fighting experience.

An essential aspect of fire fighting is to provide access for emergency vehicles. Similarly, within a building, hose reels and portable fire extinguishers are provided as a manual first aid means of controlling fire spread. Automatic sprinkler systems are sometimes used in buildings; these are also for controlling the spread of fire and are regarded as a first aid. The latest generation of sprinkler releases a mist or fog rather than spraying large quantities of water, however large storage tanks or fail-safe water supplies are still required. Such systems are designed to control fires while people escape from the building rather than extinguishing the fire and they are often isolated by the fire fighters when they arrive.

In larger structures, either dry or wet risers are provided to allow the hoses used by fire fighters to be connected to a water supply within the building. In taller buildings, whose height makes access by ladder unfeasible, fire fighting lifts and stairs are provided. In order for fire fighters to reach all parts of the interior, smoke clearance for each compartment should be facilitated through cross ventilation or by mechanical means.

78A

54B

54C

External cladding requires more maintenance than traditional construction techniques. This is because windows can usually be cleaned and maintained from inside the building whereas cladding panels are fixed and have few opening lights. Most cleaning and maintenance is achieved via cradles, gantries or ladders. This equipment can also be used to replace damaged panels. Cradles are platforms enclosed by balustrades in which maintenance personnel stand to access all parts of the facade. They are supported by cables from a roof-mounted hoist. Travelling gantries are used to clean glazed roofs. Vertical areas of glazing not exceeding 10 metres (33ft) are generally accessed by ladders. The choice of maintenance system is based on building height, shape, structural design and frequency of use. Maintenance equipment tends to be standardised but is often modified to suit each application.

## Cradles on jibs

Cantilevered frames, or jibs, are used to support cradles, which can be fixed either on pads or wheels which run either on rails or on a dedicated zone, or runway, on the roof. These jibs allow the cradle to be moved from one part of a roof to another and are used to swing the cradles in and out of position in front of the facade. This method has the advantage of allowing more space to be used to access other roof-mounted equipment such as air handling plant.

## Cradles on trolley systems

Trolleys supporting access cradles are mounted on either rails or wheels. Rails are mounted on either the parapet or the supporting roof deck, and allow the loads from the trolley to be transferred to specific points on the building structure, such as immediately above columns. In common with cradles on jibs, this system can be either hand operated with a winch or be electrically powered. Buildings taller than three storeys usually require an electrically driven motor.

Details
1. Cleaning cradle
2. Motor and counterweight
3. Hydraulically operated arm
4. Steel cable
5. External wall parapet
6. Roof surface
7. Wheels or guide rails

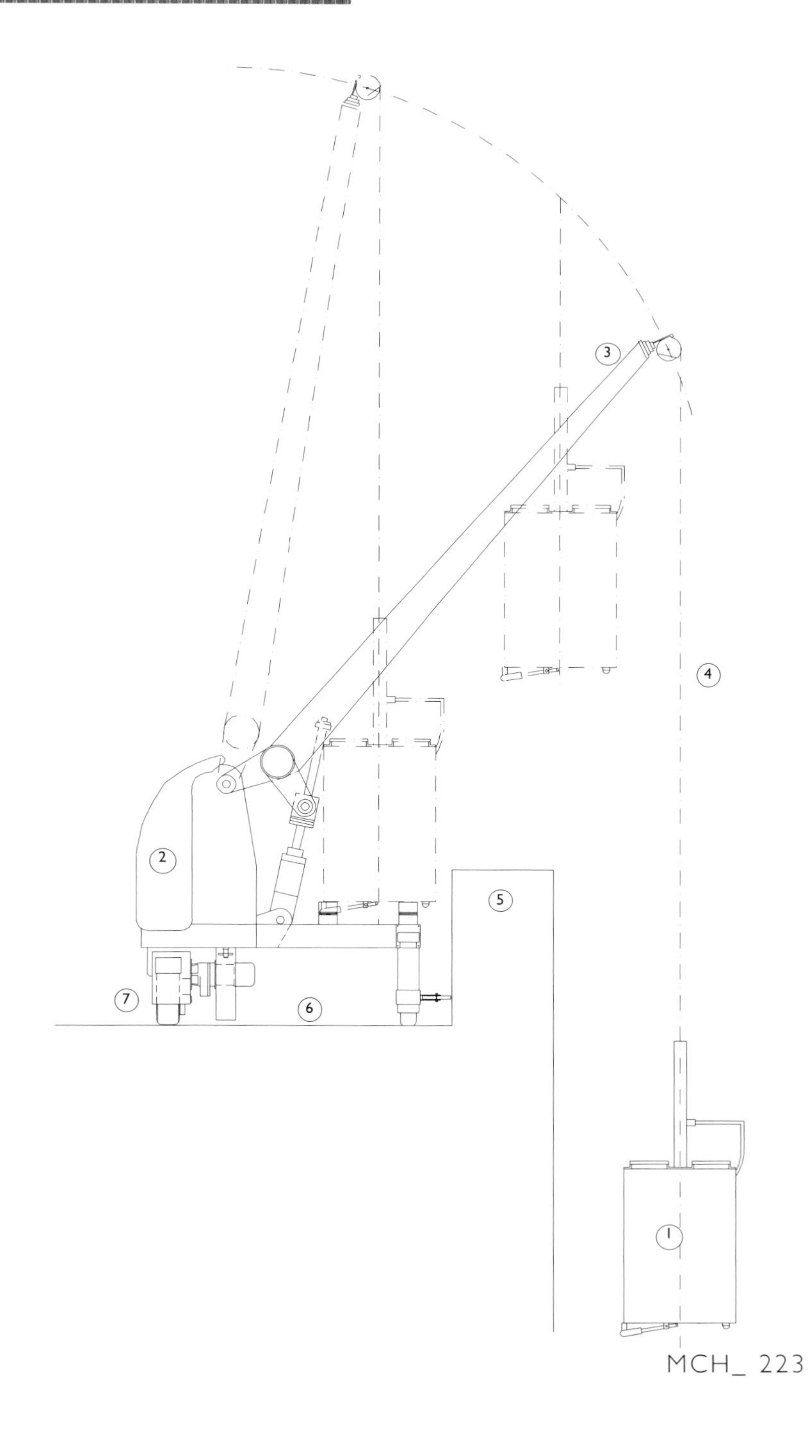


## Cradles on external rails

In this system, cradles are suspended directly from rails that are positioned at the top of the wall, directly above the cradles, avoiding the need for jibs. Additionally, the cradles can be guided on vertical rails that are integrated into the window mullions.

## Travelling gantries and ladders

Gantries run on rails set at each end of the travelling platform. The platform is enclosed with balustrades to provide a safe working area and may house additional fixed ladders to reach different parts of the external walls. Travelling ladders are supported by a top rail, and the bottom wheels usually rest against a continuous metal strip fixed to the external wall. The safety of the people using the equipment is paramount, and all designs include safety harnesses and fall-arrest cables.

# 6

## FITTINGS

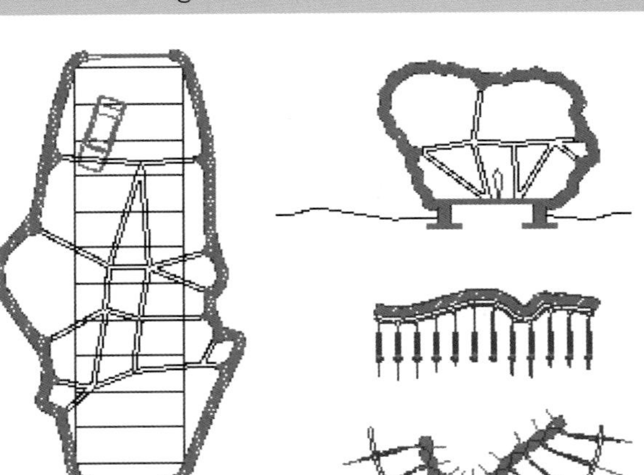

21A

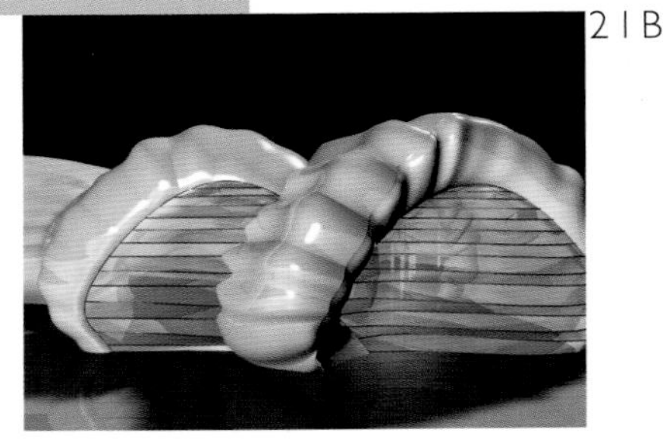

21B

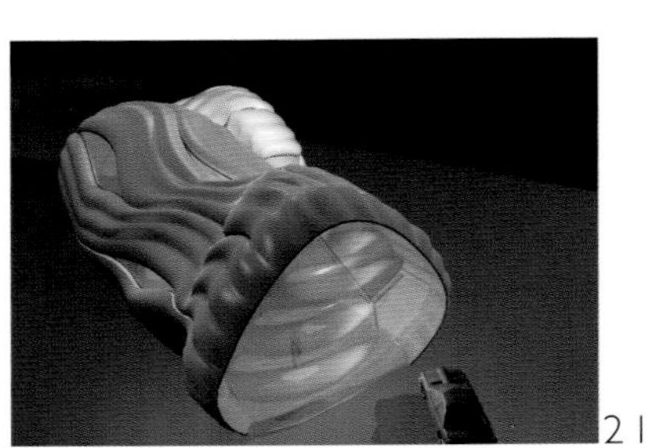

21C

105

105

Historically, traditional construction was performed mostly on site working from 'raw' building materials, where windows, doors, sanitary fittings and related servicing items were the only prefabricated assemblies delivered from a factory. There has been a slow but steady move away from site-based construction towards prefabrication in workshops. Prefabrication is preferred for the increased speed of construction and improved quality control over that available in site-based construction. These qualities help prefabrication to take on the demands of the ever-increasing technical complexity of construction projects from small to large scale. The degree of prefabrication used ranges from small-scale components such as pre-formed toilet cubicle components to fully pre-assembled toilet modules.

An essential difference between site-based work and prefabrication is that work on site is progressively refined by successive operations whereas prefabricated components usually have to fit correctly as soon as they are installed. For example, in site based construction, the imperfections in walls are covered over by plaster, while the local cracking at edges in plaster is concealed by skirtings; the imperfect fit of doors into openings is concealed by architraves, and so on. These successive operations use lapped joints to provide increased refinement. In contrast, prefabricated items, such as toilet modules, air handling units and bathroom units for hotels, cannot be lapped by other components. When fitted into the building on site they make use of butt jointing where completed components are set effectively side-by-side. The need for components to fit properly the first time on site has led to greater co-ordination of building operations before construction actually starts, both between designers and between manufacturers. This trend is set to continue with the ever-greater emphasis on higher quality of building construction combined with the driving down of construction costs. While both prefabricated and site-based forms of building construction involve collaboration between manufacturers and designers, this process is critical to the success of more prefabrication-based projects. The design team must become conversant with the constraints of mass production as well as with its possibilities. This has the benefit of moving away from the method of manufacturers offering 'standard' ranges with very limited possibilities of modification, to a wider dialogue between the design team and industry. This change is leading to the greater use of prototypes for testing building components and away from the emphasis on the external wall as the only component that is typically prototyped and tested on projects.

## Prefabricated modules

Prefabricated modules for toilets and bathrooms have been in use since the 1960's, (and are occasionally expressed as distinct components within a built form, such as in the Lloyds Building, London or the Hong Kong and Shanghai Bank in Hong Kong), but they are often concealed within the construction, as in bathrooms for hotels and housing projects. Modules are constructed in factories and arrive on site with everything pre-installed. As with other prefabricated assemblies, the use of off-site production allows components to be constructed to a very high standard and helps to reduce construction time on site. Clues to the future development of these factory-built items can be found in transport containers, portable offices for building sites and in prefabricated modules for offshore oil production platforms.

6A

76B

6C

76D

## Transport containers

The manufacture of containers is an extremely competitive market. Floors must be capable of carrying a high payload, be as light and rigid as possible, and be manufactured at minimum cost. For this reason cheaper cold rolled C- and J-sections are used in preference to hot rolled I-beams and channels. This structure is used to support a steel sheet or laminated softwood deck. The floor beams are welded to the walls of the container, which, together with the roof, form a rigid box. The walls and roof of a typical container consist of 1.6mm (1/16 in) thick profiled steel sheet. The walls must resist a portion of the payload, since the load can be partially transferred to the walls when the container ship rolls at sea. Due to its high thickness, the profiled sheet is made in a brake press rather than passing through rollers used to make profiled cladding for buildings. A typical container floor consists of 125mm (5in) cold rolled channel sections, spanning 2.5 metres (8ft 4in) at 300mm (1ft) centres. Mild steel checker plate is fixed to the beams as decking. Alternatively, J-shaped sections are used to avoid seawater collecting on the bottom flange, which would cause corrosion. The floor structure is treated with a zinc-based paint to minimise the corrosive effects of seawater. The steel floor is fixed to a frame of channel sections that form the container structure. The wood floor decking is not rigidly fixed in order to accommodate the amount of rotation, or roll, in the floor expected at sea.

## Portable offices

Portable office modules are commonly used as temporary accommodation for staff on building sites where they provide self-contained highly serviced spaces. The structural design is usually based essentially on that of the transport container. Profiled steel sheet is used on the walls and roof to clad a cold-formed structural steel frame. Steel sheet is used on the internal face of the wall. Thermal insulation, in either rigid sheet or flexible quilt form, is provided within the depth of the wall. Most proprietary designs have a half-hour fire resistance. Offices are constructed on a modular basis so that they can be connected to form two- or three-storey buildings without structural modification. Offices are provided as a kit of parts, comprising access systems of steel staircases and walkways, and are delivered to site by road and lifted into position on site by crane.

## Modules for offshore oil production platforms

Production platforms are constructed from either steel or concrete, and are designed specifically for each location. Design varies with factors such as the distance offshore, the proximity to other drilling locations, the depth of water, the number of wells to be drilled, the weather pattern and the method of secondary recovery. Modules for production platforms are individually fabricated to suit each particular application of different offshore conditions. They are prefabricated and delivered by barge to the platform where they are transferred into their final position by crane and welded in place. Separate modules provide generators, living quarters, helicopter landing pads, mud tanks and chemical stores, production decks for the wellheads and air conditioning plant. Prefabricated modules are most commonly made as a welded or bolted space frame. In common with the manufacture of toilet modules, a single contractor is often responsible for their design, fabrication and delivery.

1C

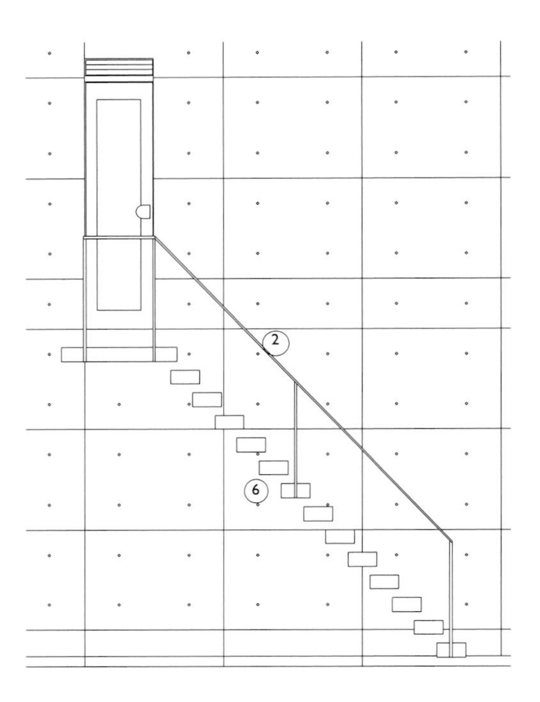


Part elevation

Vertical section

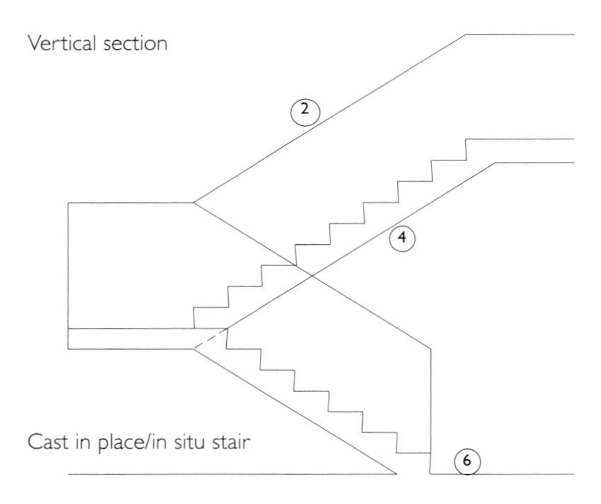


Cast in place/in situ stair

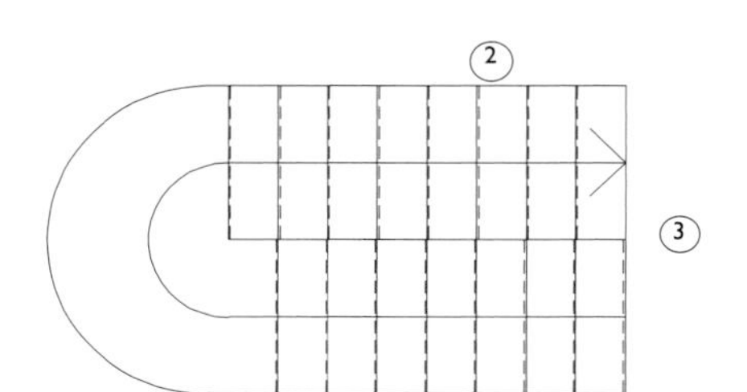


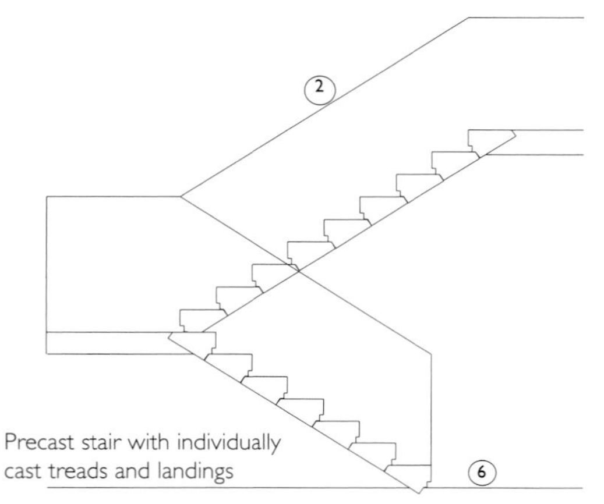


Precast stair with individually cast treads and landings

Concrete stairs have the advantage of good fire resistance combined with the ability to absorb impact sound. Concrete can be used in conjunction with both steel and concrete structures. Stairs can be of cast-in-place or precast types, but the type used must be compatible with the overall type of construction used.

## Cast-in-place concrete

Cast-in-place stairs are made by setting steel reinforcement into reusable steel or timber formwork. A screed is added afterwards to provide a smoother and more exacting finish to the exposed surfaces. The screed can be used as a self-finish, when used internally, but the dusty nature of concrete walking surfaces often leads to the use of floor paint, or polishing the concrete, or inserting treads in another material such as timber. Where the screed is used as a self-finish, anti-slip nosings are added. They can be recessed or surface fixed. Where the stair is exposed to view, the formwork into which the concrete is cast has to be designed and built carefully to reflect the quality of the finish demanded.

The junction at stair landings is often designed to create a single arris line across the soffit. This allows handrails on different flights to be properly aligned, achieved by offsetting flights by a distance equal to one tread width.

## Precast concrete

Precast stairs are manufactured either as complete flights, sometimes with a landing attached to one end, or as individual treads which are fixed together on site. Precast stairs are used primarily where there is a large number of stairs of the same design used in a single project, and where a shorter construction time is an important factor. (These stairs also assist in the construction process itself by providing convenient access.) This is particularly important if the design of the stair is complex, where the cost of precast staircases can be considerably more than the cast-in-place type.

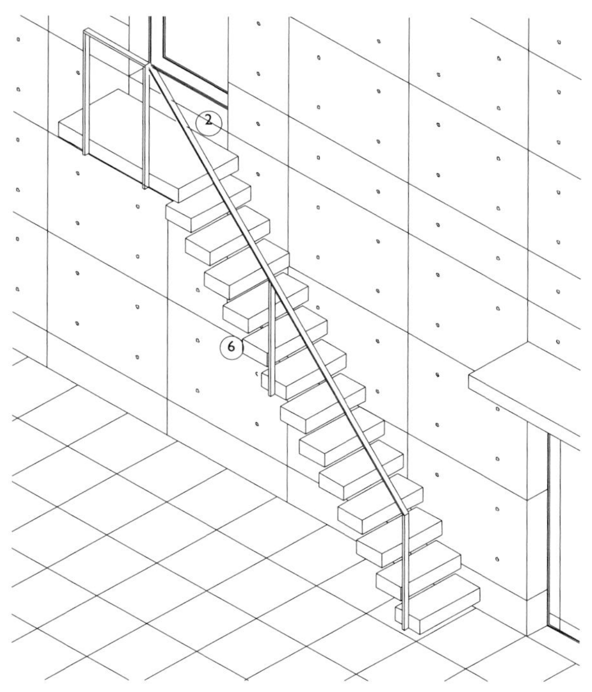

10A

## Guardrails

Guardrails on concrete stairs are most often pre-fabricated in parts, typically in steel, and then grouted into slots or holes drilled into the sides or treads of the stair. Alternatively, they can be fixed to the side of the stair with steel brackets. Handrails in metal are either integral with the guardrails or fixed to an adjacent wall.

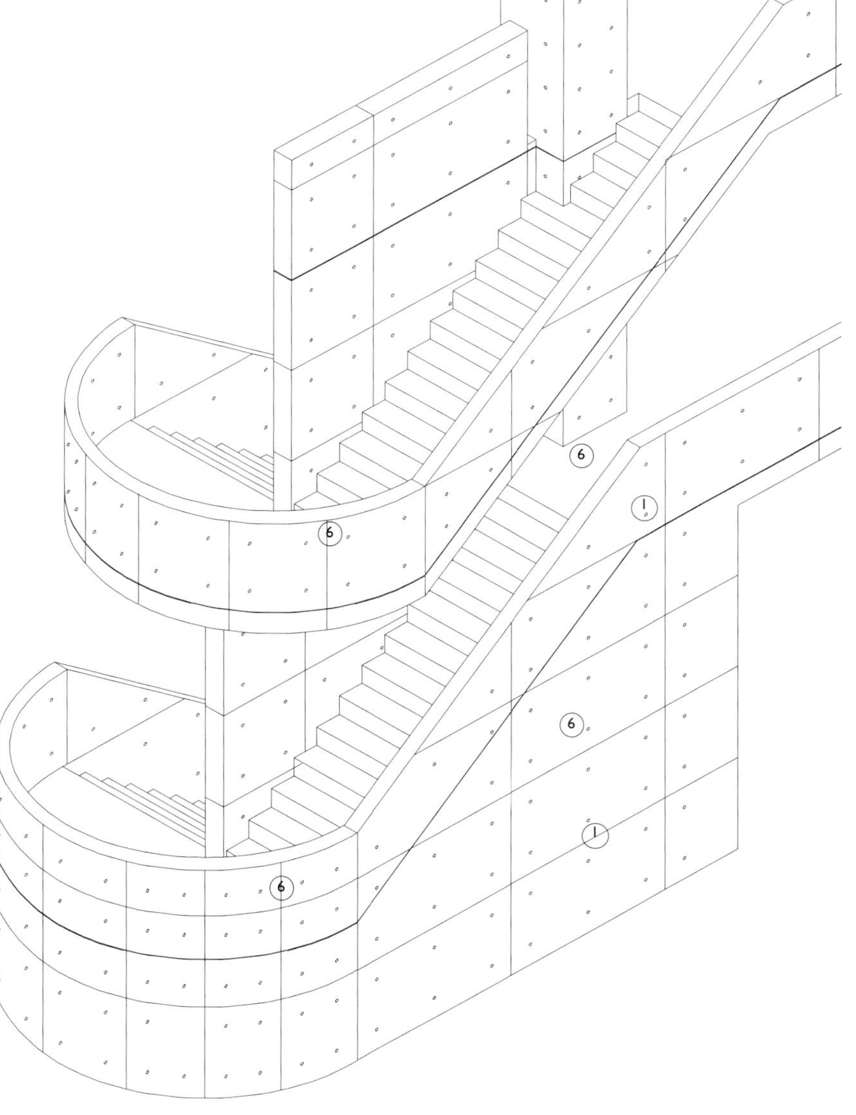


Details
1. Guarding
2. Handrail
3. Staircase
4. Treads projecting from concrete wall
5. Cast-in-place staircase
6. Precast staircase

91C

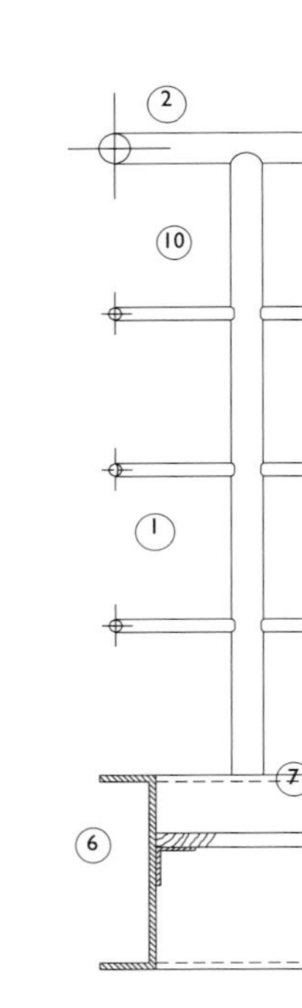


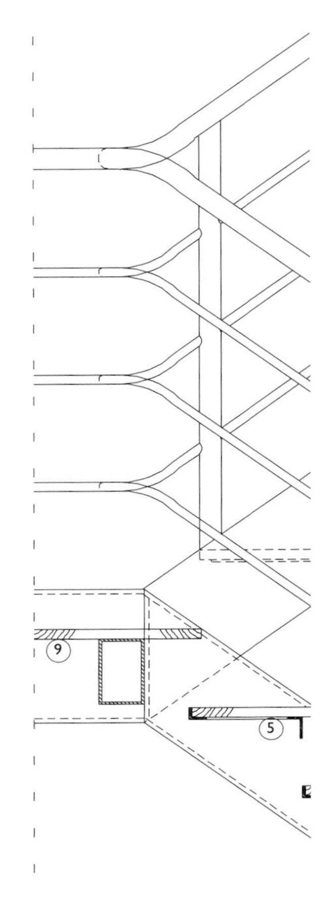

91D

Steel stairs have the advantage of being lighter in weight, allowing them to be prefabricated and delivered to site as completed sections of flights and landings. They can be lifted into place more easily than concrete stairs, but lack the high fire resistance associated with that material. There are two generic types, with either treads as plates set between stringers, or with folded sheet set between stringers. In addition, spiral stairs in cast iron and steel are available as proprietary products in a range of standard sizes. These comprise a central post to which radiating treads are fixed.

## Flat plate type

Stairs with plate set between stringers are made by bolting or welding treads formed from either smooth- or checker-plate. Stringers,

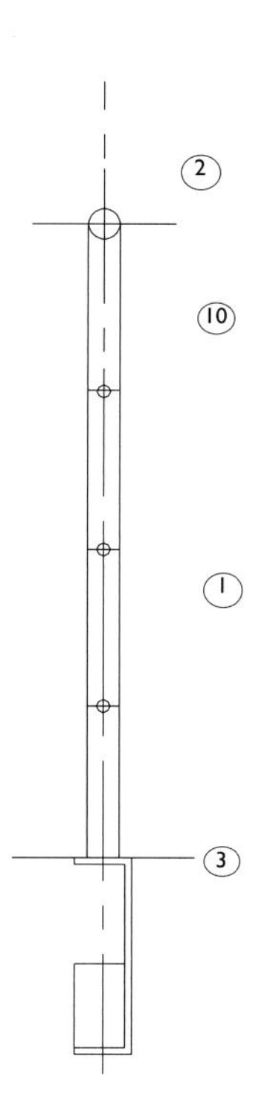


Vertical section

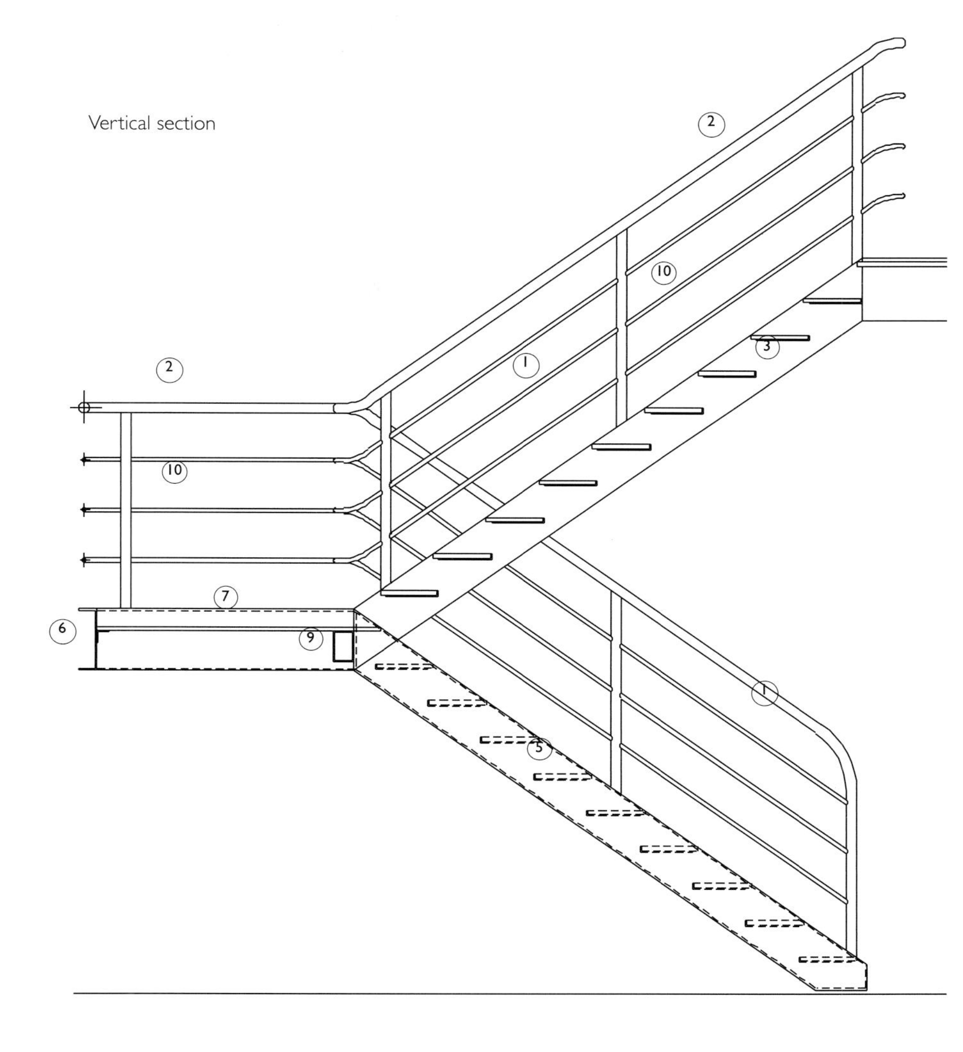


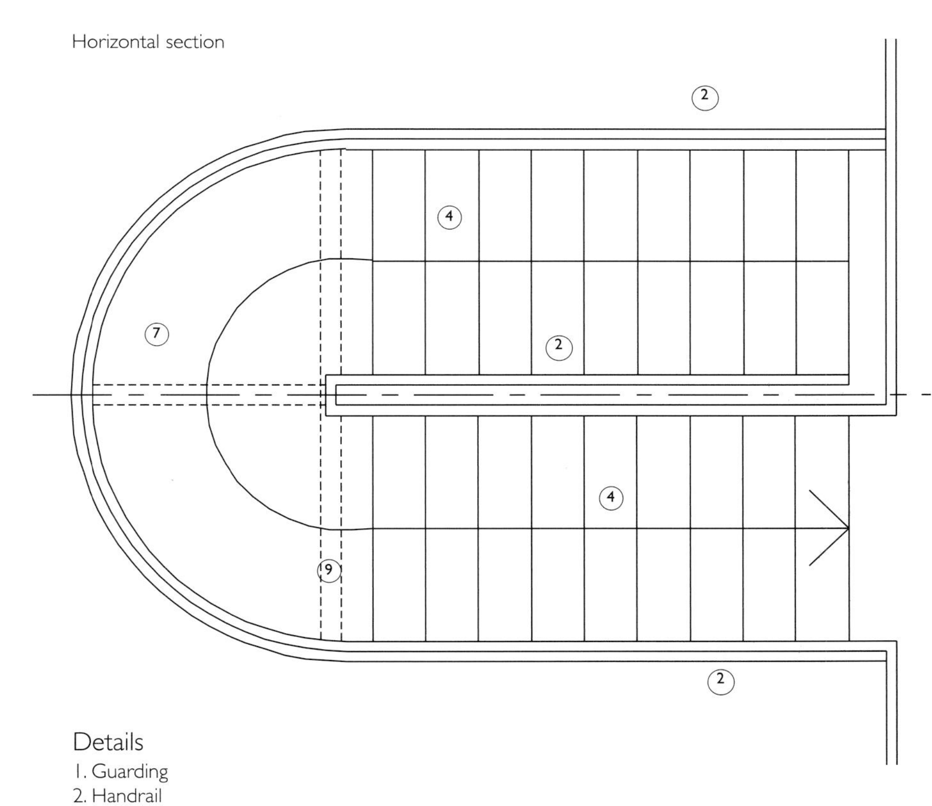


which form the beams at the sides of the stair, are usually made from steel channel, which provides a straight vertical face to which the treads can be fixed. The position of the treads within the depth of the stringer is critical in order that stringers can be aligned neatly where they intersect at landings. Bolted or welded connections are made to be sufficiently stiff so that the stair does not rattle or experience any significant movement while in use. Depending on their width, treads may need to have either their edges folded to provide stiffness which can accommodate an additional finish such as a decorative timber tread, or formed as a composite steel tray with a concrete fill. An alternative method of stiffening treads is to weld a vertical steel plate to their underside to form a T-section.

Details

1. Guarding
2. Handrail
3. Steel stringer
4. Tread
5. Inserts, typically timber or concrete, set onto steel plate
6. Steel channel
7. Landing in steel plate
8. Steel plate deck
9. RHS box section
10. Balustrade

100B

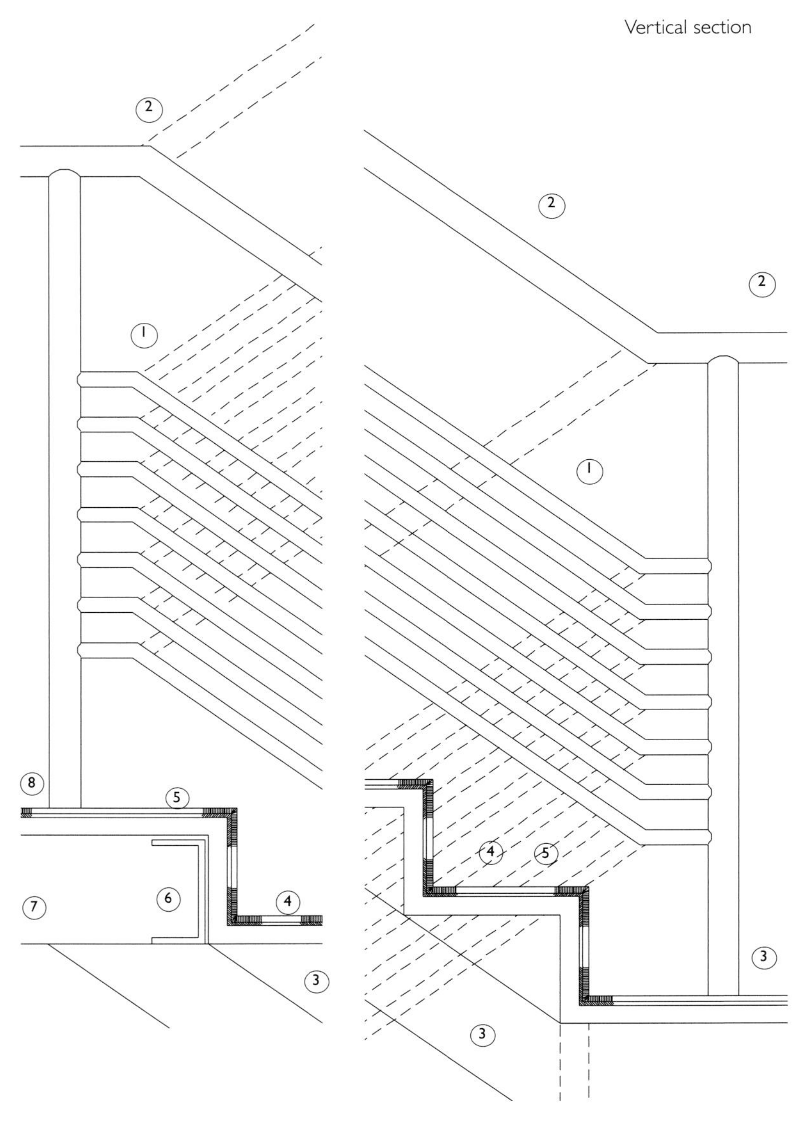


103F

Stairs can be formed by folding a steel sheet and supporting it, either on stringers set at the sides or by a single central stringer set beneath the plate. The inherent rigidity of the folded sheet allows a wide range of economic solutions for the arrangement of the stringers. As with the flat plate type, the assembly can be either bolted or welded, and decorative inserts in other materials such as timber can be added. Both stair types can be finished in a variety of coatings from galvanizing to painting to polyester powder coating. Softer coatings such as PVDF are rarely used due to their poorer wearing qualities. In addition to visual considerations, the choice of finish is determined by the required degree of durability and appearance.

## Guardrails

Steel guardrails are prefabricated but are not often fixed to staircases before delivery to site in order to make the stair both easier to install and to make it easier to align the guardrails with adjacent walkways or enclosing walls. Guardrails are usually finished before delivery to site. If a paint finish is used the guardrail will at least be prepared and primed before arriving on site where finish coats can be applied after its installation.

## Details

1. Guarding
2. Handrail
3. Steel stringer
4. Tread
5. Inserts, typically timber or concrete, set onto steel plate
6. Steel channel
7. Landing in steel plate
8. Steel plate deck
9. Folded steel plate to form continuous tread and riser
10. One stringer with projecting arms to support treads/risers or two stringers set near edges of stair

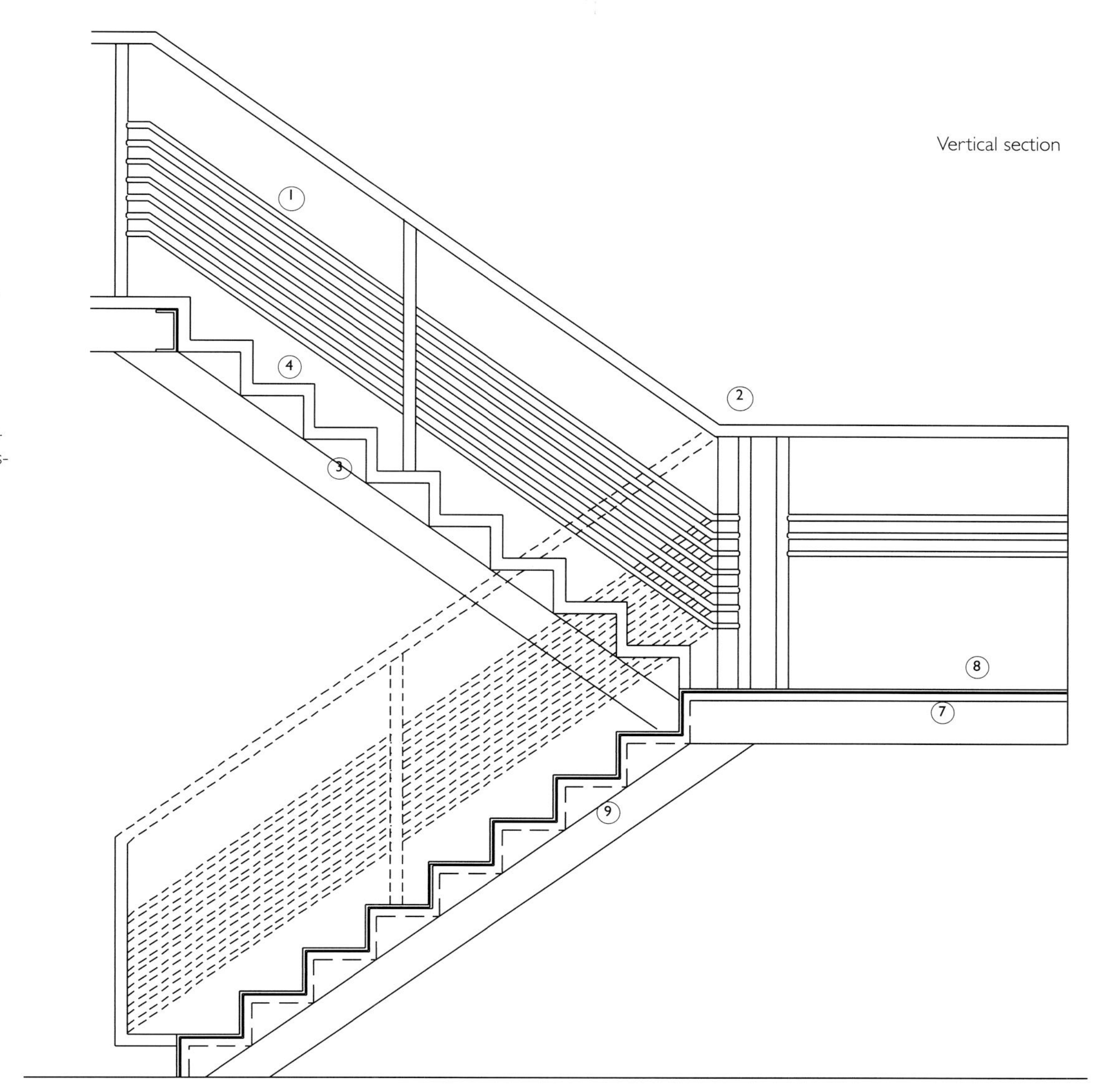


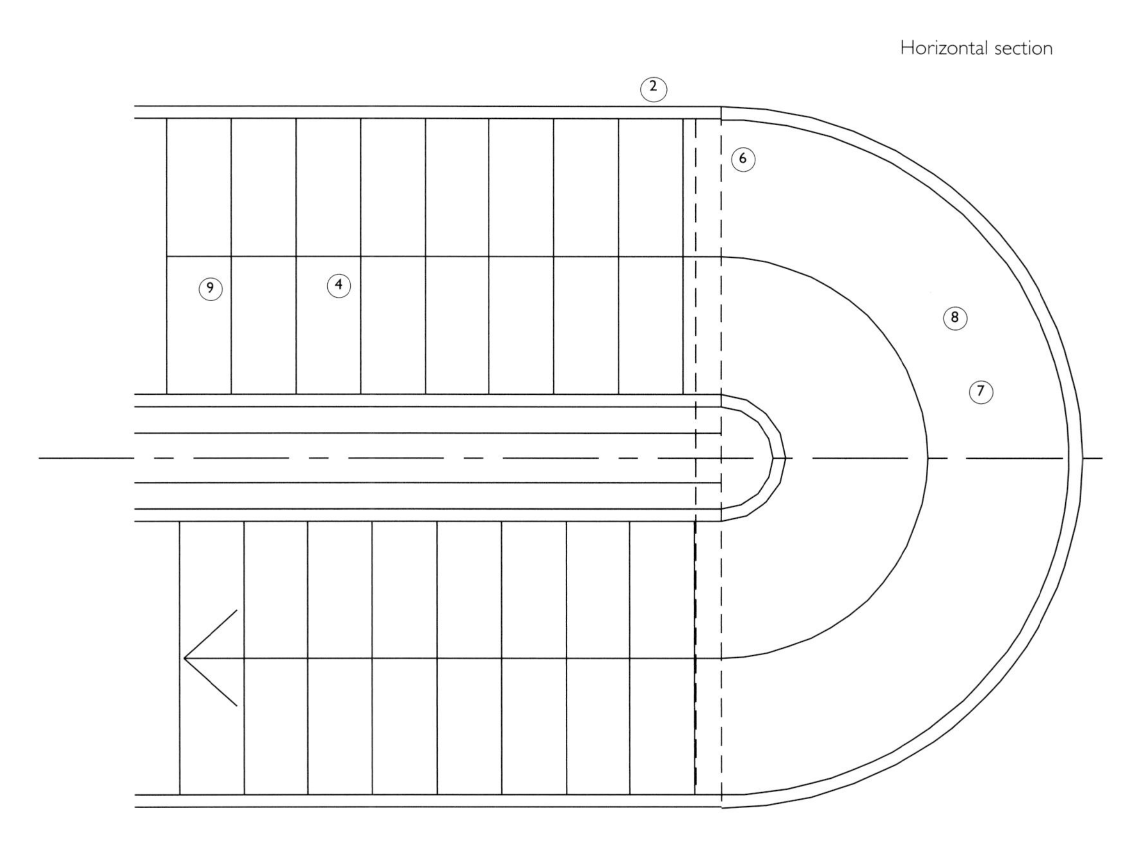

42D

42E

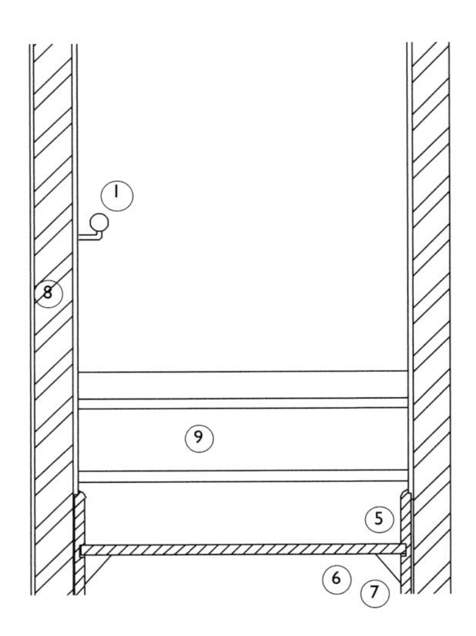


42C

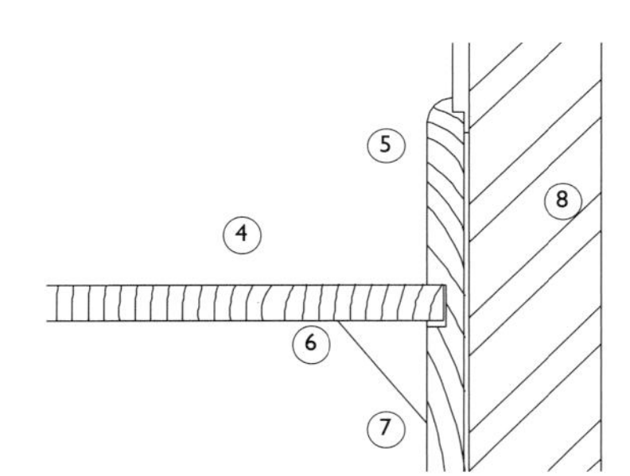


22D

An advantage of timber staircases is that they can be integrated easily into adjacent timber construction and can be modified easily on site in a way that is very difficult to achieve in an equivalent steel or concrete construction. Timber stairs consist of stringers or carriages, which support the stair, to which treads and risers are fixed. Several carriages may be positioned between stringers as loading demands. The stair is generally delivered to site as a complete structure secured with timber wedges, which avoids the need for visible fixings. Relatively small pieces of timber are used. The thin timber sections are prone to shrinkage and creep unless they are locked together. Because timber stairs have poor fire resistance, they are often restricted to residential use.

Treads and risers are fixed together with tongue-and-groove (or rabbeted) joints that provide a tight fit and avoid creaking when the stair is in use. Both treads and risers are fitted into rebated slots cut into the stringers where they are wedged to provide the correct alignment of treads. Trimmer beams are sometimes added at the top and bottom of the staircase to stabilise the staircase and provide fixing points to the adjacent floors, which are typically also of timber construction.

An alternative approach is to use heavy timber sections to form staircases that resemble those in steel. Steel brackets that are bolted into the timber sections connect the stringers and separate treads together. They can easily accommodate steel guardrails which are bolted through the large stringer sections.

## Guardrails

Timber guardrails follow the traditional use of balusters at close centres, typically set 100mm (4in) apart. The lack of large structural members in timber stairs makes the use of balusters at wider centres, as used in steel or concrete construction, less suitable. Timber connections are more fragile than those in steel; they must also accommodate more movement due to moisture. Balusters at close centres allow imposed loads on the guardrail to be spread evenly along the length of the stair stringers.

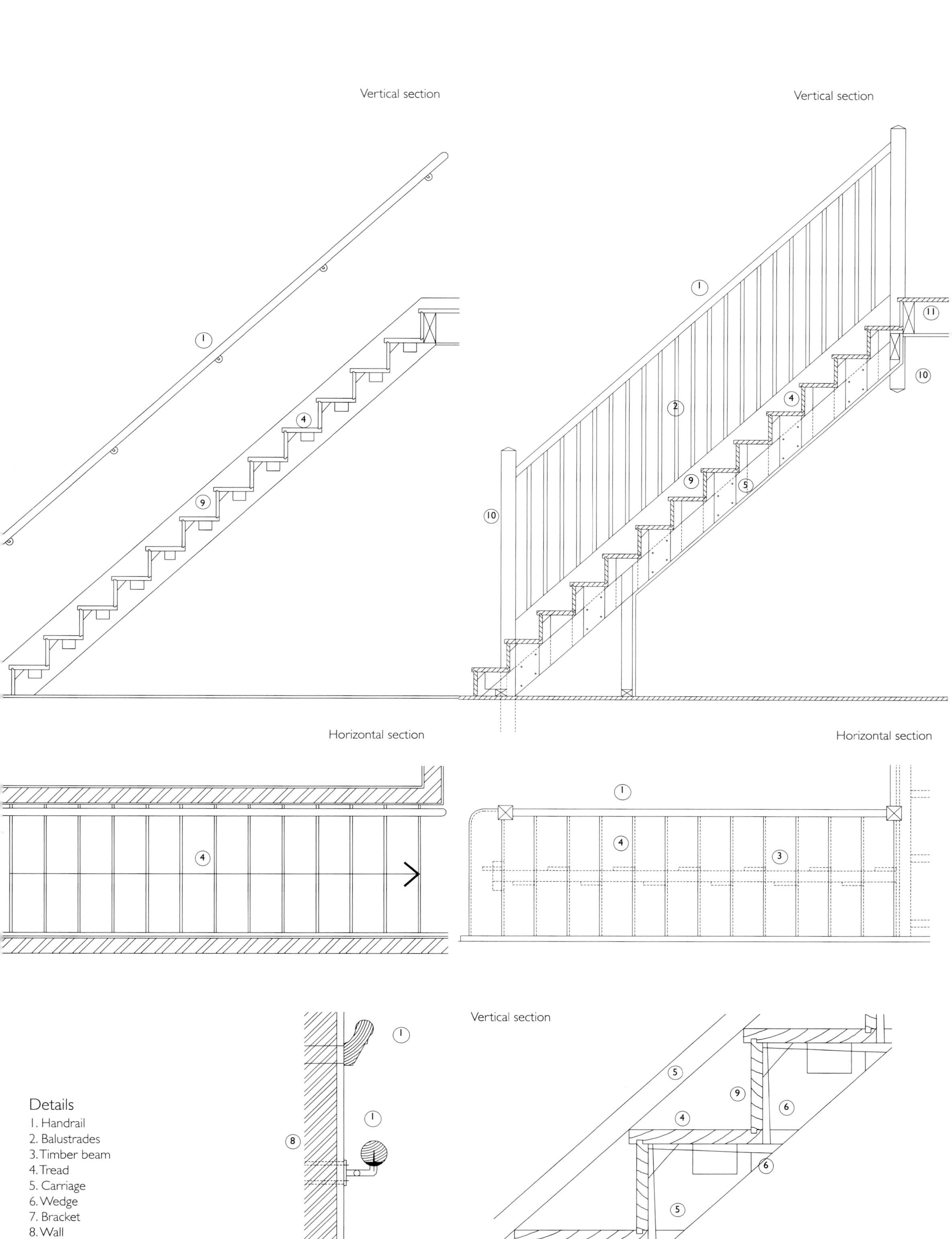
Vertical section
Vertical section
Horizontal section
Horizontal section
Vertical section
Details
1. Handrail
2. Balustrades
3. Timber beam
4. Tread
5. Carriage
6. Wedge
7. Bracket
8. Wall
9. Riser
10. Newel post
11. Trimmer joist

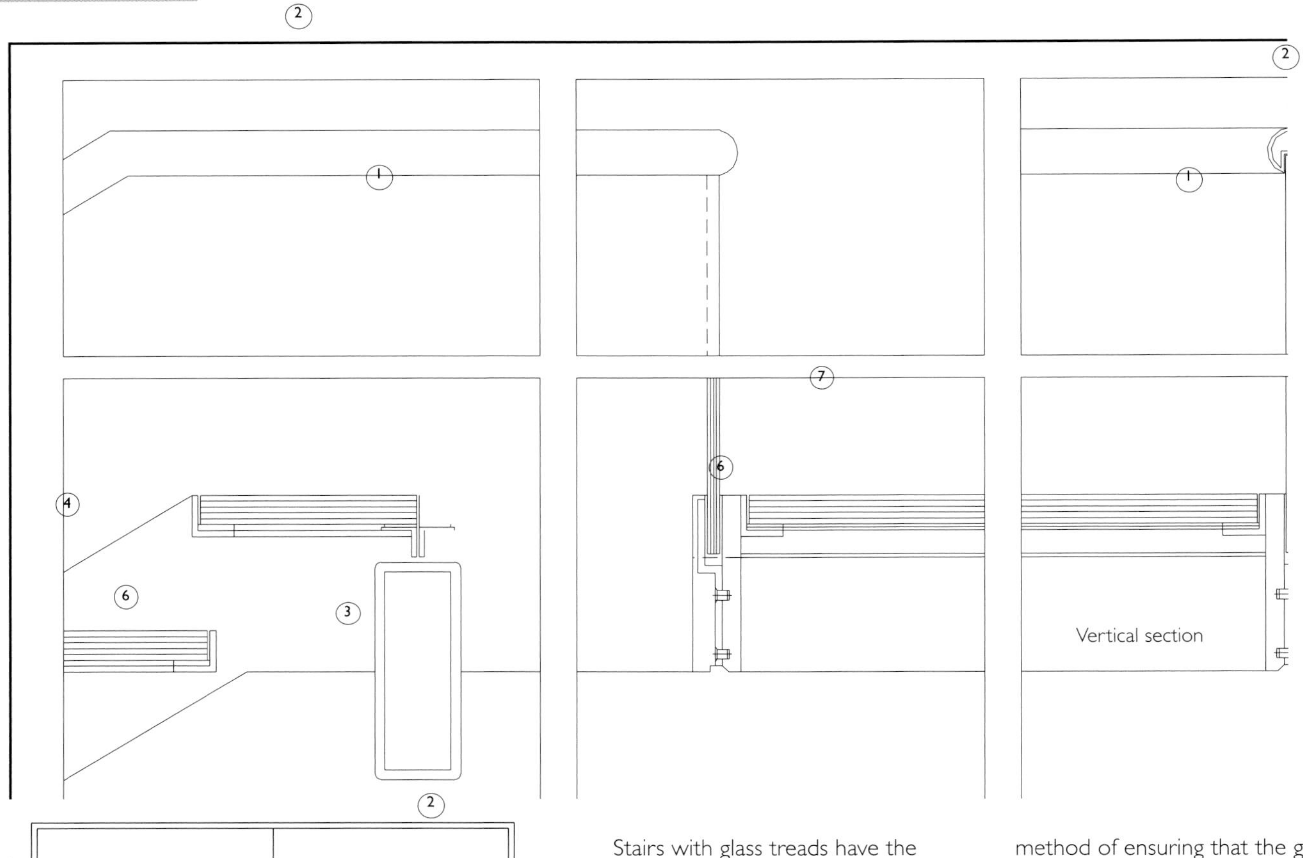


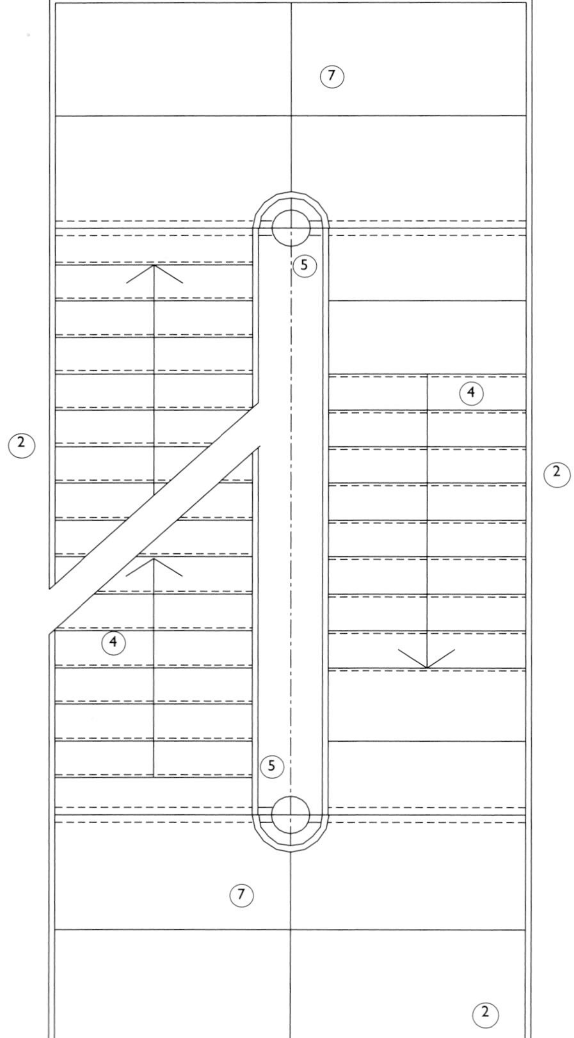
Horizontal section

Stairs with glass treads have the advantage of transparency, allowing light to penetrate down the stair enclosure. The glass used is usually a thick laminated glass which can be made of two sheets of float glass with a thin interlayer, or toughened glass with a more robust interlayer that stays in place if the two toughened sheets are broken. Glass treads can be supported either within a steel tray which provides support on all edges, or be two edge supported. An alternative is to bolt fix the glass using techniques taken from their primary application in glazed walls. This method allows the stair to be suspended from cables, a technique still in the early stages of use. The tread assemblies are then usually supported by steel stringers, though concrete can also be used.

Where glass treads are set within a steel supporting tray, the laminated glass sheet is set directly onto a silicone-based bedding. In addition to holding the glass in place, the bedding provides both a cushion and a method of ensuring that the glass is evenly supported along all the edges and is fixed level. An additional weather seal is used on the sides of the glass between the glass and frame, if the stair is to be used in external conditions. Treads often have a surface treatment to provide slip resistance. Sandblasting, etching or the addition of a carborundum coating, typically in strips, is used for this purpose.

## Guardrails

All-glass, or structurally glazed, balustrades can be constructed with sheets of toughened or laminated glass and used with a stair built in another material. They comprise sheets of toughened or laminated glass fixed at floor level with either a clamped plate secured by bolts or by bolt fixings directly through the glass. The glass must be sufficiently strong and rigid to span vertically without additional vertical support. A 12mm (1/2 in) thick sheet is typically used. A handrail can be added by introducing

45D

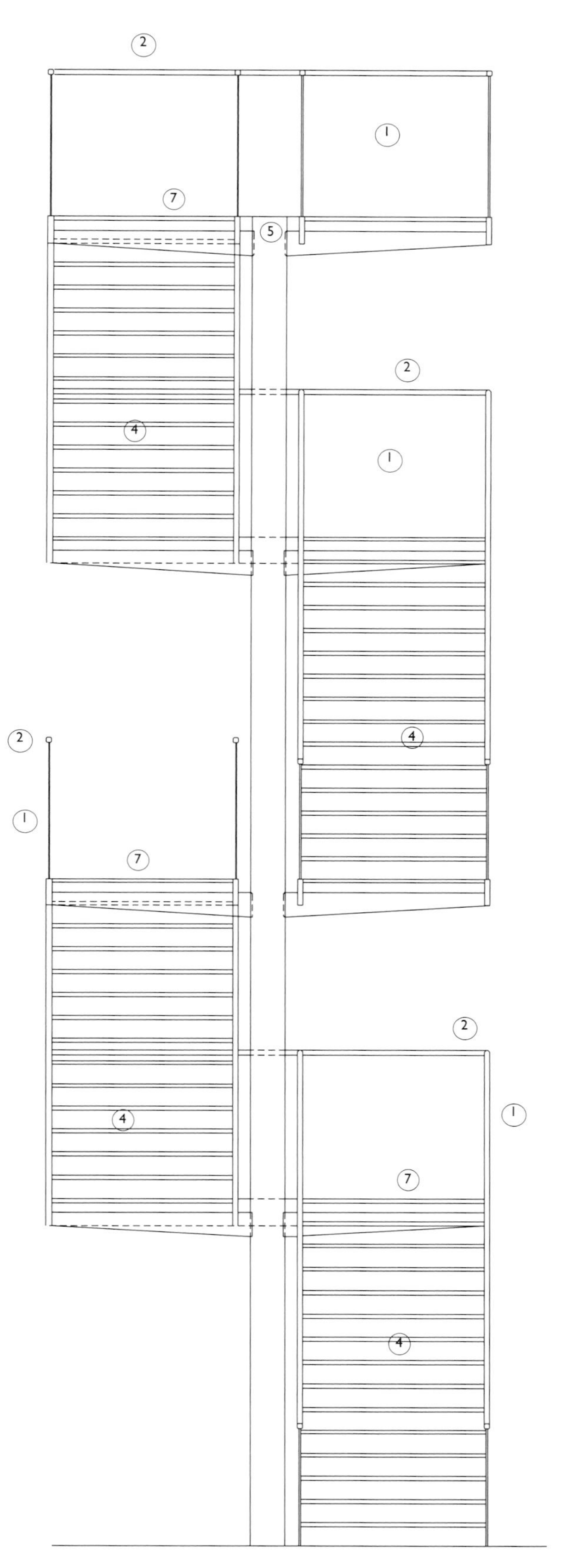

Vertical section

a rebate into the handrail section and setting it directly onto the top of the glass guardrail. An alternative method is to form a guardrail from steel posts that support a handrail, with laminated or toughened glass sheets used as infill panels. The glass can be fixed with clamps or be bolt-fixed back to the posts. Glass sheets are set with a vertical gap of around 10mm between them which is filled with a translucent or transparent silicone seal.

Details

1. Glass balustrade fixed at base
2. Stainless steel handrail
3. Steel stringer
4. Glass treads, typically sandblasted to provide friction
5. Steel tube support frame
6.. Steel angle
7. Glass landing in steel plate

84

94B

Traction elevators are electrically powered. Cables, running over pulleys and balanced by counterweights in order to minimise the efforts of the elevator motor, support cars. Both car and counterweight move in vertical steel guide rails. Traction elevators have advantages over hydraulic elevators with their higher speed and hardwearing qualities. Speeds vary from 0.5 metres per second for smaller installations to 6.0 to 7.0 metres per second for tall buildings.

Traction elevators have an overrun space at the top of the shaft. This protects anyone working on top of the elevator should the car travel to the top floor accidentally. The overrun also allows cables to run over the pulley to the motor. The overrun space also accommodates the cable pulleys. Motor rooms house the elevator motors and ancillary equipment that drives the cars. These are usually located at the top of the shaft, adjacent to the overrun. The motor room can also be positioned to one side of the elevator pit at the bottom of the shaft, but this requires more sophisticated cabling. Motor rooms require ventilation because of heat generated by the machinery.

A pit is located at the bottom of the elevator shaft. It is usually about two metres (6ft) deeper than the lowest floor served in order to accommodate the underside of the elevator car. The top of the elevator shaft is vented to the outside to ensure that air is drawn up the shaft and out of the building in the event of fire.

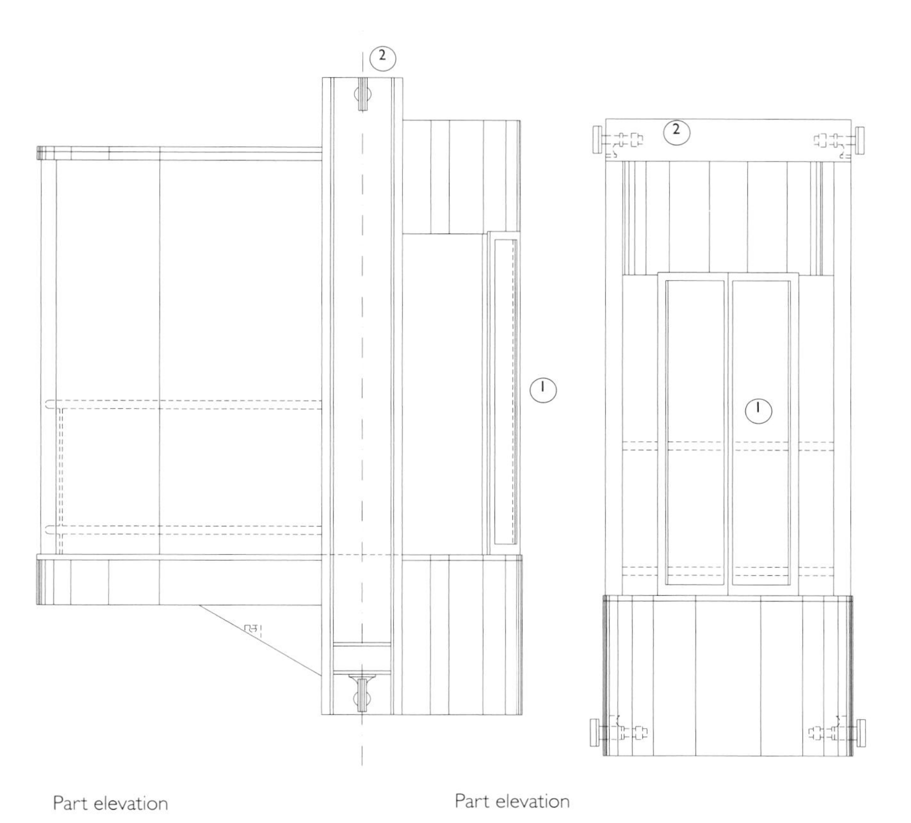


Part elevation

Part elevation

Details
1. Glass fronted doors
2. Runner

94A

106A

106D

94C

106B

106C

113B

113A

Details
1. Clear toughened glass
2. Roller guides
3. CHS (steel tube) framing supporting 'structural deck' of lift car
4. Lighting trough
5. Metal plate supporting light fittings
6. Reflector luminaire
7. Extract fan
8. Metal access panel
9. Bolt fixed glass enclosure
10. Angled downlighters
11. Flexible mild steel electrical conduits
12. Removable maintenance platform hung from steel tube frame
13. Floor finish

113C

Vertical section

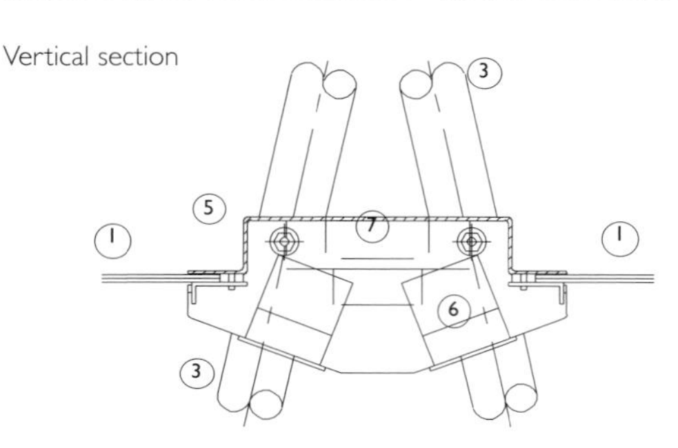


A hydraulic elevator consists of a piston cylinder set into a well beneath the elevator pit, which accommodates the underside of the elevator car at the lowest floor level. The piston can be a single or telescopic tube. If it is a single tube, the well must be deep. The hydraulic piston is fixed to the bottom of the shaft and connected to the elevator car. The ram is activated by hydraulic oil which is forced into the cylinder well under high pressure, which then moves the elevator car. Hydraulic elevator mechanisms are suited to installations serving up to five floors. They are usually slower than traction elevators, with speeds ranging from 0.63 to 1.00 metres per second. The motor room can be located a short distance away from the elevator shaft, rather than next to it, either at the bottom or the top, which is an advantage in small buildings. Hydraulic elevators are also useful where an overrun cannot be accommodated easily at the top of the elevator shaft.

The machinery room can be independent of the elevator shaft, and may be located up to five metres (16ft) away. Machine rooms are smaller than those of traction elevators, and contain pumps, motors, fluid storage tanks and control equipment. Machinery rooms require ventilation due to heat generated by the equipment. Hydraulic elevators have an overrun at the top of the elevator to protect anyone working on the top of the car, in case it accidentally rises to the top of the shaft. Like traction elevators, the top of the shaft is externally vented to ensure that air is drawn up and out of the building in the event of fire.

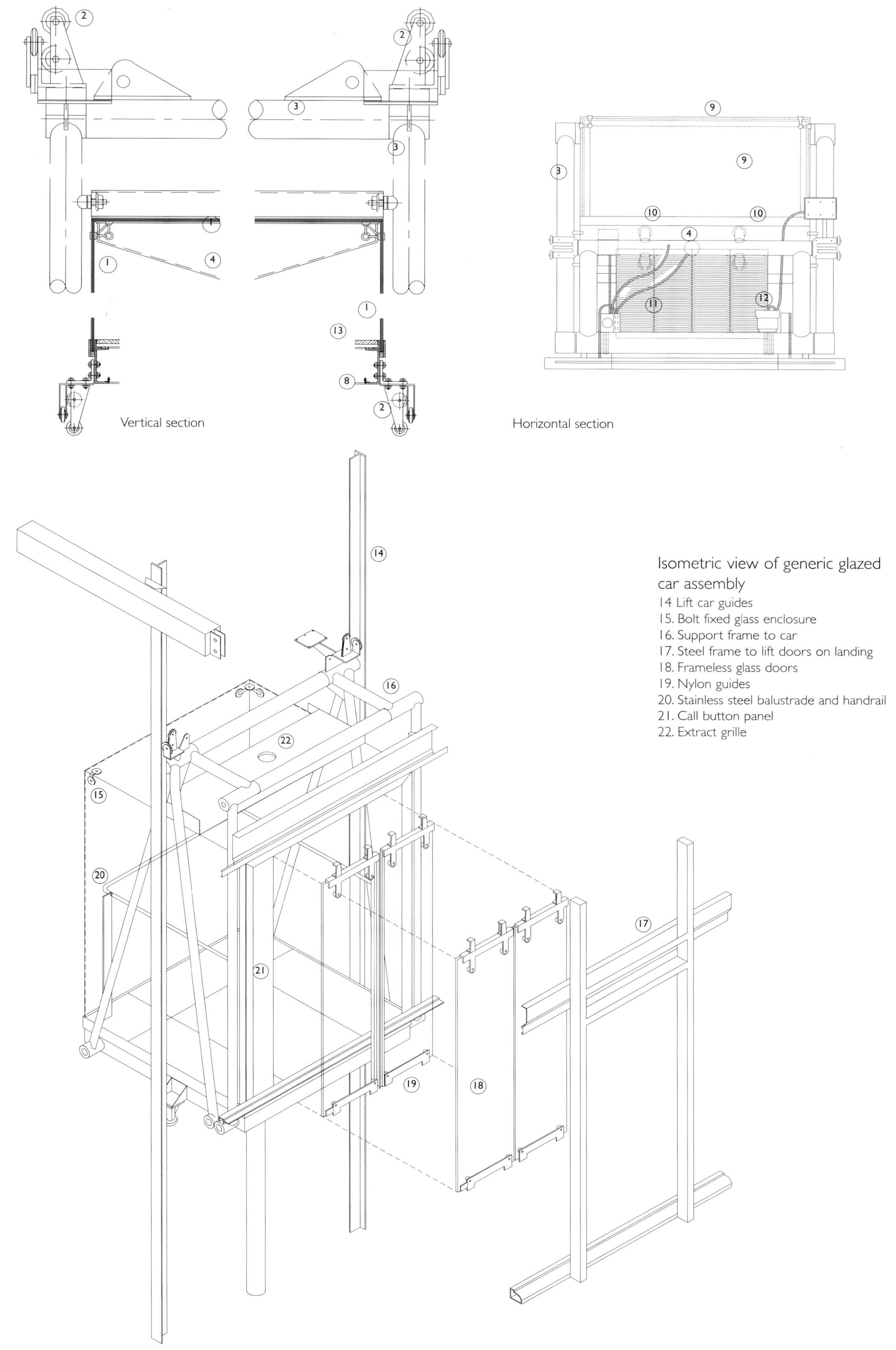

Vertical section

Horizontal section

Isometric view of generic glazed car assembly

14 Lift car guides
15. Bolt fixed glass enclosure
16. Support frame to car
17. Steel frame to lift doors on landing
18. Frameless glass doors
19. Nylon guides
20. Stainless steel balustrade and handrail
21. Call button panel
22. Extract grille

Vertical section

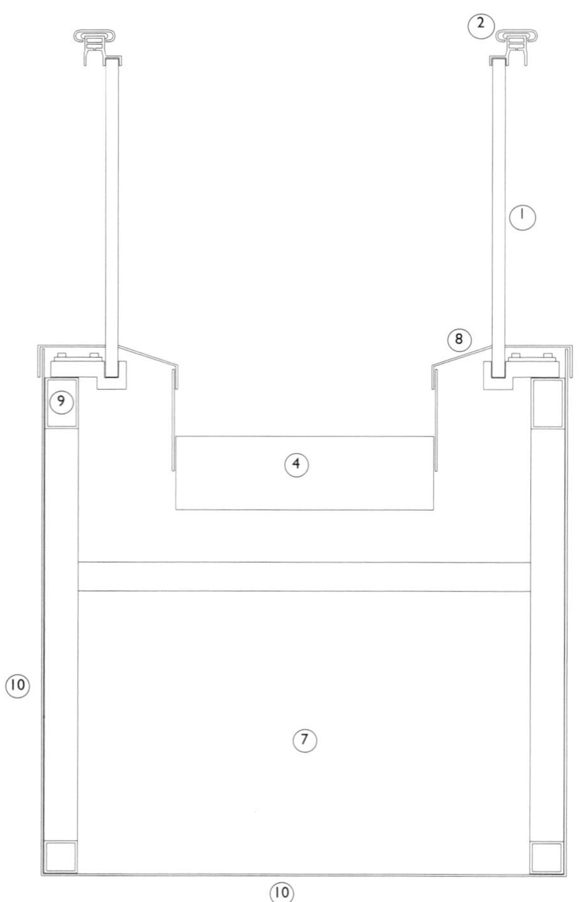


Escalators consist of a set of treads connected together in a closed chain. The complete mechanism of chain, motor, treads and handrail is supported on trusses or girders that span from the top of the escalator to the bottom. Long spans sometimes require an intermediary support to the supporting structure. Casings, which enclose the mechanisms, can be constructed from a range of materials, including glass and steel sheet. The balustrades supporting the continuous handrail belts are often constructed in structural glass. Escalators are manufactured and delivered to site as complete assemblies, and are often one of the last elements to be fitted. It is therefore essential to ensure that spaces are large enough for the completed unit to be brought into the building and that there are sufficient fixing points to assist the support of the escalator during its installation. Escalators are fixed directly to the building structure because of their considerable weight. Although a standard range of sizes is available from manufacturers, escalators are typically adapted for each application and a certain amount of flexibility exists in the overall length and width of components.

## Travelators

Travelators, or walkways, are similar to escalators but transport the user along horizontal or gently inclined planes up to a ten-degree slope. The same mechanisms are used in both devices. Their maximum span is currently around 32 metres (106ft).

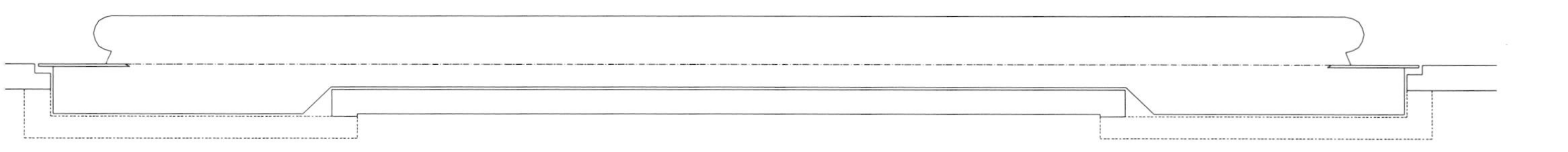

Part elevation
Travelator

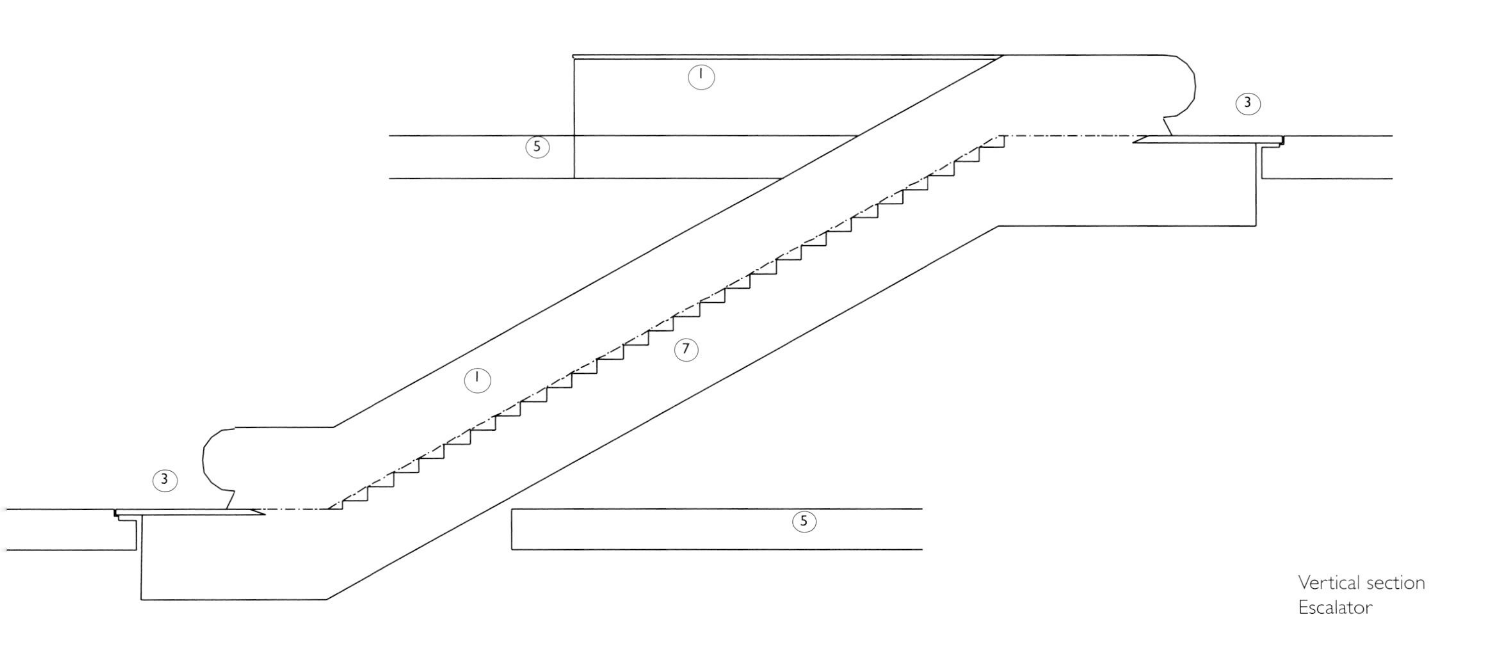


Vertical section
Escalator

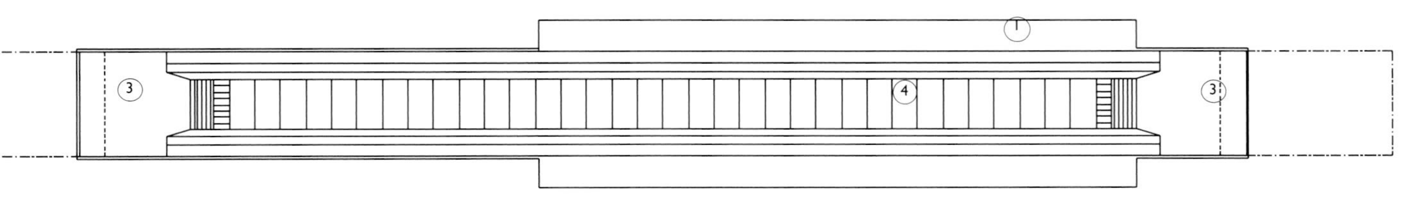


Horizontal elevation
Escalator

Data

Escalators

| | |
|---|---|
| Gradients: | 30° to 35° |
| Speed: | 0.5 to 0.65 metres/second |
| Standard tread widths:: | 600mm (2ft), 800mm (2ft8in), 1000mm (3ft3in), 1400mm (4ft8in) |

Travelators

| | |
|---|---|
| Gradients: | 10° to 12° |
| Speed: | 0.5 to 0.65 metres/second |
| Standard tread widths: | 800mm (2ft8in), 1000mm (3ft3in) |

Details

1. Balustrade (shown glazed)
2. Moving handrail
3. Metal tread plate
4. Treads
5. Adjacent floor slab
6. Treads rise to fixed position
7. Mechanism housing
8. Pressed steel cover plates
9. Structural steel hollow section forming frame to support escalator
10. Casing, typically metal, opaque glass

96B

96A

Plastered walls provide a smooth, continuous finish that is usually painted. Plastering is a traditional method that has been adapted for use on different wall backgrounds. An advantage of plaster is that complex shapes and edges can be formed. Dry premixed plaster in powder form is used, in order to ensure a consistency of mix, and is mixed with water on site. Plasters are relatively soft and can have slots cut into the material easily for the passage of electrical wiring. This ease of use makes it a very practical wall finish which can be cut and patched to accommodate changes throughout the life of a building. Plastering is a labour-intensive operation that is carried out entirely on site and the quality of finish is very much dependant upon the skill of the individual plasterer. For this reason it is important to ask for sample areas of plastering to be provided on site to ensure that the required quality can be achieved. Like pouring concrete, plastering is a wet trade, which involves mixing the material with water and allowing it to dry out. This drying process can slow down other building operations, particularly where construction time is an important factor.

A wide range of plasters is used in different countries as a result of the availability of minerals varying between regions. Manufacturers should be consulted to advise on appropriate mixes for different applications. All plasters use either a sand-cement mix, a gypsum plaster (gypsum is a naturally occurring mineral) or a lightweight plaster made from minerals such as perlite or vermiculite. Plasters are usually applied in two coats, as an undercoat and finish. The finish coat has a finer texture in order to achieve a smooth surface by using a mix that would not be strong enough to be used as an undercoat. Plasters have different levels of surface hardness. For example, soft plasters might be used for exhibition spaces where paintings are frequently moved; acoustic plasters are used to increase sound insulation.

When detailing edges, junctions and openings in plastered walls, joints

97A

97B

are needed where there is either a change in background material or a change in structure, such as where a wall meets a column. This is to accommodate thermal and structural movement at the junction between two materials. These junctions are treated either by a hairline joint, which is mostly concealed, or by a recessed joint that forms a shadow gap. The hairline joints use a layer of expanded metal lathe that is nailed or screwed to the background and which spans the joint to form a continuous background. A breather membrane is set behind the metal lathe to isolate the plaster if structural movement is expected. The plaster is then continued across the joint and is keyed into the lathe where a different plaster mix, called isolating plaster, is sometimes used. Shadow gaps are often formed with galvanized steel trims to create a crisp line between the different backgrounds. The joint is revealed as a continuous groove down the joint. Both plaster and trim may be painted to create a homogeneous appearance.

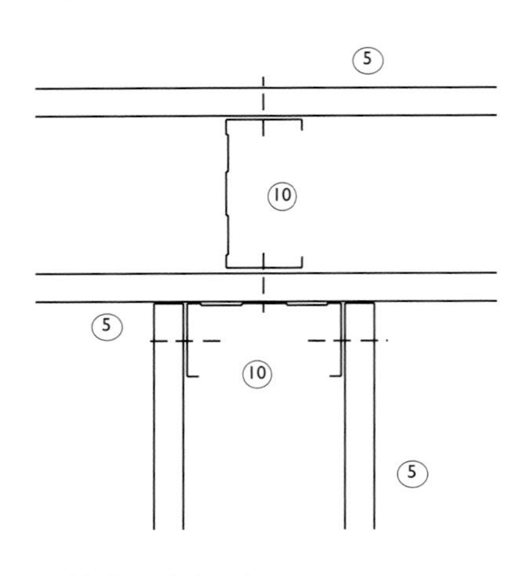

Horizontal elevation

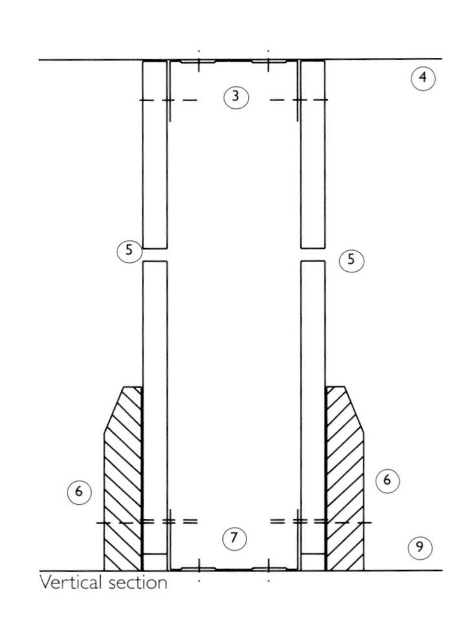


Vertical section

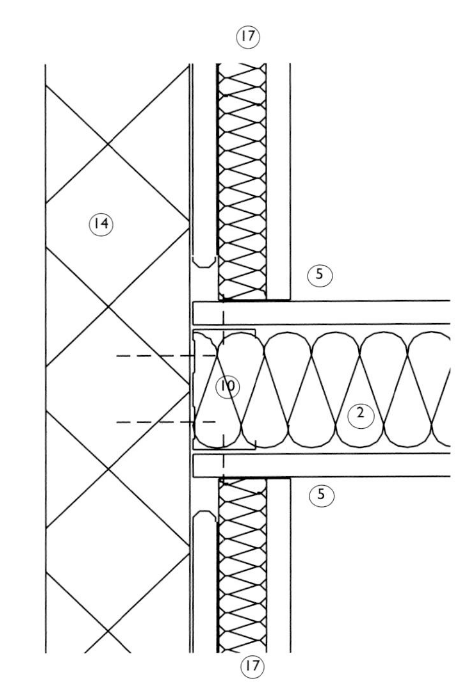


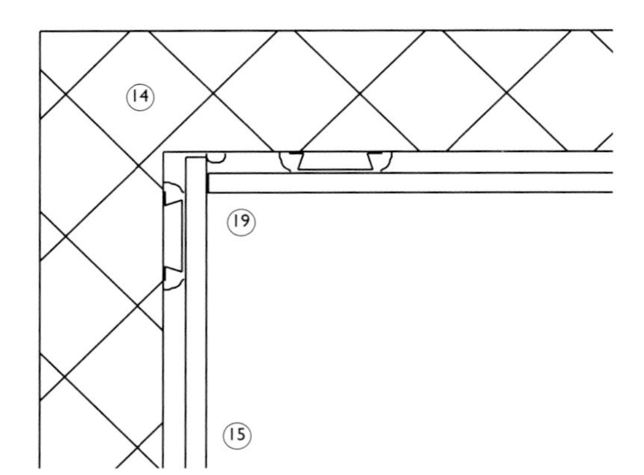


Horizontal elevation

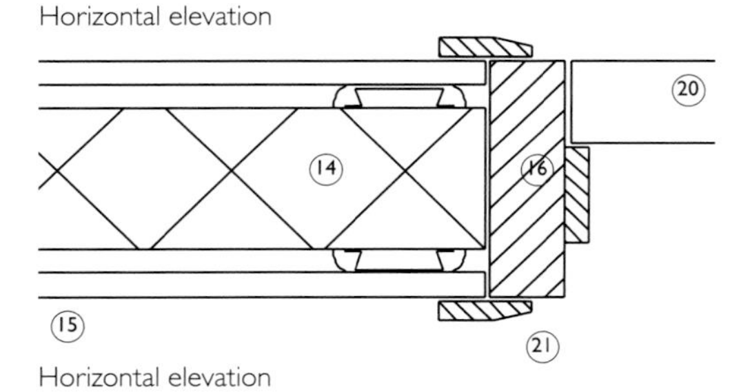


Horizontal elevation

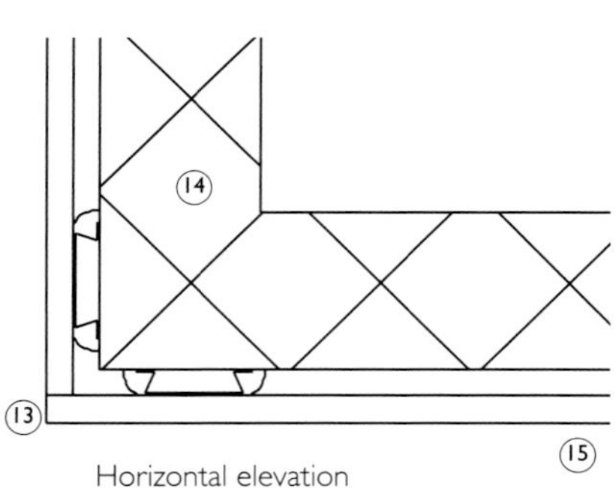


Horizontal elevation

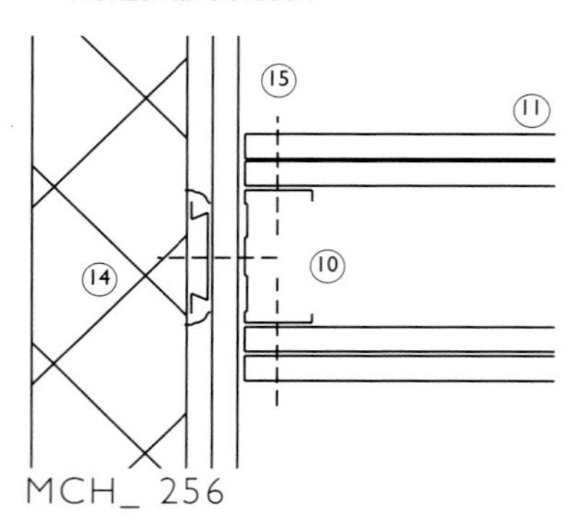


Wallboard systems use gypsum plasterboard sheet to give a plastered wall finish to a variety of backgrounds. A thin skim coat finish over the plasterboard provides a continuous, dry fixed partition system that is fast to build. It is particularly useful on timber- or metal-framed partitions where its use avoids the need for the more expensive traditional technique of plaster and lathe. An alternative use of dry lining is to fix plasterboard sheets back to a masonry wall using either steel or timber battens which are screwed back to the background wall, or alternatively on plaster dabs which are literally dabs of plaster. A variety of gypsum plasterboard types is used; glass-fibre reinforced boards can be curved to form radiused corners. High-impact resistant boards are used where a hardwearing plaster finish is required, and fire-resistant boards are used in fire protection, particularly to structural steel frames.

Wallboard systems are finished with either a 2 to 3mm (1/16 to 3/32 in) thick skim coat or a full 15mm (9/16in) plaster coat. In practice, the full plaster coat is not often used except, for example, on curved dry lining partitions in order to provide a very smooth finish. An alternative to a full skim is to use tapered edge plasterboard sheets to allow joints to be covered with paper tape and filled in order to avoid cracking. The wall is then coated to provide a surface ready for applied finishes such as painting. This almost 'dry' process allows following trades to start work sooner than with plaster, which requires much longer to dry.

Wallboard covered stud partitions are assembled on site and can be easily modified either during or after construction. It is difficult to re-use plasterboard sheets although the studs can be recycled. Wallboard covered partitions often perform better acoustically than an equivalent blockwork partition. Partitions can be formed using three or more layers of plasterboard sheet without the use of studs to a maximum height of approximately three metres (10ft).

Timber stud partitions are limited to around 3.5 metres (11ft 8in) in height, since cut timber sections of greater length are more difficult to obtain and are expensive. Metal studs provide most flexibility in terms of partition thickness and height, spanning up to around eight metres (26ft 8in) without need for intermediary support. With longer spans, partition thickness can be reduced to around

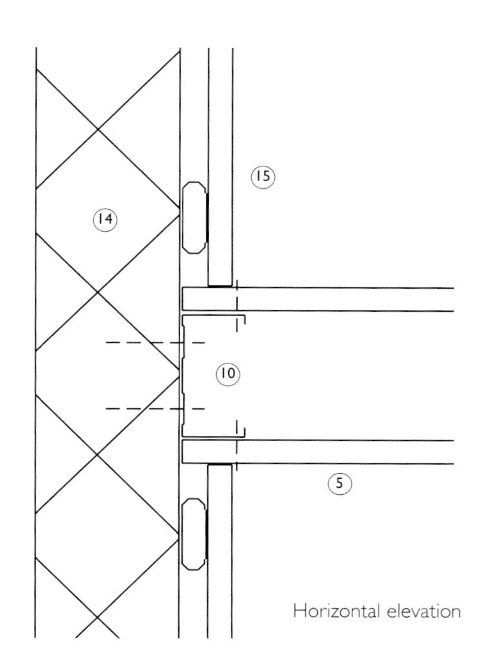

Horizontal elevation

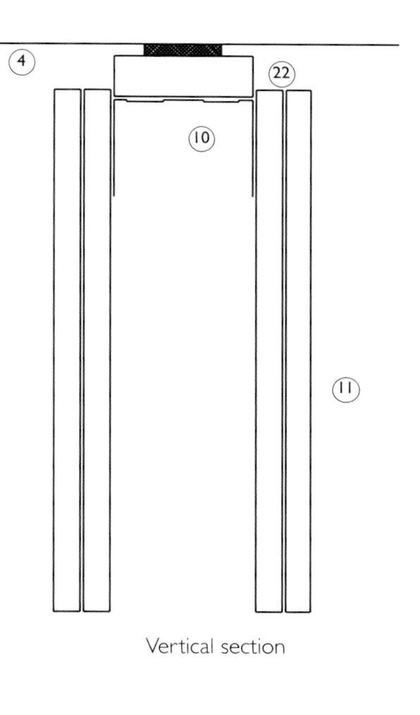

Vertical section

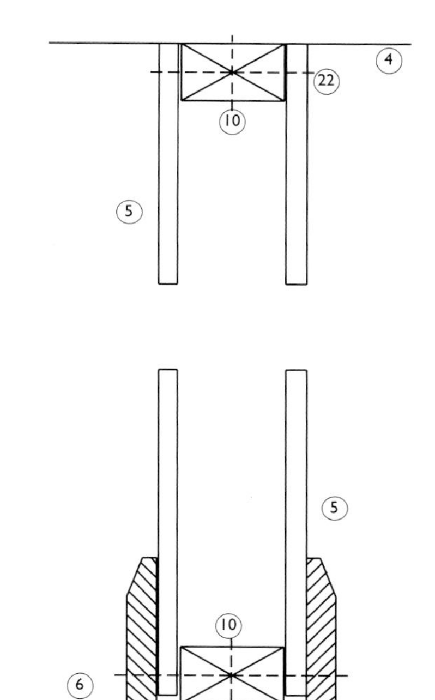

Vertical section

150mm (6in) by setting studs at closer centres, typically 300mm (1ft) and by changing the stud type. Short vertical spans, up to around 3 metres (10ft) use cold formed pressed steel channel sections, while longer spans up to approximately 5 metres (16ft8in) use the same channels set face to face to form a box section, while spans up to 8 metres (26ft) use cold formed back-to-back channels that form an I-section. These studs are also available in varying depths to provide different overall partition thicknesses to accommodate the layering of sheets for varying amounts of stiffness, fire resistance and acoustic performance.

In common with plastered walls, expansion joints are required where there is a change in construction in order to accommodate thermal or structural movement at the junction of two different backgrounds. This typically occurs at the junction of a partition and a structural column. Junctions can be continuous if constructed using a latex-based plaster. Alternatively, a shadow gap is formed which uses galvanized steel trims to create the shape of the joint in a way that imitates a plaster recessed joint.

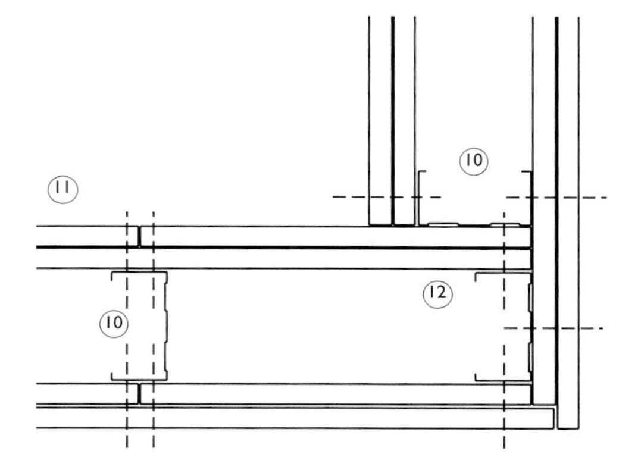

Horizontal elevation

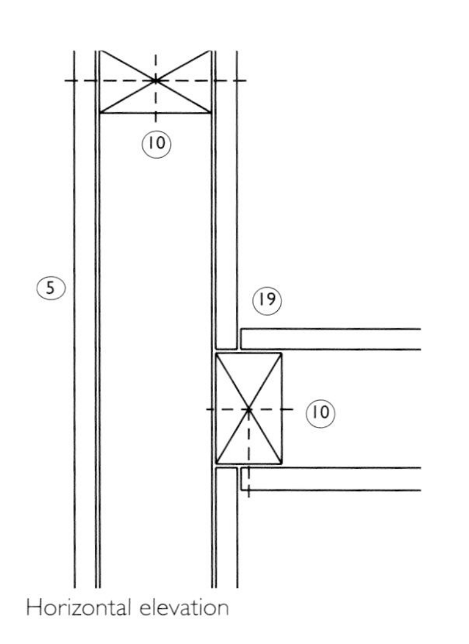

Horizontal elevation

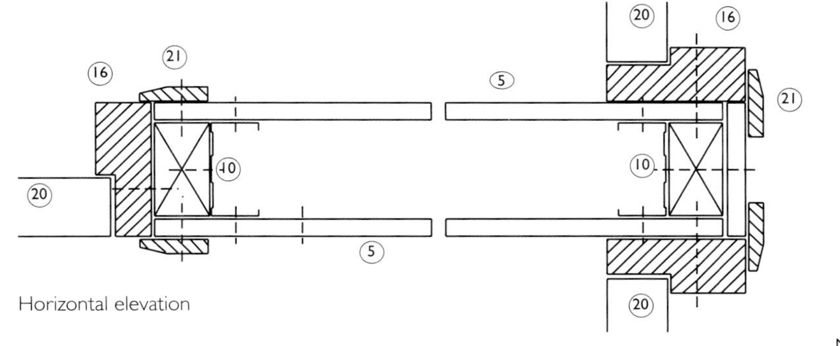

Horizontal elevation

Details
1. Movement joint
2. Quilt insulation
3. Top rail in timber or metal (pressed steel shown)
4. Ceiling level
5. Plasterboard
6. Softwood skirting
7. Pressed steel bottom rail
8. Plasterboard
9. Floor level
10. Stud in timber or metal (pressed steel shown)
11. Two layers plasterboard/drywall
12. Plasterboard/drywall laps adjacent run of partition
13. Plasterboard/drywall butt jointed at edge
14. Wall in different material (concrete block shown)
15. Plasterboard/drywall on battens or dabs
16. Internal timber door frame
17. Insulated board
18. Mineral quilt providing acoustic insulation
19. Plasterboard/drywall butt jointed at internal angle
20. Door leaf
21. Architrave
22. Packer

76G

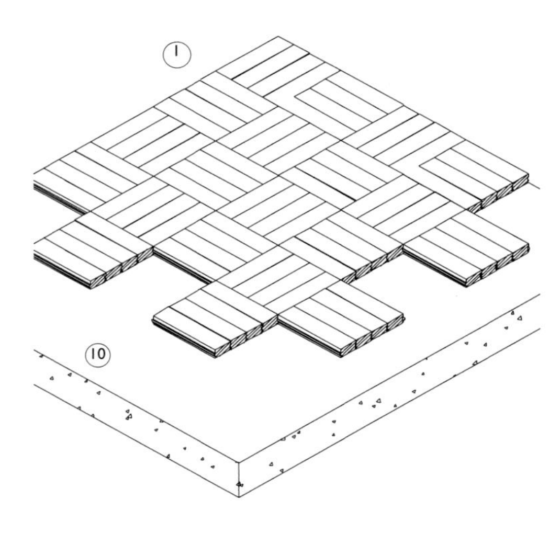


Fixed floor finishes are those which are not intended to be accessible or demountable and form a permanent part of the building. Floor finishes are usually harder and more resilient than the floor structure, or substrate, beneath. Since the properties of finishes and substrates are usually different, the finish used must be laid in a way that allows both it and the substrate to move together or be separated in order to avoid cracking in the floor finish. Where there are movement joints in the floor structure, the floor finish is usually broken to allow expansion and contraction. As a result, the finish can influence the design of the substrate. For example, a cast-in-place concrete slab may require contraction joints at centres which do not correspond to the module of a proposed tiling layout but which should be adjusted to suit the design of the floor finish.

## Concrete

Concrete can be used as a self-finish to a floor slab, with its surface sometimes polished to provide a smooth, dust-free finish, either as a screed or as a power-floated finish to a floor deck. Grains of carborundum can be added to improve the wear of the floor. Concrete floors can be coated with floor paints which have improved enormously in recent years to provide fairly hardwearing surfaces that will last up to 5 years but are not suitable for very heavy use which can cause the paint to wear away quickly. An epoxy coating is used where a harder surface finish is needed for heavy foot traffic, particularly for industrial applications. This polymer coating has excellent resistance to abrasion but is a very hard surface to walk on. A softer surface is provided by rubber-based compounds which are used in indoor sports facilities, but they have less abrasion resistance. Both epoxy and latex coatings, together with variations such as polyester resins, are referred to as poured floor finishes due to their ability to be poured in place to provide a self-levelling joint-free floor finish. They are poured over large floor areas to produce thicknesses of up to 6mm (1/4 in). Made by mixing a resin with a curing agent, they provide a hard, smooth finish which can be coloured and is resistant to chemical attack.

Terrazzo is a hardwearing floor finish that is applied to a concrete substrate. It consists of crushed marble aggregate mixed with cement. The material is applied as wet mix to a thickness of between 15 and 25mm (5/8in to 1in) depending on whether it is bonded directly to the screed or concrete slab beneath. Terrazzo is bonded by laying it on the concrete substrate while it is curing (but hard enough to walk on). It is laid in bays formed by movement joints (expansion and contraction joints) in the concrete substrate. The bays are separated by stainless steel, brass or bronze angles in a very similar way to ceramic tiles. Terrazzo is finished by grinding and polishing.

## Stone

Stone is used as a fixed floor finish in the form of paving slabs that are bedded in a sand-cement screed and can be used both internally and externally. The thickness used depends on the strength and thickness of the stone and the size to which the stone can practically be cut

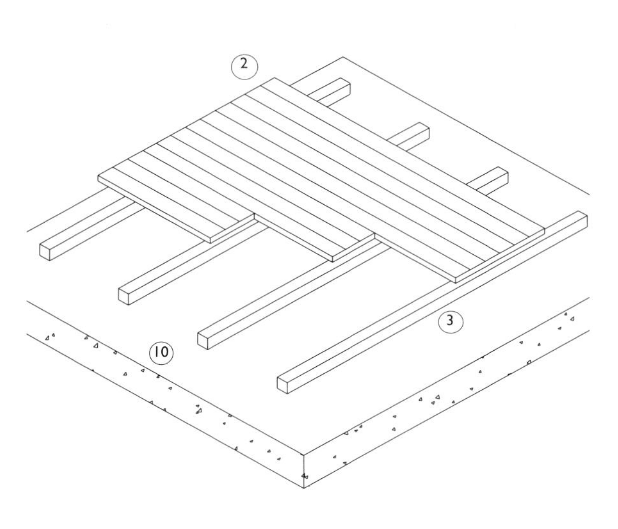

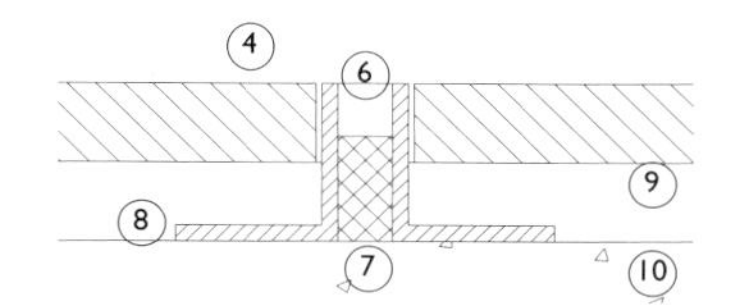


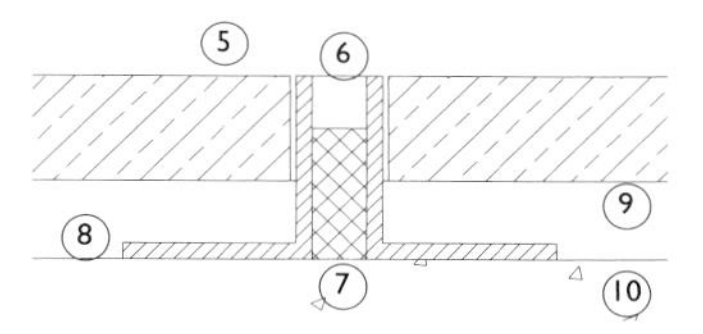


Details
1. Timber flooring block
2. Timber boards
3. Timber battens
4. Ceramic tile
5. Stone
6. Compressible seal
7. Compressible backing
8. Stainless steel angle
9. Bedding compound
10. Concrete floor

from blocks. The most common thickness is 20-30mm (1in to 1 1/2in) with slab sizes adjusted to allow it to be carried by hand and laid in place. A common size is a module of 500x500mm (20x20in) or 600x600mm (2ftx2ft) slabs, which allows 10mm for the joint between them. For small applications, stone can be ordered from choosing a sample, but for large areas, the material is cut to order and the Architect should visit the quarry to choose the material. In common with other natural materials, it is difficult to define precise visual qualities since its natural formation produces an enormous variety of appearance. The usual method is to define 'extremes' of quality between the most and least marked, and the lightest and darkest in a set of sample slabs. Because stone has little strength in tension, the substrate must be firm, with very little structural movement or deflection.

Stone floors are ground before being laid to provide different surface finishes. They can be polished, using fine sanding wheels, or honed, using rougher grinding wheels. Floor sealers are often not recommended for some stones since some slabs will be more absorbent than others, resulting in exaggerated colour differences across a floor surface.

## Timber

Timber flooring can be used as a floor finish on concrete, timber or cold-formed steel framed floors. In all three instances, the substrate is sealed with a vapour barrier to avoid the timber rotting. The timber is either loose-laid as solid strip flooring, fixed on bearers resting on the substrate, or bedded in mastic or bitumen as block flooring. Strip flooring has traditionally been used with timber strips less than 100mm (4") wide, while boards indicate anything wider up to a limit of around 150mm (6"). Nowadays strip flooring refers to all sizes of hardwood tongued and grooved board. Thicknesses range from 9 to 38mm (3/8in to 1 1/2in). Almost all types are proprietary systems that are either fixed to supporting battens as a sprung floor, typically used in sports halls, or are held together by pressed steel clips as a continuous material that rests on the substrate. Beech and maple are the most common hardwoods used, which have a light coloured appearance. A gap between the floor and the wall is left to accommodate movement in the wood due to changes in temperature and humidity. This gap is typically 10mm (3/8in) for a four metre (13ft) wide bay of strip flooring.

Wood block flooring consists of small hardwood blocks that are bonded with bitumen to a concrete floor. Block sizes vary from 25 to100mm (1 to 4 in) wide and 150 to 300mm (6 to 12 in) long. Depths vary from 19 to 38mm (3/4 to 1 1/2in). Blocks are bonded directly to the concrete with a bitumen-latex adhesive. Simple rectangular patterns or herringbone patterns are the most common ways of laying blocks. Both strip flooring and block flooring require a surface seal, such as polyurethane sealer, to avoid dirt being trodden into the grain. This is re-applied every few years depending on the amount of wear experienced by the floor.

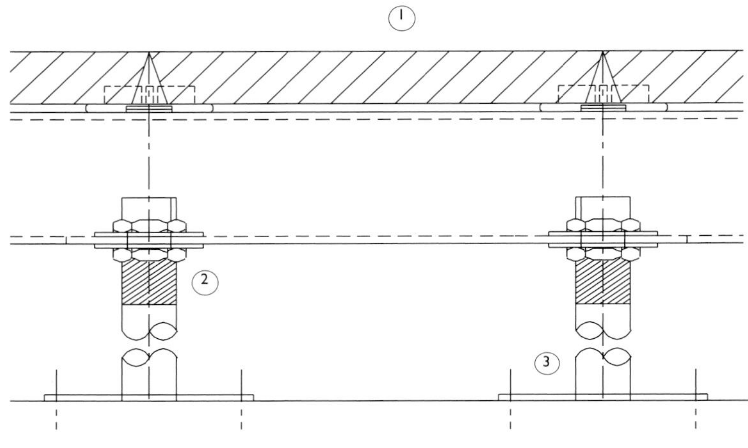


Vertical section

Raised floors were first used as computer floors to provide a zone for electrical cables and air handling ducts to large computers housed in a separate room. They are now used throughout office buildings as a means of providing a zone for mechanical and electrical services including ventilation, as either a complement or replacement for those used in suspended ceilings. Raised floors are used as a method of bringing electrical cables under the floor to a large number of points while allowing tiles to be moved to accommodate changing servicing requirements. Completely open voids can also be used as an air plenum to supply or extract air in a mechanical ventilation system.

Raised floors are manufactured as proprietary systems with different loading capabilities, designed to be either fully accessible or semi-accessible, depending on the ease of use and frequency of access required to the floor void beneath. Fully accessible types have a much greater range of void depth, ranging from 100mm (4in) to around 2000mm (6ft). Semi-accessible types are restricted to low floor voids of around 150mm (6in). The fully accessible types are generally steel composite panels with a concrete-based infill, supported on variable height pedestals. Semi-accessible types are generally made from timber composite panels supported on concrete pads or timber battens. The two types have varying degrees of rigidity in their framework to suit the degree of accessibility required. In some systems, structural stability of the frame is lost if too many modules are removed for maintenance access.

Fully accessible floors are made with a variety of construction methods. Some are made as a support framework with legs beneath. The panels fit into the frame to be supported on all four sides. Others have self-supporting panels supported on an adjustable leg in each corner. Semi-accessible systems vary even further in their design. Some comprise a single precast concrete tray with integral legs. These are laid side by side directly onto the floor slab. Others are very similar to the timber sprung floors used in sports halls, consisting of a rectilinear grid of softwood battens resting on acoustic pads to prevent sound transmission. Timber composite boards are screwed down to the frame.

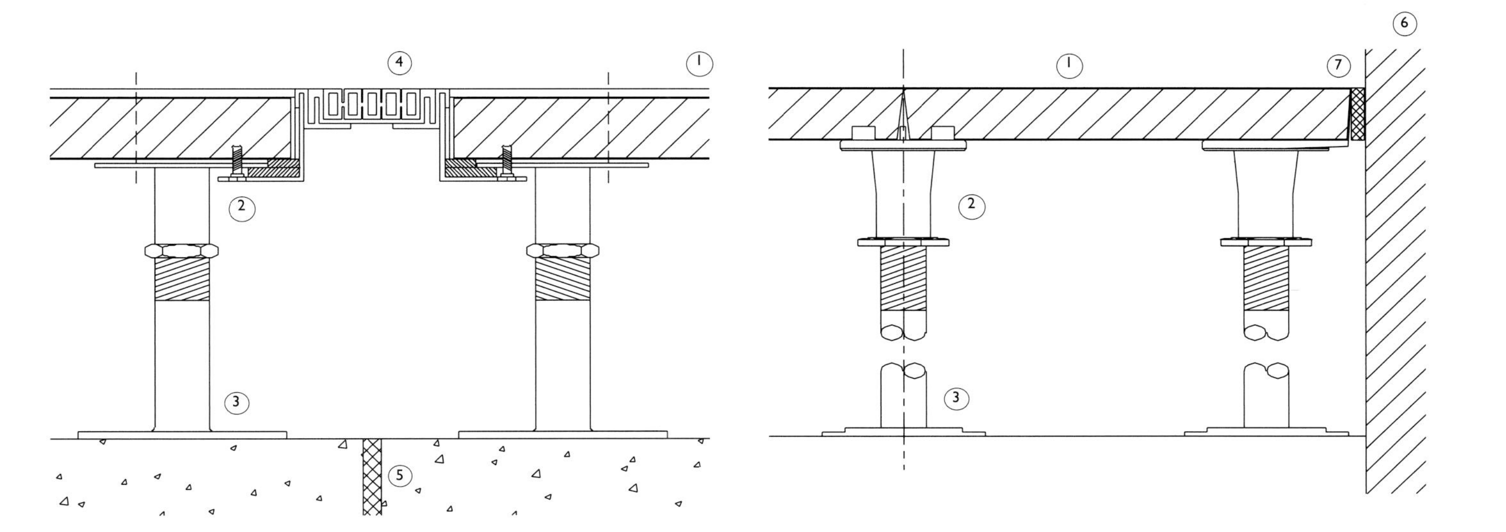


Ventilation grilles and electrical socket boxes can be incorporated into all types. Floors typically have carpet tiles loose laid on top, to provide both a comfortable walking surface and for acoustic purposes. Carpet can be separately bonded to each tile or be loose laid off the grid to conceal joints in the floor beneath.

Details
1. Floor panel
2. Pedestal
3. Base fixed to floor slab
4. Compressible expansion joint
5. Expansion joint in floor structure
6. Adjacent wall or partition
7. Rubber-based seal

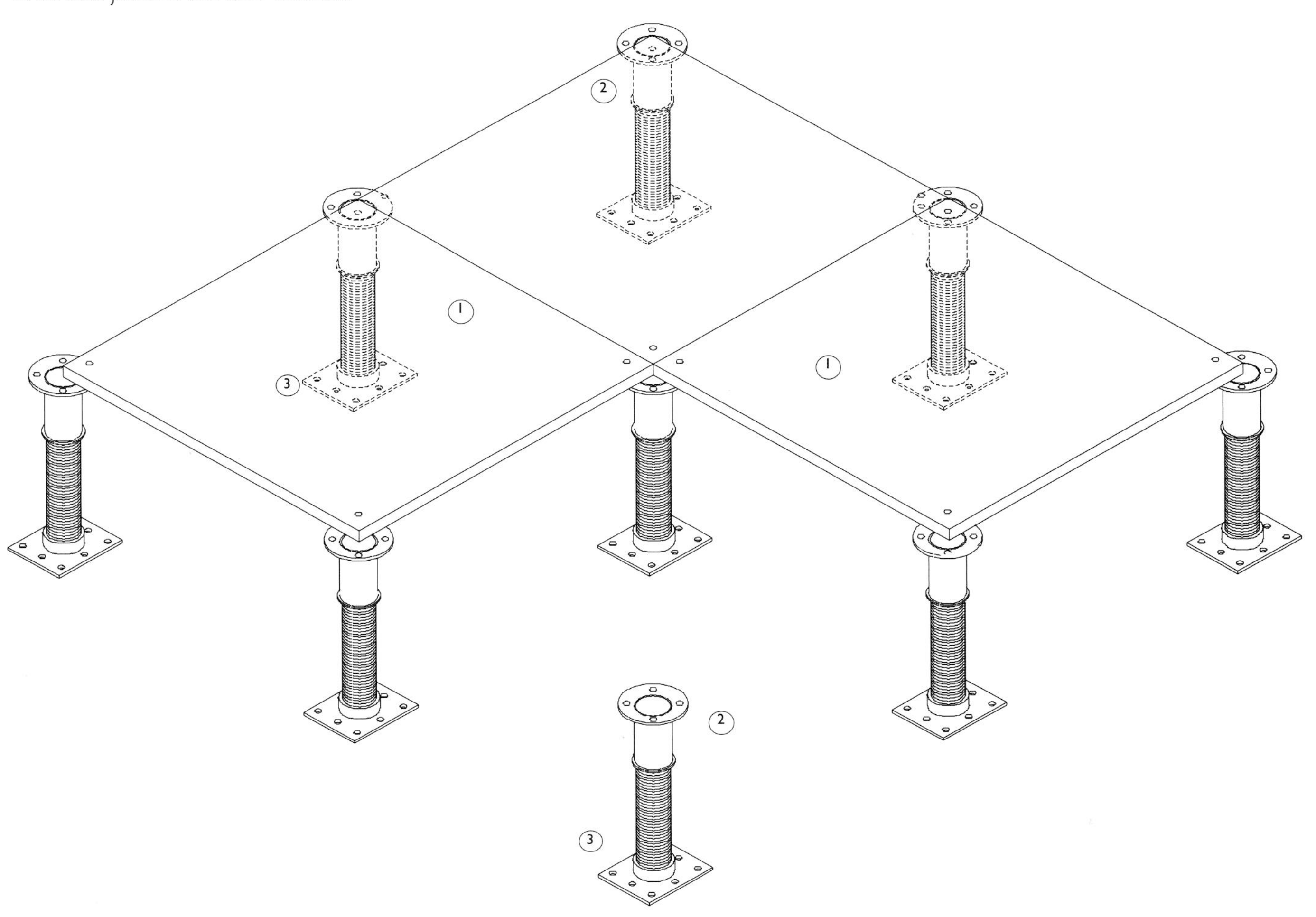

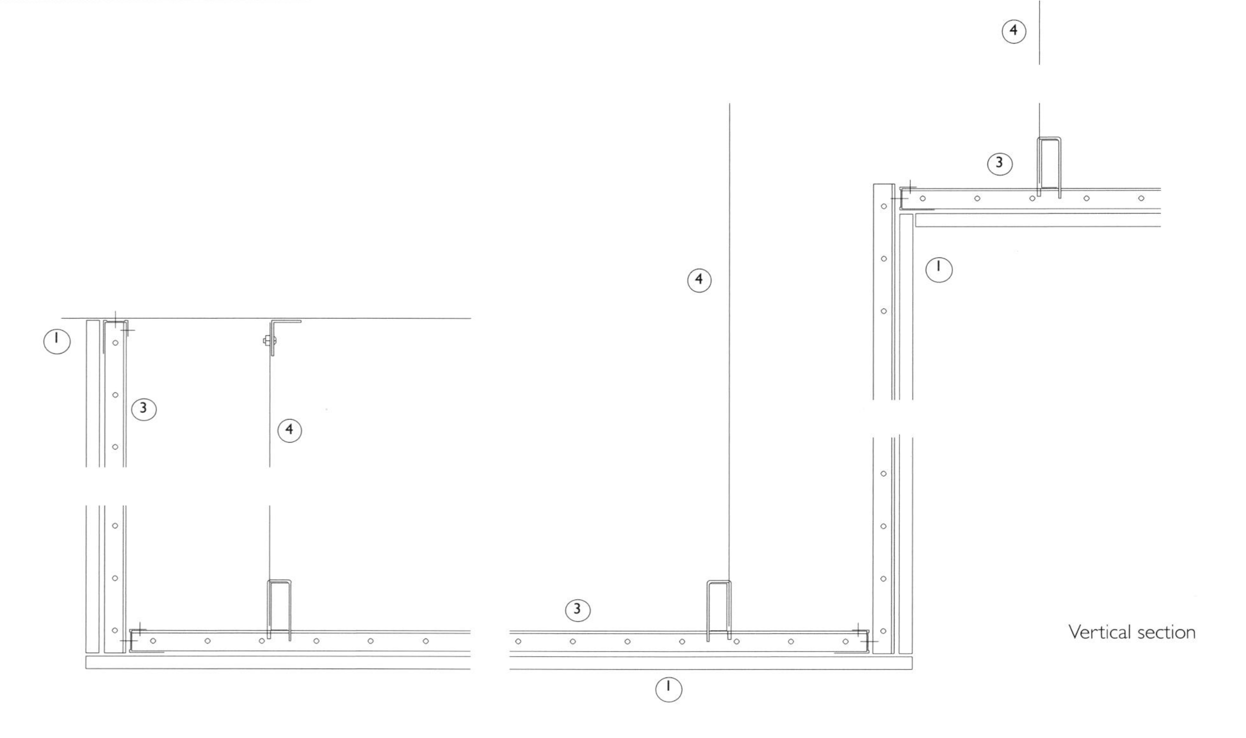


103E

81

Suspended ceilings are used primarily to provide a service void between a ceiling plane and the underside (soffit) of the structural slab above. This zone is used to house recessed light fittings, ducts for mechanical ventilation and associated equipment. There are two generic types of suspended ceiling. The fixed version is used where a continuous plastered surface is required and where there is no need to access the ceiling void from below. Accessible types are used primarily in office buildings where they integrate with a modular layout of partitions. They are designed to suit both cellular office layouts and open office areas requiring individual lighting and mechanical ventilation.

Fixed suspended ceilings consist of either layers of plasterboard sheets or wet-applied plaster on metal lathes which are supported on a frame suspended from the soffit of a structural slab or floor on either wires or galvanized steel strips. They can be designed as a simple timber or metal supporting frame or be specified as proprietary manufacturers' systems.

Fixed ceilings create a smooth, continuous soffit for recessed lighting, and a ceiling plenum for ducts that do not require access. They can also provide a fire-resistive layer where this is not provided by the supporting floor structure. Access hatches can be used but are difficult to conceal. Fixed suspended ceilings can be used to form single direction curves by bending plasterboard around a modelled frame.

Complex shapes can be formed with metal lathes. Fibrous plaster is often used, which provides a smooth surface that is easy to work but lacks the strength of other plasters. As a result, it is applied onto a reinforcing mesh in a very similar way to ferro-cement. Fibrous plaster is used where curves in two directions are required as well as in repetitive decoration, where items can be made in a workshop and later fixed to the ceiling.

8F

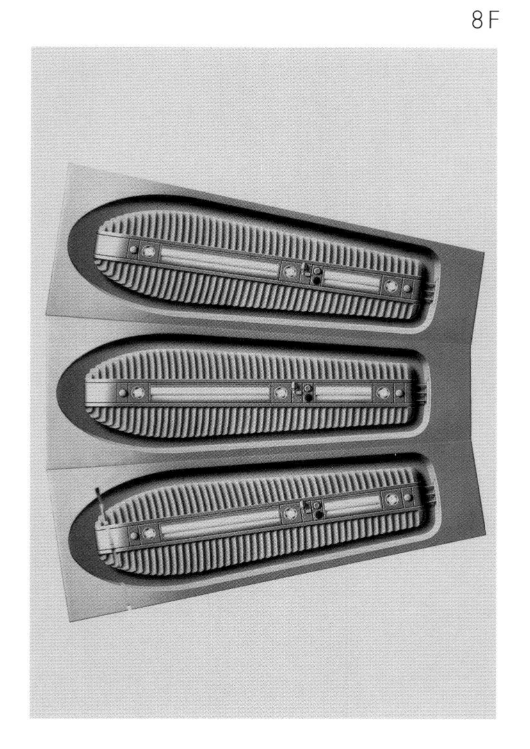

98F

Details
1. Plasterboard/drywall
2. Metal hanger
3. Fixing rails
4. Suspension rod, wire or pressed metal strip

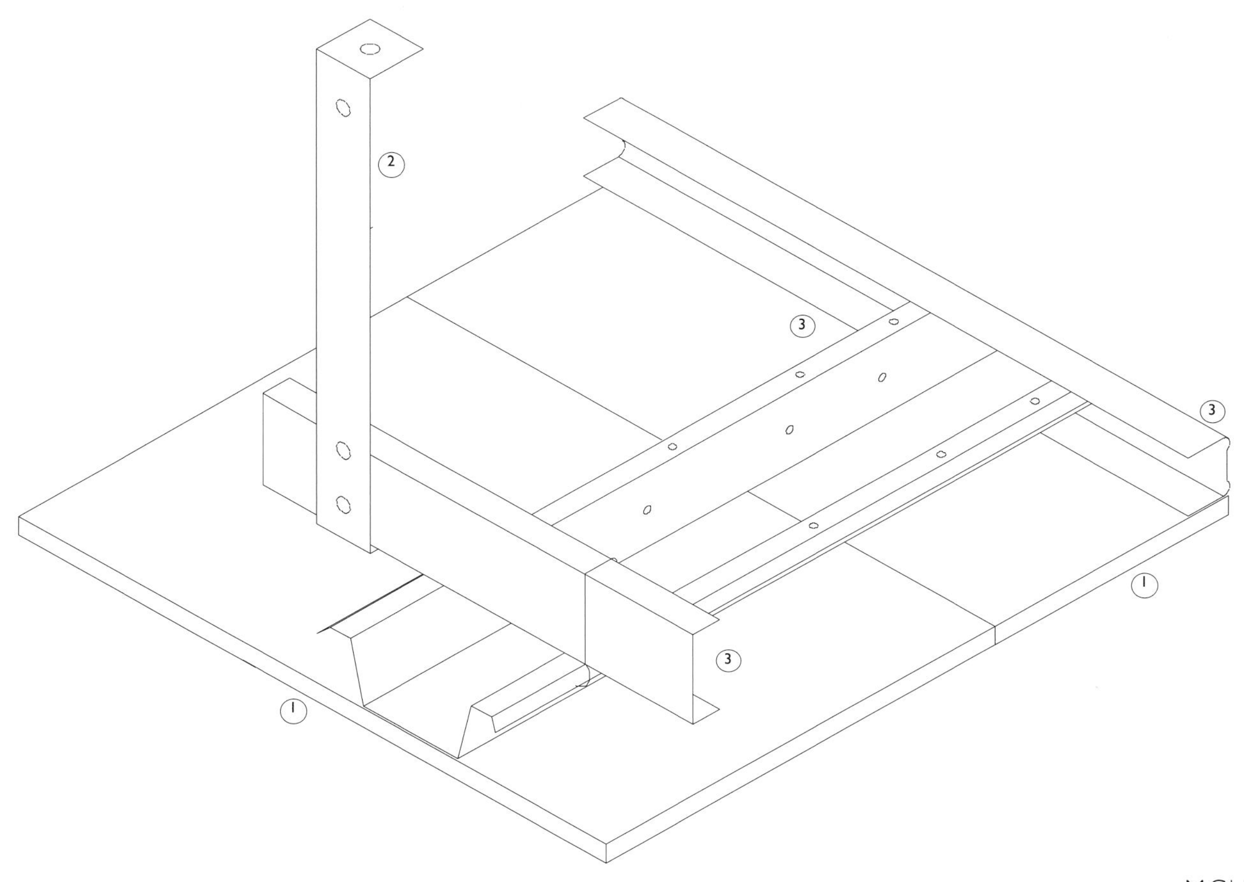

Vertical section

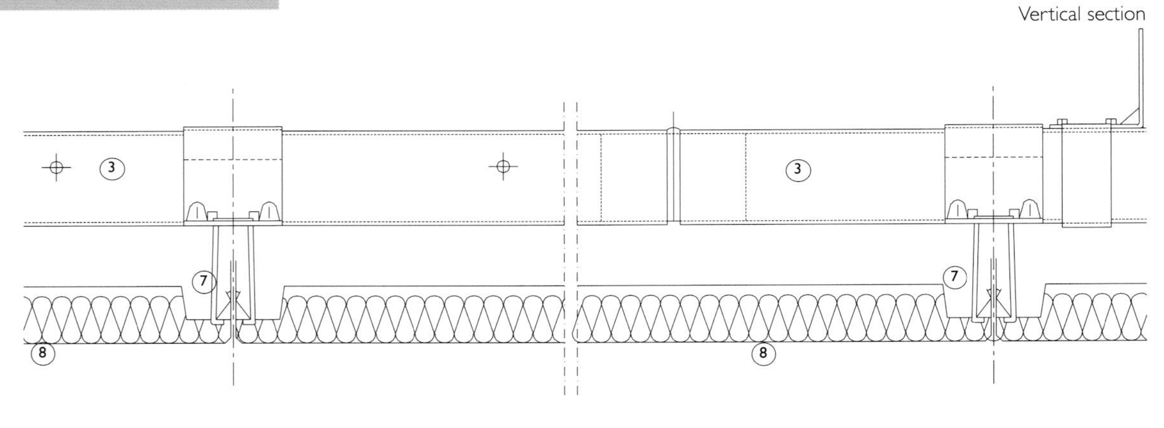


8K

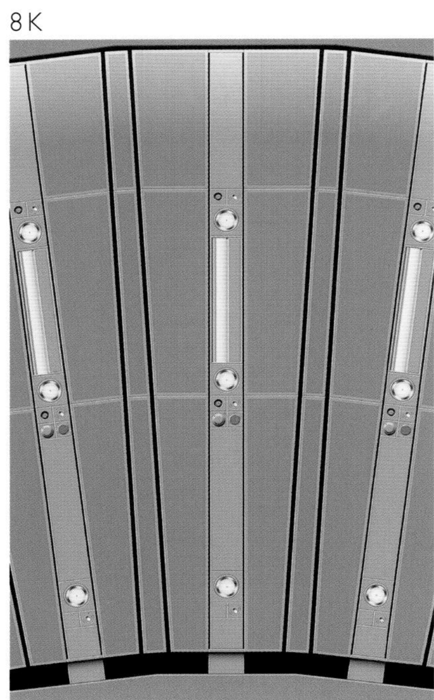

98F

Details
1. Plasterboard/drywall
2. Metal hanger
3. Fixing rails
4. Suspension rod, wire or pressed metal strip
5. Lamp fitting
6. Luminaire
7. Clip to secure panel
8. Ceiling panel (perforated metal panel with acoustic lining shown)
9. Support rail

Accessible ceilings use a supporting grid in steel or aluminium. The two most commonly used types are the T-section and the spring clip. Both are designed to be as light as possible with varying degrees of strength and rigidity, which are defined by the loads imposed on the ceiling by services from above and by partitions fixed to it from beneath. In some manufacturers' systems, dimensional stability of the frame is lost if too many tiles are removed during maintenance access. The T-section uses an inverted T-shaped aluminium extrusion which holds the ceiling tile in place. The tile either sits directly on the section, creating an exposed grid, or is set hanging partially below the frame in a semi-concealed grid. The spring clip system allows the tile to be slotted into the support grid from beneath in a concealed system. The support grid can also be concealed by fixing the tile onto a T-section grid from beneath. This is done where aluminium or steel sheet is countersunk screwed to the support frame. Noise control within a room, or between adjacent spaces, is achieved by setting a sound-absorbent board or quilt onto the upper face of the ceiling tiles. The tiles are usually perforated to allow sound to pass through to the insulation behind.

The most common support grids are the one-way grid and the two-way grid. The one-way system has identical sections set parallel, at centres corresponding to the width of the tiles. Tiles are supported on two sides and span between the sections of the grid. Partitions beneath can be supported continuously in one direction but are supported only at points in the other direction. The reduced flexibility makes the system economical. A two-way system uses either metal cross tees that span between the main runners or by using the same sections set perpendicular to one another to create a full grid. The latter type is made by half-jointing the main runners where they cross. The increased flexibility of the two-way system makes it more expensive. The frame is fixed back to the soffit with members of varying rigidity: wires, rods and angles. Suspension wires are

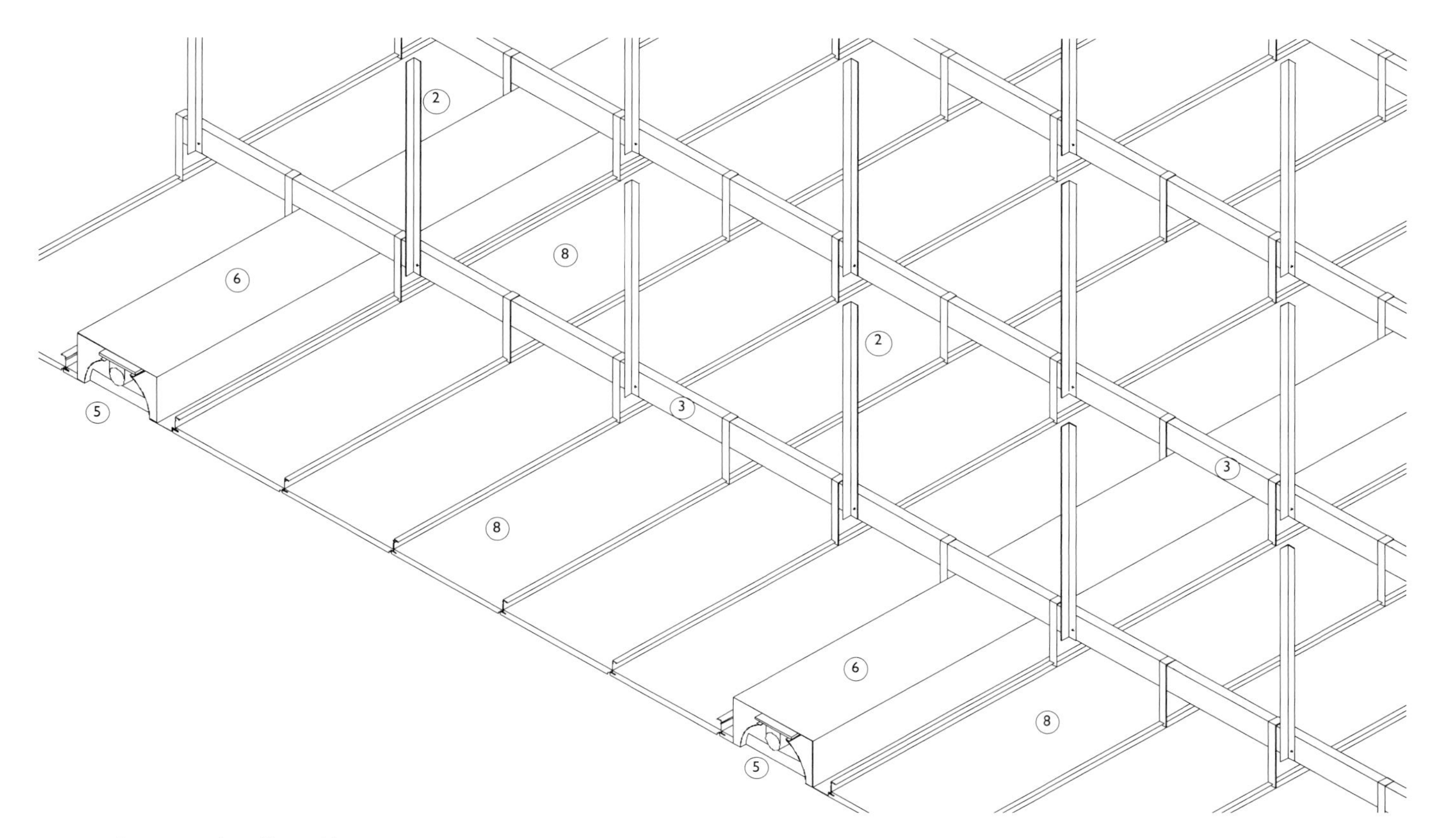


not usually appropriate if partitions are fixed in such a way as to exert pressure from below.

Like their supporting grids, ceiling tiles are designed to be lightweight. The most economical generic tile is mineral fibreboard, which provides high acoustic insulation but is limited to smaller spans due to its lack of rigidity. Greater thicknesses of board add considerably to the weight of the ceiling. The typical grid size in this material is 600x600mm (2ft x 2ft). Larger tiles are made from perforated steel trays. Since steel is a poor absorber of sound, it is perforated to enhance its acoustic performance, and in addition can be lined on the upper face of panels with either thin mineral quilt or an acoustic pad. Ceiling grids up to around 2000x3000mm (6ft 6in x 9ft 10in) are possible. Aluminium sheet can also be used but it requires a greater depth of vertical edging to attain the spanning capability of steel. This increases the overall depth of the ceiling.

103A

18E

18A

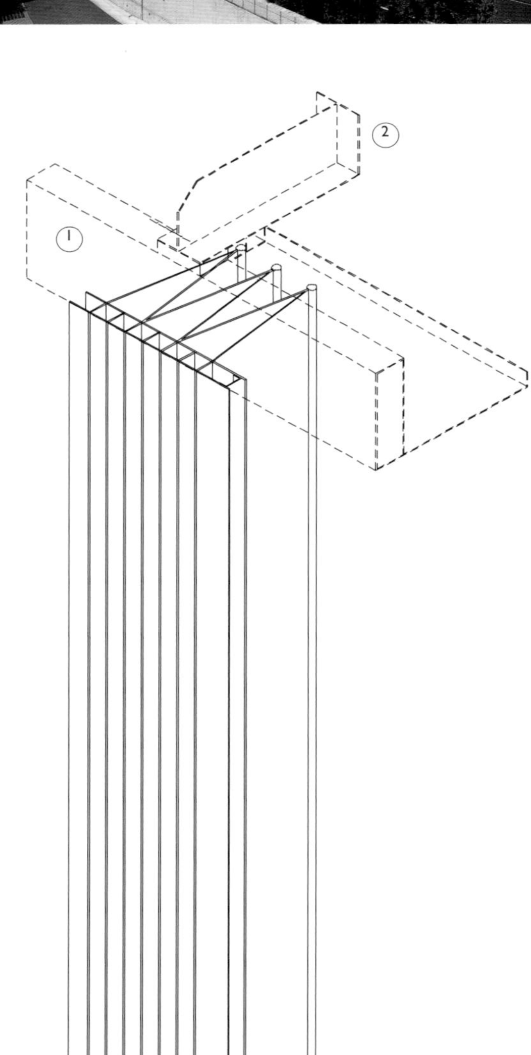


Folding doors are made as a series of metal or timber panels connected together vertically. Small folding doors are made in steel, aluminium and timber. Large doors are usually made from steel due to the high strength and rigidity required. Thermal insulation is often included within a metal door, forming a sandwich panel. Due to the complexity of their construction, large doors are not usually insulated.

## Construction

Each leaf is supported at the top by a wheel or ball bearing pulley running in a horizontal track. The track supports the full weight of the door assembly, unless the beam above the door opening cannot be made sufficiently strong. In this case, the doors sit on wheels in a bottom track, but large bottom-rolling doors are rarely used due to the risk of the wheels being blocked by debris. For top-rolling doors, the track is fixed back along its length to a structural support above. The bottom of the door is held in a guide rail set into the depth of the floor. The doors are completely suspended from the top rail and no load is exerted on the bottom rail. Each leaf can be supported centrally, where it hangs evenly, or at the end of the leaf, where it is always subjected to a torsional force when open. In large doors, pairs of hot rolled steel channels forming vertical members are connected by rows of lattice bars that run between them. Door leaves are made from thin steel sheet, connected together along their vertical edges by a continuous hinge that forms a watertight joint. Pairs of leaves are connected to the vertical channels. Manually operated doors have a series of handles on the verticals of the folding frame to move the door. Electrically operated doors have a geared motor fitted to the top track that moves the pulleys by a continuous chain.

## Details

1. Rail to support top hung doors
2. Support bracket fixed to beam or external wall
3. Motor driving chain which operates doors
4. Inside
5. Outside
6. Guide rail set into floor

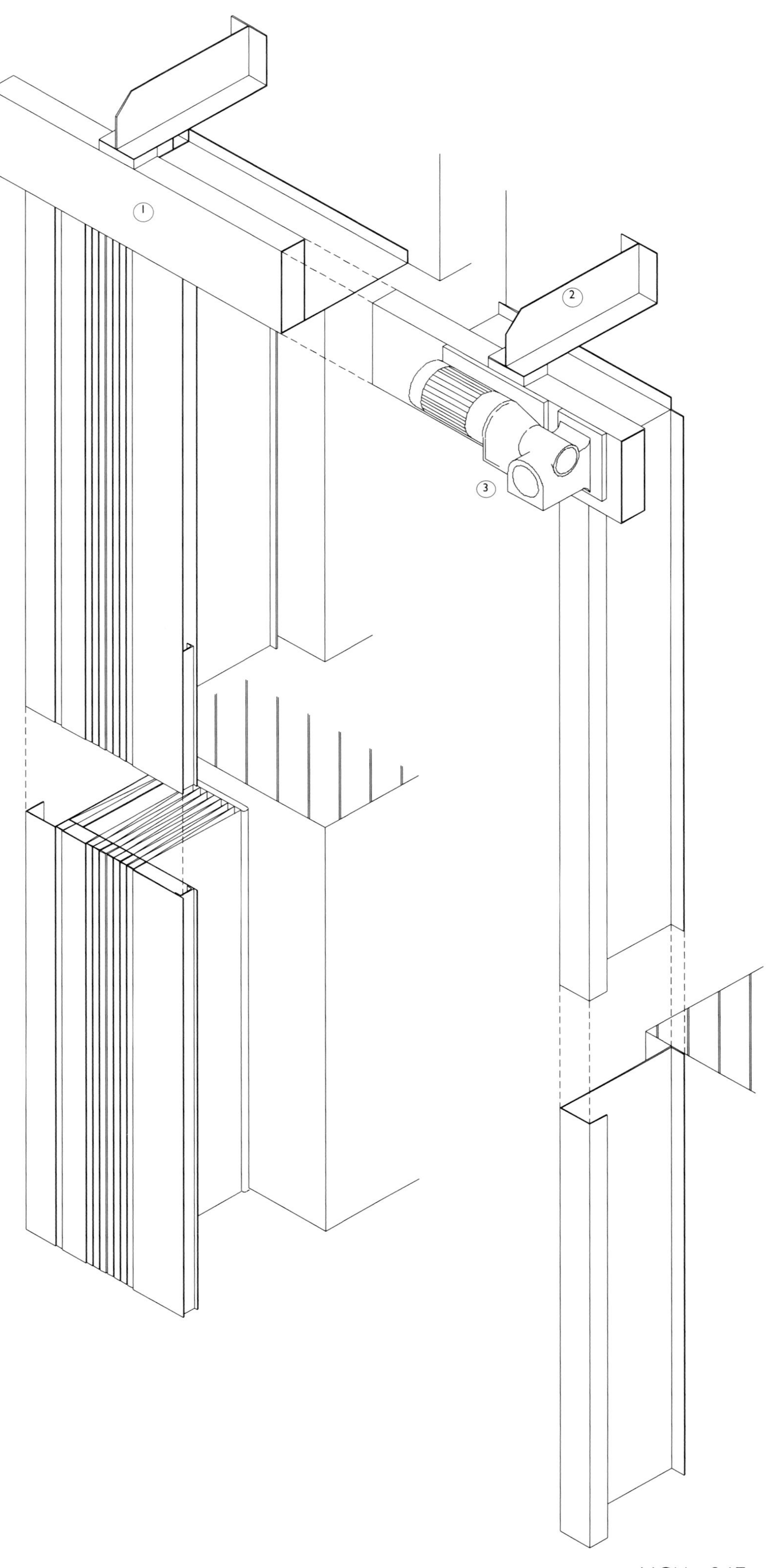


## Weatherproofing

Leaves are usually manufactured to be watertight. The supporting frame for the door is fixed directly to the adjacent structure and sealed with either a neoprene or EPDM gasket or a silicone seal. The track at the top of the door is enclosed in an open box section to exclude rain. This section is fixed either directly to the supporting structure or to a cantilevered bracket. The same type of weather seal is used on the vertical sides. The bottom track usually consists of a steel rail cast into concrete, and the ground in front of the door is laid to a minimum fall of 2 per cent to drain water away from the door opening. The slot is not usually drained.

9A

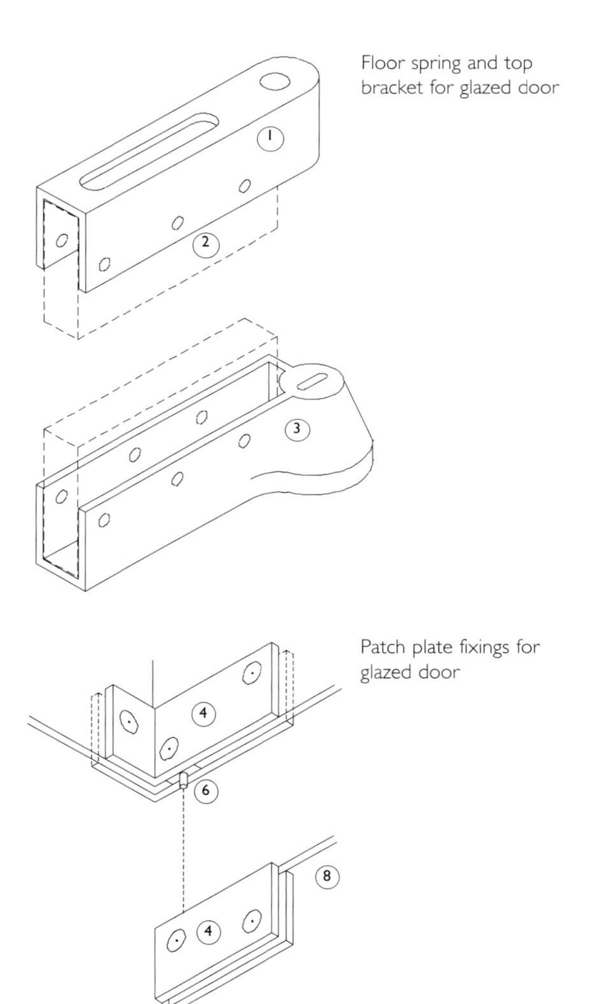


Floor spring and top bracket for glazed door

Patch plate fixings for glazed door

Ironmongery is the hardware that allow windows and doors to be opened, closed and locked. The term also includes coat hooks and related items. A single range of manufacturer's ironmongery tends to be used throughout a building, often using a common size and style of door handle, hinge and lock. Ironmongery needs to be hardwearing and sturdy since it is often the part of the building users actually touch and use constantly.

Historically, the most common locks in use, the rim lock, combined with a latch, and the spring-loaded mortise lock, were developed by the end of the 19th century. Subsequent changes were primarily in the refined design of the mechanism and appearance. Throughout the 20th century, architects have been involved in designing ironmongery, with complete ironmongery systems unified by a common design theme available since the 1960's.

The traditional materials for ironmongery are iron, brass and bronze, using manufacturing techniques of forging and sand casting. Aluminium, stainless steel and plastic have become popular in the last 50 years since they are less expensive than brass and bronze and require no additional finish. Surface finishes are an important consideration in designing or specifying ironmongery, since lacquers and paint wear away very quickly. Aluminium, stainless steel and plastics allow fittings to be cast, pressed and extruded using large-scale industrial techniques.

## Door hinges

Most hinges for timber doors are mortised, a technique that involves their being set into the depth of the door and frame. Steel doors often have surface-mounted, non-mortised, hinges. Frameless glass doors can be supported on patch-plate hinges similar to the fittings used for bolt-fixed glazing. Only the pin and knuckles are visible when the door is closed. Larger projecting hinges are used where the door is required to open through 180 degrees. Projecting hinges called parliament or H-hinges, narrow down to use a shorter pin, making them lighter in appearance. Although the pin can be removed in all hinge types to lift the door off, some are made with a short pin in each knuckle to make the door easy to remove. These lift-off hinges are used where doors must be removed regularly. Where there is a minor variation in floor height from one side of the door to the other, such as carpet on one side of the door, a rising butt type is used to lift the door as it is opened. The number of hinges per leaf depends on the weight and service requirements of the door. Door manufacturers will recommend where they should be fixed and how many butts are required.

## Door pivots

Doors can be pivoted at the top and bottom as an alternative to side hinges. They are mainly used on frameless glass doors where hinges would be too visible. Pivots are usually located slightly away from the edge

Common hinge types

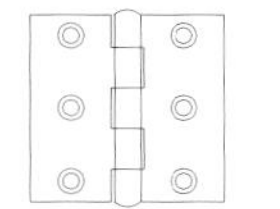

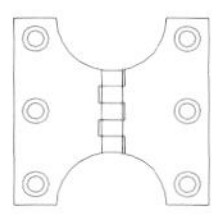

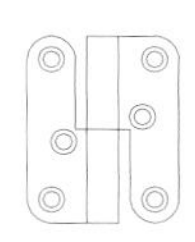

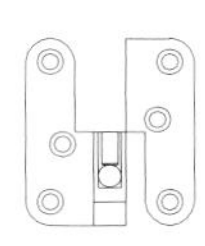

of the leaf and can be hung from the head or supported at floor level. Floor springs are commonly used where the door is required to remain in the open position or return to the closed position. They are recessed into the depth of the floor and support the weight of the leaf. The mechanism is housed in a box up to 75mm (3in) deep with a cover plate on top which is set at floor level. Floor springs are often used as an alternative to door closers in frameless glass doors.

## Door closers

Door closers are used to ensure that a door returns to the closed position after being opened. They are used primarily in escape routes to prevent the spread of fire and smoke during an emergency. A concealed type, called a spring closer, is used mainly in residential buildings where a powerful pull is not required to close the door. Types used elsewhere are either surface-mounted at the top of the door leaf or recessed into the ceiling. These closers comprise a box containing a hydraulic mechanism that pulls an arm to close the door.

## Handles and push plates

Door handles are either lever, knob or pull type. Most have a cover plate to conceal the junction between the handle and the door. Push plates are used on the opposite side of doors with pull handles. A large range of handles is available.

## Locks and latches

Locks and latches can be either recessed into a door, surface-mounted or be integral with the door handle. The most common type of door lock used is the cylinder rim night latch (deadbolt lock). This is often used with a separate mortise, or recessed, dead lock (mortise lock). A third type, called lock and latch sets, combine door handle, lock and latch in a single mechanism. Sets of locks are grouped to reduce the number of keys needed to open all the doors of the building, but to ensure that individual keyholders have access only to a few locks with a single key. A system of key suiting (keying systems) is used where a master key will open all doors, but servant keys will open only a few. An intermediary level of key, called a submaster, will open a group of servant locks.

## Door bolts

Bolts are used to secure closed doors without the need for a key. The most common type is the espagnolette bolt which may be furnished with an additional key. Espagnolettes provide a simple way of securing bolts at the top and bottom of the door as well as at the sides if required. Panic bolts, used on fire escape doors, have a horizontal bar which is pushed to release bolts in the top and bottom of the door. As with other types of ironmongery, bolts can be either mortised or surface-mounted.

9

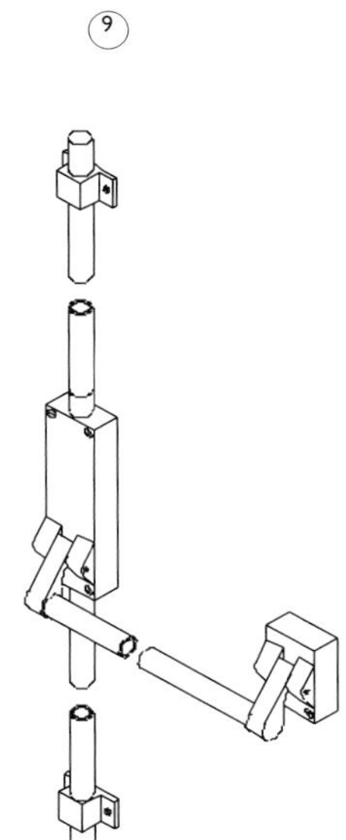

Details
1. Top bracket
2. Door leaf
3. Bottom bracket fixed to floor spring assembly
4. Stainless steel patch plates (shown secured to glass fin)
5. Top hinge
6. Steel pin
7. Flush bolt
8. Top of glass door leaf
9. Isometric view of panic bolt assembly
10. Isometric view of espagnolette bolt assembly
11. Isometric view of door bolt assembly

10

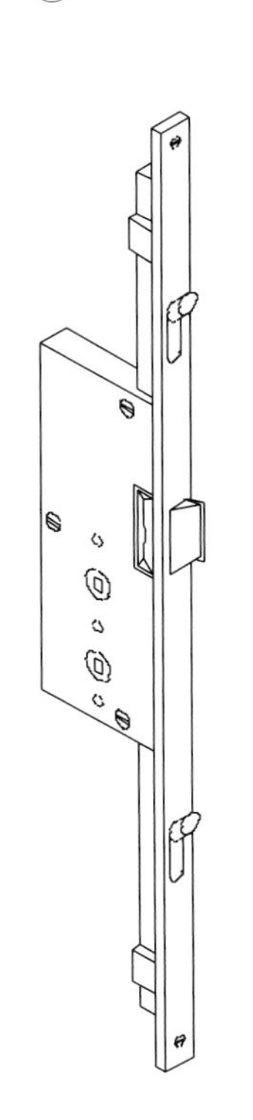

11

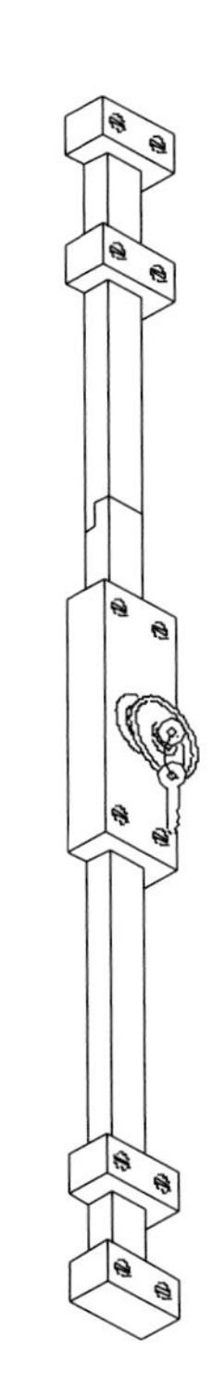

# REFERENCES

Project references and photographer contacts

Projects illustrated in the text are referenced in terms of

- Project title
- Photographer of images shown
- References to associated articles in architecture magazines
- Brief description of the building

Rokko Housing, Kobe, Japan.
Tadao Ando Architect & Associates.
Japan Architect Spring 1994.
Photographers: 1a. Mitsuo Matsuoka.
1b, 1c, 1d, Tadao Ando.

This housing project is set on a slope at the foot of the Rokko Mountains in Kobe and comprises three separate projects. The reinforced concrete structure is composed of sets of standard size building units which follow the irregular natural topography of the site, creating an enormous variety of spatial experience.

Toto Seminar House, nr Kobe, Japan.
Tadao Ando Architect & Associates.
Architectural Review October 1999.
Architectural Record February 1999.
Photographer: 1e, 1f, Mitsuo Matsuoka.

This building is both a retreat for company employees and a training and seminar facility. The built volumes are constructed in reinforced concrete, each set in a different orientation to one another. There is a wide range of spatial experience as one walks around the building which still uses an economic form of construction.

Telecommunications Authority Building, Oporto, Portugal.
João Álvaro Rocha and
José Manuel Gigante.
Domus March 1997.
Photographer: 5a, 5b, Luís Ferreira Alves.

The composite panels used to clad this building have the same metal finish on the inside as the outside. The panels, which are 60mm thick, are insulated but are arranged in order to reveal the form of the reinforced concrete frame behind. Tall, narrow panels are used in front of columns and thin, long panels are used to reveal the thickness of the floor slab. Joints are staggered to avoid the gridded appearance of other composite panel walls that often give few clues as to the spatial organisation of the building behind.

Arene de Nimes, France.
Labfac / Finn Geipel, Nicolas Michelin.
Techniques et Architecture May 1989.
Photographer: 6a, 6b, 6c, 6d, 6e, 6f, J. Denerdaud + P. Blot.

This inflatable roof structure is partly a balloon, and partly a tent, providing both shelter and thermal insulation in colder months. The form of the roof is one which suggests that it is growing from the arena structure and engaging with it rather than being a temporary addition that might have hovered above it.

Freshwater Pavilion, Neeltje Jans, Holland.
NOX Architects.
Domus September 1997.
Photographs: 7a, 7b, 7c, 7d, 7e, courtesy of Lars Spuybroek, NOX Architects.

The Freshwater Pavilion is connected to the Saltwater Pavilion discussed in the Structures chapter. The building is not placed on the ground in the traditional manner but is set like another dune within the landscape. The metal covering both provides a fluid geometry in sympathy with that of the dunes and is set apart as an enclosure to the multimedia exhibition within the pavilion. The pavilion design is based on the interaction between architecture and events created by water, light and sound.

Leipzig Exhibition Centre, Germany.
Von Gerkan Marg und Partner.
Domus June 1996.
Photographers: 2a. Punctum/H-Ch. Schink.
2b. Busam/Richter.
2c. H G Esch.

The Central Hall, designed with Ian Ritchie, is constructed as a steel framed arch structure in the form of a vault. The transparency of the structure is maximised by the use of bolt fixed glazing.

Tateyama Museum of Toyama, Japan.
Arata Isozaki & Associates.
Domus October 1996.
Photographer: 3a, 3b, 3c, Yasuhiro Ishimoto.

The roof of this museum building is in the form of an upturned boat hull, which forms part of the symbolism of the design. This form is covered with slates, which are small enough to follow the form of the roof without distorting the shape. The overall appearance is one of a continuous surface rather than the strong lines associated with sheet metal roofs. The supporting roof structure in timber is partly revealed inside the building, adding to the sense of the roof as a boat hull.

Uji-an Tea House, Kila-shinagawa, Tokyo, Japan. Arata Isozaki & Associates.
Domus October 1996.
Photographer: 4a, 4b, Katsuaki Furudate.

This tea-room, used for tea ceremonies, has a shallow domed lead-finished roof, covering a traditional interior. The roof covering is shaped to provide a joint pattern that reinforces the circular form within the constraints of a narrow sheet width of metals. The underside of the roof is covered with stainless steel to give the roof a completely independent form from that of the spaces below.

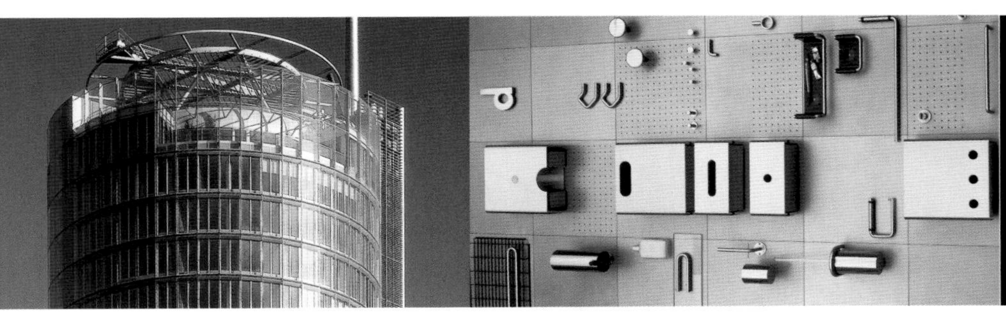

› 10A – 19

RWE AG Essen, Germany.
Ingenhoven, Overdiek und Partner.
Architectural Review June 1997.
Architectural Record June 1997.
Photographs: 8a, 8f, 8j, 8k, 8l, 8m, 8n, 8o, 8p, 8s, courtesy of Overdiek und Partner.
Photographers: 8b, 8c, 8h, 8i, Holger Knauf, Dusseldorf.
8d, Damian Heinisch, Essen.
8e, 8t, 8u, Hochtief Hauptverwaltung.
8g, HG Esch.
8q, 8r, Graziano Canzian.

This office building is one where every element of the construction is thought through to a very high level. The innovation and craftsmanship can be seen from the external double wall through the walls and ceilings to the roof structure.

D-Line ironmongery (hardware).
Photograph 9a, courtesy of Allgood plc.

The D-Line system was designed by Knud Holscher and first manufactured by Petersen in Denmark in 1972 as a fully co-ordinated range of ironmongery (hardware) from door handles to toilet roll holders unified by a common appearance and manufacturing method. An earlier system by Knud Holscher and Alan Tye was marketed by the British company Allgood in 1965 as the Modric system, an abbreviation of 'modular geometric'.

› 10A – 19

Expo 2000, Hannover, Germany.
Halle 8/9. von Gerkan, Marg und Partner.
Photographer: 10a, 10b, 10c, 10d, 10e, 10f, 10g, Heiner Leiska.

This exhibition hall has a roof constructed in the manner of a suspension bridge with reinforced concrete elements spanning between supporting cables.

Centre for Nature Information, Fetsund, Norway.
Askim Lantto.
Domus December 1998.
Photographs: 11a, 11b, 11c courtesy of Askim Lantto.

This building, set in woodland, is constructed as a timber frame with timber infill panels which are partly demountable to encourage a changing relationship to exhibits both inside and outside the building.

Bridgewatchers House, Rotterdam, Holland.
Bolles and Wilson.
Domus April 1999.
Photographs: 15a, 15b, 15c, 15d, 15e, courtesy of Bolles and Wilson.

This building houses a control cabin for river and bridge traffic, and is part of a wider project of the integration of the docks into the urban landscape of Rotterdam. The Bridgewatchers House is constructed as an assembly of planar surfaces built in different materials, which contrast both with each other and with the tubular steel legs. The metal rainscreen wall is used to create a smooth surface which contrasts with the other elements of the design.

Lerner Hall Student Center, Columbia University, New York, USA.
Bernard Tschumi / Gruzen Samton, Associated Architects.
Domus June 1999.
Photographers: 16a, Peter Mauss/Esto. 16b, Lydia Gould.

This building comprises an 1100 to 1500 seat auditorium, assembly hall, dining room, lounge and bookstore. The glazed external wall and glass walkways maximise the effect of transparency that allows views through the wall from the outside to the spaces inside.

Sun Tower Office Building, Seoul, South Korea.
Morphosis.
Domus November 1999.
Photograph 17, Young-Il Kim.

This ten storey tower in central Seoul is built in two parts and partially wrapped in perforated metal sheet set onto a supporting frame. The perforated metal is fixed 200mm (8in) from the reinforced concrete frame and forms an outer skin to the building, giving the offices greater scale and significance by the addition of this outer layer of 'clothing'. The blurred appearance of the perforated metal during the day contrasts with the material being back-lit at night.

Centre for American Studies, University of Oxford, England.
Kohn Pedersen Fox (International).
Photographs: 12a, 12b, 12c, Jim Dunster, courtesy of Kohn Pedersen Fox (International).

This building comprises a reading room and associated book storage for a research and teaching facility within Oxford University. The reading room has a large south facing glazed wall with a lightweight metal roof that projects beyond it to provide solar shading. A screen of fritted glass louvres is set forward of the glazing to provide protection both from solar gain and glare. The section was conceived with the aim of encouraging natural ventilation in the reading room throughout the year and has been designed to avoid the need for any fans or mechanical ventilation

Olympic Velodrome and Swimming Pool, Berlin, Germany.
Dominique Perrault.
Domus February 1999.
Photographers: 13a, Werner Huthmacher.
13b, 13c, ADAGP/G. Fessy.

Both the velodrome and swimming pool buildings are set below ground, and as a result the velodrome roof can be seen from the street. It is covered with panels of stainless steel mesh which contrast with the trees that create the adjacent landscaping. The design of the roof is concerned with the texture of the roof covering rather than the joints and connections. The material is fixed in a very straightforward way. The visual effects created by the mesh covering vary with the weather, from a sparkling luminescence in sunshine to a fine woven appearance in cloudy weather.

Meteorite Exhibition Centre, Essen, Germany.
propeller z.
Domus February 1999.
Photographer: 14a, 14b, 14c, 14d, 14f, 14g, Margherita Spiluttini.

This building comprises three parts; an underground exhibition space, an aluminium-clad tube housing a café and service spaces, and a glazed central space which links the pieces together. From the constructional point of view, the metal tube exploits the fact that profiled metal can be curved in one direction to a small radius.

› 20A – 29A

Nine Square Grids House, Japan.
Shigeru Ban.
Domus December 1999.
Photographer: 18a, 18b, 18c, 18d, 18e, 18f, Hiroyuki Hirai.

This house, located on a hillside in the Kanagawa region of Japan, is set out in a square plan divided into nine squares without any partitions. Wall panels slide to open up the external walls. The reflective quality of the floor tiles and the smooth surface of the ceiling reinforce the adaptability of the space rather than the enclosing walls. The structural frame is made from light gauge steel sections.

Port buildings, Chipiona, Cadiz, Spain.
Cruz y Ortiz.
Photograph 19 courtesy of Cruz y Ortiz.

The precast concrete panels in this building are used to reinforce the linear and refined nature of this composition rather than give the vertical emphasis often associated with precast wall construction.

› 20A – 29A

Kew House, Melbourne, Australia.
Sean Godsell Architects.
Domus December 1999.
Photographers: 20a, Trevor Mein.
20b, 20e, Earl Carter.

This house is formed by a single volume which is 9 metres (29ft 6in) wide x 18 metres (59ft) long x 3 metres (9ft 10in) high. The volume projects 5.5 metres (18ft) beyond the edge of a slope. The structural steel frame has been oxidised and sealed with a clear primer. The long north (sun facing) facade and west facade are screened with operable oxidised steel shutters. The straightforward form of construction reinforces the texture of the wall, floor and ceiling surfaces in a strong design for the contemporary Australian home.

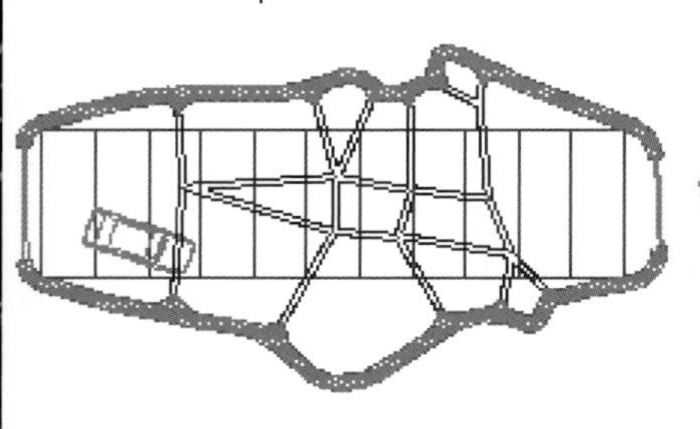

Project for '5 Speed' Prefabricated housing near Eindhoven, Holland.
NOX Architects.
Domus January 2000.
Photographer: 21a, 21b, 21c, 21d, courtesy o
NOX Architects.

This project for prefabricated homes is conceived around five inter-related machines, or machine-like conditions, which are linked metaphorically as in a gearbox. In constructional terms, built forms are created by a machine comprising a set of pistons acting o flexible ribbings connected to a synthetic hood forming a mould which can be shaped as desired. Into this hood is injected high density polyurethane which can form a shell of complex geometry. The shell is then coated with waterproofing layers and finishes. The overall concept is innovative and points to a more flexible form of construction than that used in more conventional forms.

Santa Justa Railway Station, Seville.
Cruz y Ortiz.
Architectural Review June 1992.
Photograph 25 courtesy of Cruz y Ortiz.

The curved entrance wall to this building is a dramatic use of brick. The internal spaces also have long brick walls which act as filters for the strong sunlight and manipulate the daylight across their surfaces. The long, low form of this building is given a strong presence by the use of this material.

Glyndebourne Opera House, Glyndebourne, England.
Michael Hopkins and Partners.
Architectural Review June 1994.
Photographer: 26a, 26b, 26c, 26d, 26e, 26f, 26g, Jim Dunster.

The innovative use of loadbearing brickwork has produced a very strong form that does not require the movement joints associated with modern cavity wall construction.

Kansai International Airport, Japan.
Renzo Piano Building Workshop,
N. Okabe, associate architect.
Architectural Review November 1994.
Photographers: 27a, S. Ishida © Renzo Piano Building Workshop.
27b, Gianni Berengo Gardin © Renzo Piano Building Workshop.
27c, Noriaki Okabe © Renzo Piano Building Workshop.
27d, Yoshio Mata © Renzo Piano Building Workshop.

The airport terminal building has arched trusses supporting the roof structure. Due to the large scale of the project, the need for repetition was important in both the structure and the cladding in order to remain economic while maintaining a high quality of construction.

House in Carreço, Viana do Castelo, Portugal.
João Álvaro Rocha.
Photographer: 22a, 22c, 22d, 22e, 22f, Luís Ferreira Alves.

This house has an interplay of concrete volumes and timber surfaces in an innovative and highly crafted construction.

Rehabilitation Center for Handicapped Children, Perbál, Hungary.
Péter Janesch - Tamás Karácsony.
Domus February 2000.
Photographs 23a, 23b, 23c, courtesy of Péter Janesch - Tamás Karácsony.

This home, run by a foundation, has a very caring attitude towards handicapped children and aims to provide a secure home in a rural setting. The buildings make use of materials which reveal their construction. Children are encouraged to touch these exposed materials. The use of loadbearing brick enhances the sense of straightforwardness of construction.

Conference Centre, San Sebastian, Spain.
Rafael Moneo.
Domus March 2000.
Photographs 24a, 24b, 24c, courtesy of Rafael Moneo.

The conference centre comprises two buildings; an auditorium and a congress hall. An outer glass box encloses the two inner volumes. The outer glass wall is a double skin wall. The outer skin of curved glass is separated from the inner skin of flat glass by a supporting steel structure. The overall effect of the glass is one of white-coloured translucence.

› 30 – 38C

Glass bridge, Rotterdam, Holland.
Dirk Jan Postel, Kraaijvanger o Urbis.
Architectural Review February 1995.
Photograph 28, H.G. Esch.

This all-glass structure is discussed in the Structures chapter.

Design Centre, Linz, Austria.
Thomas Herzog + Partner.
Architectural Review May 1995.
Photograph 29a courtesy of Herzog + Partner.

The loadbearing structure which supports the glazed outer skin consists of a flat arch in steel which spans 76 metres (249ft). The arches are set at 7.2 metre centres (23ft 7in). Secondary beams in the longitudinal direction are set at 2.7 metres (8ft 10in). All these steel components were specially fabricated from cut and welded steel sheet and plate. Longitudinal bracing is provided by rods set diagonally into the curved plane of the roof.

Palau D'Esports Sant Jordi, Barcelona, Spain.
Arata Isozaki and Associates.
Photographer: 30, Yasuhiro Ishimoto.

This sports stadium has a space frame roof with a very elegant overall form.

Bibliotheque Nationale, Paris, France.
Dominique Perrault.
Architectural Review July 1995.
Photographers: 31a, Georges Fessy.
31b, Michel Denance.
31c, Dominique Perrault/ADAGP.

This building consists of a rectangular base building with an L-shaped tower in each of its four corners, where books are stored. The towers have an outer glass wall. Timber screens are set internally to provide solar shading and to protect the book collection. Perforated metal is used to enclose external staircases.

Crawford Municipal Art Gallery, Cork, Ireland.
(EEA) Erick van Egeraat associated architects.
Photograph 35a, Christian Richters.

Hogeschool, Rotterdam, Holland.
(EEA) Erick van Egeraat associated architects.
Photographer: 35b, 35c, Christian Richters.

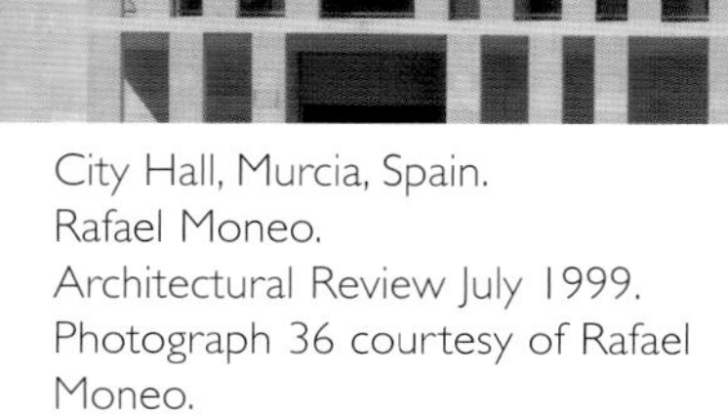

City Hall, Murcia, Spain.
Rafael Moneo.
Architectural Review July 1999.
Photograph 36 courtesy of Rafael Moneo.

The sandstone on the facade facing the city square is used in two ways. Firstly, the sandstone facing the columns is coursed in a way that reveals its thickness at the corners. This contributes to the massive quality of the stone blocks. In contrast, the wall immediately behind the columns does not reveal its thickness around windows. The stone in this wall is used as a textured surface, with the large-scale visual grain that sandstone can provide. This dual use of stone as 'building block' and as 'texture' is used well in this building.

Glass Museum, England.
Design Antenna.
Architectural Review August 1995.
Photographer: 32a, 32b, Dennis Gilbert / View.

This building is discussed in the Structure chapter.

National Laboratory of Veterinary Investigation, Vairão, Portugal.
João Álvaro Rocha, José Manuel Gigante.
Photographer: 33a, 33b, 33c, Luís Ferreira Alves.

The monolithic use of concrete, timber and glass is very much in evidence in this elegant project.

Henley River and Rowing Museum, Henley on Thames, England.
David Chipperfield Architects.
Architectural Review January 1997.
Photograph 34a courtesy of Henley River and Rowing Museum.
Photograph 34c courtesy of David Chipperfield Architects.

In this building, oak is used to form both a timber rainscreen wall and a deck one side of the building. The rainscreen configuration allows the use of large timber sections to be used, where their mass can be seen and touched. This sets a visual contrast with the flatter, thinner appearance of traditional lapped board construction.

Ecotopia, Holland.
Dirk Jan Postel, Kraaijvanger o Urbis.
Photographer: 37a, 37b, Christian Richters.

Cinema, Rotterdam, Holland.
Koen van Velsen Architects.
Architectural Review May 1998.
Photographer: 38a, 38b, 38c, Kim Zwarts.

The profiled polycarbonate sheet used to enclose this building is translucent, allowing the building to glow softly at night when the building is, perhaps, most in use. The lightweight quality of this material allows it to be used in a visually dramatic way with less structural support than would be required for an equivalent design in glass. The profiled polycarbonate sheet is fixed either side of a supporting frame, providing a smooth, continuous surface both inside and outside the building.

› 39A – 47C

Wall-less House, Japan.
Shigeru Ban.
Japan Architect Summer 1998.
Photographer: 39a, 39b, Hiroyuki Hirai.

The floor becomes the wall on one side of this house. The internal fittings have been reduced to an absolute minimum. The bath is reduced to an elegant tub, the toilet is reduced to a single pan. As with the Nine Square Grids House, also by Shigeru Ban, the freedom of the floor plan is central to this striking design.

Educatorium, Utrecht University, Holland.
Rem Koolhaas, Office for Metropolitan Architecture.
Architectural Review March 1999.
Photographer: 40a, 40b, 40c, Hans Werleman, Hectic Pictures.

This building houses two lecture theatres, three examination halls, a canteen and performance area. A large internal glazed wall is secured with bolt fixings that use a simple circular disc instead of a countersunk bolt, giving it greater visual presence. The long sloping glazed external wall is fixed with patch plates giving the wall a delicate and resilient appearance.

Jewish Museum, Berlin, Germany.
Daniel Libeskind.
Architectural Review June 1999.
Photographs 44a, 44b courtesy of Daniel Libeskind.
Photographs 44c, 44d, 44e, 44f, 44g, 44h, courtesy of Jewish Museum, Berlin.

The reinforced concrete frame to this exquisite building is wrapped on the outside with thermal insulation which is in turn covered by zinc sheet. The zinc is fixed back to profiled steel sheeting to provide both a rigid substrate and a zone for ventilating the back of the zinc sheet.

Bank Extension (2 phases), Budapest, Hungary.
(EEA) Erick van Egeraat associated architects.
First phase. Architectural Review July 1995.
Photographer: 45a, 45b, 45c, 45d, Christian Richters.

An existing building has an all-glass roof (glass panels and glass beams) to allow light to penetrate down through the floors. A whale-like form containing a single main space is set into this new roof. A staircase with glass treads beneath the roof aids the passage of daylight down the building.

Bank Extension (2 phases), Budapest, Hungary.
(EEA) Erick van Egeraat associated architects.
Second phase. Architectural Review June 1999.
Photographer: 91a, 91b, 91c, 91d, Christian Richters.

A unitised glazing system is used as the outer layer of a double skin wall. An inner wall of sliding panels provides a zone into which blinds are set and as a passage for return air. Some of the glass is screen printed in a way that is reminiscent of the stone of adjacent buildings.

The British School in the Netherlands, The Hague, Holland.
Dirk Jan Postel, Kraaijvanger o Urbis.
Architectural Review March 1999.
Photograph 41, Jan Derwig

Each classroom in this school has a view of the garden. The south facade is clad in timber panels with a lightweight frame supporting metal louvres set in front of this wall to provide solar shading. The combination of the tactile timber windows and delicate solar shading is very successful in this building.

Patio houses, Esposende, Portugal.
João Álvaro Rocha.
Photographer: 42a, 42b, 42c, 42d, 42e, Luís Ferreira Alves.

The use of aluminium windows in this project is both straightforward and elegant. The sliding doors are used as an elemental part of the facade rather than being an element that fills an opening in a wall.

Photovoltaic Installation, Northumberland Building, University of Northumbria, England.
Architects Journal April 13 1994.
Photograph 43, courtesy of University of Northumbria.

This installation is discussed in the Structure chapter.

› 47D – 77E

Glenlyn Medical Centre, East Molesey, England.
Jim Dunster Architect.
46a, 46b, 46c, 46d, 46e, 46f, courtesy of Jim Dunster Architect.

This building uses a laminated timber frame that provides both the highly crafted finish associated with off-site manufacturing and the warmth of timber construction within a modest budget. The result is a very elegant building.

Olympia Fields Park and Community Center, Olympia Fields, Illinois, USA.
Weiss / Manfredi Architects.
Photographs: 47a, 47b, 47c, courtesy of Weiss / Manfredi Architects.

An existing farm building has been renovated with a new cladding and with additions to serve as a community centre. This has been done in a way that accepts the existing framed structure but has openings carefully cut into the fabric. This is a fine example of using modern building techniques in an existing building to enhance the old construction while working skilfully with the new construction.

Women's Memorial and Education Center, Washington DC. , USA. Weiss/ Manfredi Architects.
Architectural Review August 1999.
Photographer: 47d, 47e, courtesy of Weiss / Manfredi Architects.

The spaces in this building do not reveal themselves to the street as they are set behind an existing wall. Instead, the building is lit from rooflights which also serve as a memorial place for inscriptions. Text is inscribed on glass sheets, which serve as tablets, set slightly above a bolt fixed glazed roof, which can also be accessed from the terrace at roof level. The glazing is fixed to trussed rafters which have a more delicate feel than would an equivalent solid member. Shadows from these rafters wash across the walls in sunlight in an ever-changing pattern. This is a very good use of bolt fixed glazing as an important constructional component.

Rifkind House, Long Island, USA.
Todd Williams, Billie Tsien and Associates.
Architectural Review September 1999.
Photographer: 48, Michael Moran.

This building is based on a timber frame but has glazed corners in order to dissolve the box-like quality of the traditional platform frame. This mixture of techniques allows the building to become a composition of volumes in different materials using durable construction techniques.

House in Fuente del Fresno, Spain.
Paredes/Pedrosa.
Photographer: 52a, Luis Asín.

This house is constructed as a white coloured reinforced concrete structure with oak windows that fill the openings. Natural stone is used to envelop the lower floor. The roof is finished in zinc. Each element of the construction is well-crafted, with the overall construction created by fitting these independent pieces together in a straightforward, elegant way.

Town Hall, Valdemqueda, Spain.
Paredes/Pedrosa.
Architectural Review December 1999.
Photographer: 52b, 52c, 52d Baltanás / Sánchez.

Like the house in Fuente del Fresno, this building is composed of separate, and well detailed, elements which are carefully fitted together. The structure is reinforced concrete, slate and grey painted steel are used on the south facade, with concrete and iroko hardwood used on the north facade.

House in Kensington, London, England.
Porphyrios Associates.
Photograph 53 courtesy of Porphyrios Associates.

This elegantly proportioned example is worthy of the long tradition of this form of window construction. The use of well-constructed timber sash windows continues in northern Europe, where traditional forms of construction continue to flourish. Since these windows are opened by hand, the warm, tactile quality of timber continues to dominate in residential projects.

Government Training Centre, Herne-Sodingen, Germany.
ourda Perraudin.
Architectural Review October 1999.
Photographer: 49a, 49b, 49c, 49d, 49e, 49f, 49g, Christian Richters.

The essential concept for this building is of a glazed box which modifies the external climate, and into which other buildings are inserted. The glazed roof of the 'outer' building is covered with 10,000m$^2$ of photovoltaic (PV) panels. The PV panels are arranged in different densities to create varied lighting effects beneath and also provide solar shading to the glass.

Pump House, London.
Richard Rogers Partnership.
Photographer: 50, Matthew Antrobus.

Housing, Tilburg, Holland.
(EEA) Erick van Egeraat associated architects.
Architectural Review November 1999.
Photographer: 51a, 51b, Christian Richters.

The use of sliding timber panels, made from clear glass and western red cedar, provide a gentle transition between inside and outside, encouraging the social interaction of residents. The timber is weathering to a graceful silver-grey colour.

Cleaning equipment. Lloyds Building, London.
Richard Rogers Partnership.
Photographer: 54a, 54b, 54c, Robert Peebles.

The new Lloyds Building integrates the maintenance and cleaning equipment into the overall architectural design, since these items will always be clearly visible from the surrounding city streets. This has provided the opportunity for this equipment to be specially designed.

Cleaning equipment. Lloyds Building, London.
Richard Rogers Partnership.
Photographer: 77a, 77b, 77c, 77d, 77e, Robert Peebles.

Stainless steel is used to 'wrap' elements in a way that gives them a durable and crisp appearance, while making them appear as elements distinct from the cladding.

Daimler Benz Design Centre, Sindelfingen (Stuttgart), Germany.
Renzo Piano Building Workshop.
Architectural Review January 2000.
Photographer: 55a, 55d, Peter Horn. © Renzo Piano Building Workshop.
55b, Gianni Berengo Gardin © Renzo Piano Building Workshop.

This design centre is composed as a series of linked halls, creating a fan-shaped plan. The roof glazing has blinds slung beneath which provide solar shading and control of daylight in these tall spaces.

House re-modelling, London, England.
Jim Dunster Architect.
Photograph 56a, 56b, 56c, courtesy of Jim Dunster.

The staircase in this project has been well designed and crafted with skills more associated with the best metal stairs. This is partly due to the arrival of economic CNC (computer numerically controlled) machines for timber which can bring the variety and precision of metalwork to that of timber construction.

Yatsushiro Municipal Museum, Kumamoto Prefecture, Japan.
Toyo Ito & Associates.
Japan Architect Summer 1993.
Photographer: 60a, 60b, courtesy of Toyo Ito & Associates.

This museum is in a park setting. The roof structure has a visual lightness that interacts with the park, created by a curved and welded stainless steel roof which has a perforated metal ceiling on its underside.

Shima Art Museum, Mie Prefecture, Japan.
Naito Architect & Associates.
Japan Architect Spring 1994.
Photographer: 61a, 61b, 61c, Kazunori Hiruta, courtesy of Naito Architect and Associates.

The free-standing reinforced concrete walls of this museum allow the roof structure to act as restraints against their outward thrust. This allows the roof structure, which is exposed internally, to have lightweight steel ties. The roof is covered externally with tiles whose undulating texture contrasts with that of the smooth concrete walls.

Shimosuwa Municipal Museum, Lake Suwa, Japan.
Toyo Ito & Associates.
Japan Architect Spring 1994.
Photographer: 62a, 62b, courtesy of Toyo Ito & Associates.

The metal rainscreen panels cast reflections on the water forming what seems like a continuity with the lake.

Library, Seville, Spain.
Cruz y Ortiz.
Photograph 57, courtesy of Cruz y Ortiz arquitectos.

The use of brickwork in this building is one which expresses the external walls as surface planes.

Bracken House, London, England.
Michael Hopkins and Partners.
Photographer: 58a, 58b, Jim Dunster.

The precast concrete frame within this building makes the structure a fundamental component in the architecture, rather than concealing it behind partitions and internal finishes.

Fuencarral Public Library, Madrid, Spain.
Andres Perea Ortega.
Photographer: 59, Javier Azurmendi.

The spaces in this library are a rich mixture of reflective surfaces, which provide a continuity between floors and glazed partitions.

› 66A – 75A

Retirement housing, Yatsushiro, Japan.
Toyo Ito & Associates.
Japan Architect Spring 1995.
Photographer: 63a, 63b, courtesy of Toyo Ito & Associates.

The flat roof is punctured by elliptical voids of varying size and is supported in a one-way spanning system by transverse roof beams. The roof beams are set into the depth of the profiled decking so as not to be visible from below. This emphasises the floating, separate nature of the roof plane.

Reuters Building, London, England.
Richard Rogers Partnership.
Photographer: 64, 65, 104a, 104b, 104c, 104d, 105a, 105b, 106a, 106b, 106c, 106d, 107, Robert Peebles.

This building has crisply assembled components, from the primary structure to the cladding, elevators and internal fittings.

Visitor Centre, Koga, Ibaraki Prefecture, Japan.
Naito Architect & Associates.
Japan Architect Autumn 1998.
Photographer: 66a, 66b, 66c, Kazunori Hiruta, courtesy of Naito Architect and Associates.

This building is timber framed with float glass windows and timber louvre screens fixed externally. All the timber is in Douglas fir. The zinc sheet covering on this roof is not used as a waterproof membrane but primarily as a finish material and as protection to a bitumen-based membrane beneath.

David Mellor Cutlery Factory, England.
Michael Hopkins and Partners.
Photograph 67, Jim Dunster.

Galleria Kameoka, Kameoka City, Japan.
Y. Ikehara and Associates.
Japan Architect Winter 2000.
Photographs 71a, 71b, 71c, 71d, courtesy of Y. Ikehara Architect and Associates.

This building is used primarily as a community education centre. The fire-resisting exposed steel columns have a striking appearance, particularly when seen against the bolt fixed glazing of the external walls.

Kitakame Canal Museum, Japan.
Kengo Kuma and Associates.
Japan Architect Winter 2000.
Photographs: 72a, 72b, 72c, courtesy of Kengo Kuma and Associates.

The grass roof of this museum blends with the adjacent landscaping. The canal footpath passes over this roof. As one approaches the building, the only element that is clearly visible is the stainless steel solar shading over the entrance.

Saitama Prefectural University, Tokyo, Japan.
Riken Yamamoto and Field Shop.
Japan Architect Winter 2000.
Photographs: 73a, 73b, 73c, 73d, Shigeru Ohno, courtesy of Riken Yamamoto and Field Shop.

This is a university for medical, health and welfare education. Laboratories are set out on the ground floor with their roofs used as a continuous planted deck.

Eco iD. Evolution Structural Furniture System.
Michael Hopkins and Partners.
Photographer: 68a, 68b, 68c, 68d, 68e, 68f, Jim Dunster.

This prefabricated kit system for furniture demonstrates the high level of precision which can be achieved with timber.

Hotel Industriel, Paris, France.
Dominique Perrault.
Photographer: 69a, 69b, Robert Peebles.

The unitised glazing system here shows how glass can be used to clad a complete facade without the need for additional infill panels between floors.

Sea-Folk Museum in Mie Prefecture, Japan.
Naito Architect & Associates.
Japan Architect Spring 1993.
Photographer: 70a, 70b, Kazunori Hiruta, courtesy of Naito Architect and Associates.

The form of this roof imitates that of an upturned boat. The central rooflight illuminates the timber roof structure which is exposed internally.

ini-House, Nerima-ku, Tokyo, Japan.
oshiharu Tsukamoto and Momoyo Kaijima
Atelier Bow-Wow, TIT Tsukamoto Lab.
pan Architect Spring 2000
hotographs: 74a, 74b, 74c, 74d, 74e, 74f, 4g, 74h, 74i, 74j, 74k, courtesy of sukamoto and Kaijima.

his innovative construction is built from a ght-gauge steel structural system. The ouse is built from partially prefabricated vall and floor units which are assembled vithout the use of columns or beams. The valls are built from steel channels set at 00mm (1ft 8in) centres. Panels are 8 etres (26ft 3in) high and 2.5 metres (8ft in) wide, fixed directly to the foundations. loor sections are 4.5 metres (14ft 9in) ong, 500mm (1ft 8in) wide and 100mm 4in) deep. No additional finishes such as ry lining (drywall) were used in this build- ng.

Hotel P, Hokkaido, Japan.
Toyo Ito & Associates.
Photograph 75a courtesy of Toyo Ito & Associates.

This building makes extensive use of glass blocks, which form a continuous external wall.

› 76A – 86

› 76A – 86

Furniture House, Yamanashi Prefecture, Japan. Shigeru Ban.
Japan Architect Spring 2000.
Photographer: 76a, 76b, 76c, 76d, 76e, 76f, 76g, 76h, 76i, Hiroyuki Hirai.

This example of innovative construction uses a system of factory-fabricated furniture units which extend from floor to ceiling. These units are used both as structural elements to support the timber roof deck and to organise the internal spaces. The high quality of the off-site fabrication method resulted in the building being assembled on site as a kit of parts. Little material is wasted on site, unlike the high wastage associated with site-based construction. Furniture units are either 700mm (2ft 3in) wide for cupboards or 450mm (1ft 6in) wide for shelving. Both units are 900mm (3ft) wide and 2.4 metres (7ft 10in) high.

B8 Offices, DaimlerChrysler Projekt, Potsdamer Platz, Berlin, Germany.
Richard Rogers Partnership.
Photograph 78, Robert Peebles.

The cleaning gantry in this building allows the underside of the roof to be cleaned and maintained without the need for a cleaning platform to be set up from below.

Cite Internationale, Lyon, France.
Renzo Piano Building Workshop.
Photographer: 82a, 82b, 82c, 82d, Michel Denancé © Renzo Piano Building Workshop.

Terracotta rainscreen facades, developed in part by the Renzo Piano Building Workshop, can be seen as a logical development of brick cavity wall construction. The terracotta exploits the possibilities of the thinness in the outer skin of a cavity wall, turning it into an open jointed set of panels that require no lintels above openings and which can be modulated free of the constraints of movement joints in traditional brickwork.

Belvedere Village, Ascot, England.
Porphyrios Associates.
Photograph 83a courtesy of Porphyrios Associates.

This set of buildings is constructed from different models or 'typologies', each based on the building's function. This follows traditional construction techniques, which are honed differently according to the building type. This approach contrasts with some aspects of modern construction, which seek to homogenise a range of functions into a single building type. In deriving the construction from the best traditional techniques, these buildings demonstrate an elegant continuity and development of the long tradition of site-based construction.

Lloyds Registry of Shipping Building, London, England.
Richard Rogers Partnership.
Photograph: 84, Robert Peebles.

The elevator cars in this building are mounted outside the building envelope. The result is an elegant addition to the external form of the building.

Housing at Nijmegen, Holland.
(EEA) Erick van Egeraat associated architects
in co-operation with Mecanoo architecten.
Photographer: 79a, 79b, 79c, Christian Richters.

The timber structure used to support the walkways contrasts beautifully with the reinforced concrete structure to which it is fixed.

Magdalen College, Oxford, England.
Porphyrios Associates.
Photographs 80a, 80b, 80c, courtesy of Porphyrios Associates.

This beautifully built set of buildings for Magdalen College are laid out in a way which emphasises their overall grouping for those walking around the college. Their presence creates spaces which are very sympathetic to the college-based life at Oxford. From a constructional point of view, the use of loadbearing masonry walls and the best of traditional construction techniques will ensure that the buildings will weather gracefully during their undoubtedly long lives.

Three Brindleyplace, Birmingham, England.
Porphyrios Associates.
Photograph 81a courtesy of Porphyrios Associates.

The loadbearing brick facade is mixed with stone at its base in a composition which also contains sculptured pieces of stone and large openings. It demonstrates that loadbearing brickwork can be used in conjunction with larger areas of glazing than has been seen in other contemporary examples and with a richness of materials.

› 87A – 96B

International Housing Exhibition, Fukuoka, Japan.
Rem Koolhaas, Office for Metropolitan Architecture.
Photographer: 85a, 85b, Kwano, courtesy of the Office for Metropolitan Architecture.

Villa dall'Ava, Saint Cloud, France.
Rem Koolhaas, Office for Metropolitan Architecture.
Photographer: 86, Hans Werleman, Hectic Pictures, courtesy of the Office for Metropolitan Architecture.

Canopy, Parc de La Villette, Paris, France.
Bernard Tschumi Architects.
Photographer: 87a, Peter Mauss/Esto.
87b, J. M. Monthiers

The curving roof forms of this canopy, set in a new Paris park, make use of standard steel-based building components and construction techniques, to provide a rich sculpture for the senses.

Saltwater Pavilion, Neeltje Jans, Holland.
Oosterhuis Associates.
Architectural Review December 1998.
Photographs: 88a, 88b, 88c, 88d, 88e, 88f, courtesy of Oosterhuis Associates.

The Saltwater Pavilion is connected to the Freshwater Pavilion discussed in the Walls chapter.
Like its counterpart, the Saltwater Pavilion sets itself in a relationship with the ground. The building houses an interactive exhibition based on water.

Okanoyama Museum of Art, Hyogo, Japan.
Arata Isozaki & Associates.
Photographer: 92e, Yasuhiro Ishimoto.

Jean-Marie-Tjibaou Cultural Centre, Nouméa New Caledonia.
Renzo Piano Building Workshop.
L'Architecture d'Aujourd'hui December 1996.
Photographers: 93a, Gollings © Renzo Piano Building Workshop.
93b, William Vassal © Renzo Piano Building Workshop.

This building is characterised by toroidal-shaped timber structures, which resemble upturned boats. The craftsmanship in their construction demonstrates future possibilities in naturally ventilated buildings.

88 Wood Street, London, England.
Richard Rogers Partnership.
Photographer: 94a, 94b, 94c, Robert Peebles.

The lifts in this building are fixed to the external wall but are enclosed in a glazed shaft.

The Millennium Dome, London, England.
Richard Rogers Partnership.
Photograph: 89A, 89B, 89C, 89D, 89E, Robert Peebles.

This tent structure is supported on a series of masts which radiate from positions around its centre. The building is both economic and dramatic, creating a roof structure that is also a city landmark.

Carter/Tucker House, Breamlea, Victoria, Australia.
Sean Godsell Architects.
Photographer: 90a, 90b, 90c, Earl Carter

This weekend home is set into a volume 12 metres (39ft 4in) long by 6 metres (19ft 8in) and is arranged on three floors. The outer walls have operable louvered screens to provide solar protection.

Museum of Contemporary Art, Los Angeles, USA.
Arata Isozaki and Associates.
Photographer: 92a, 92b, 92c, 92d, Yasuhiro Ishimoto.

This building has different rooflights to suit the differing daylighting requirements of the spaces beneath. The rooflights are a dramatic feature within each space.

The Menil Collection Museum, Houston, Texas, USA.
Piano and Fitzgerald.
Photographers: 95a, Hickey Robertson © Renzo Piano Building Workshop. 95b, Paul Hester © Renzo Piano Building Workshop.

The solar shading in this building is provided by ferro-cement louvres which are positioned inside the building.

High Museum, Atlanta, USA.
Richard Meier & Partners.
Photographer: 96a, 96b, Ezra Stoller/Esto.

The high quality of finishes in this building is reinforced by the light admitted by the central rooflight.

› 97A – 117

Museum of Contemporary Art, Barcelona, Spain.
Richard Meier & Partners.
Photographers: 97a, 97b, Richard Bryant/Esto, Scott Frances/Esto.

Faculty of Economics and Management, Utrecht, Holland.
Mecanoo.
Architectural Review March 1999.
Photographer: 98a, 98c, 98d, 98e, 98f, 98g, 98h, 98i, Christian Richters.

This building has much of its construction exposed without additional finishes, and this is achieved in a very elegant, straightforward way.

Ted Baker Building, London, England.
Matthew Priestman Architects.
Photographer: 103a, 103b, 103c, 103d, 103e, 103f, 103g, David Grandorge.

This re-modelling of an existing building has many well-crafted components, from the central rooflight to the main stair, ceilings and partitions.

Daggett Solar Farm, California, USA.
Photograph 108, G. Donald Bain, Geo-Images Project, University of California at Berkeley, USA.

This electricity-generating installation is discussed in the Environment chapter.

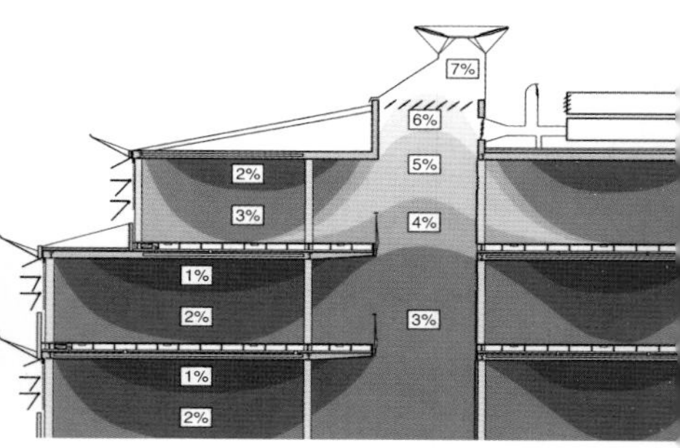


Ionica Building, Cambridge, England.
Illustrations 110a, 110b, courtesy of Battle McCarthy Engineers, London.

Bercy 2 Shopping Centre, Paris, France.
Renzo Piano Building Workshop.
Photographers: 99a, Michel Denancé © Renzo Piano Building Workshop.
99b, Gianni Berengo Gardin © Renzo Piano Building Workshop.
99c, Philippe Ruault © Renzo Piano Building Workshop.

The metal rainscreen panels in this building are used to clothe and protect an economic membrane beneath, which is set onto a curved plywood surface. The rainscreen panels are orientated so as to drain water off each panel directly onto the membrane beneath rather than allow water to run from panel to panel, where it would cause streaking.

School of Decorative Crafts, Limoges, France.
Labfac / Finn Geipel, Nicolas Michelin.
Photographer: 100a, 100b, 100c, 100d, 100e, 100f, 100g, C. Demonfaucon.

This building is treated as a single external envelope which includes the roof as the 'fifth facade'. Both wall and roof panels are interchangeable in a flexible arrangement. Many of the internal elements have similar flexibility.

B6 Apartments, DaimlerChrysler Projekt, Potsdamer Platz, Berlin.
Richard Rogers Partnership.
Photographer: 102a, 102b, 102c, Robert Peebles.

Triangle des Gares, Lille, France.
Jean Nouvel, Emmanuel Cattani et Associes.
Photographer: 112c, 112d, Yasmin Watts.

The roof configuration is discussed in the Roofs chapter.

Solarchis Solar Houses, Japan.
Architecture Studio and Maeta Concrete.
Photographer: 116A, 116B, 116C, courtesy of Solar.

Endesa Headquarters, Madrid, Spain.
Kohn Pedersen Fox Associates (International).
Photographer: 117 courtesy of Kohn Pedersen Fox Associates (International).

Luís Ferreira Alves
Rua da Alegria
Barque Habitacional Do Lima
Entrada 29
Habitacao 3A
4200 Porto
Portugal
Tel: +351 225 505 153

Esto Photographics Inc
222 Valley Place
Mamaroneck
NY 10543
USA
Tel: +1 914 698 4060
Fax: +1 914 698 1033
http://www.esto.com

Dennis Gilbert / VIEW
14 The Dove Centre
109 Bartholomew Road
London NW5 2BJ
England
Tel: +44 20 7284 2928
Fax: +44 20 7284 3617
Email: info@viewpictures.co.uk
www.viewpictures.co.uk

David Grandorge
21 Shepton Houses
Welwyn Street
London E2 0JN
mobile +44 0961 380628

Hiroyuki Hirai
2-12-16-201
Aobadai
Meguro-ku
Tokyo 153-0042
Japan
Tel: +81 3 3462 5504
Fax: +81 3 3462 5535

Holger Knauf
Düsseldorf
Tel: +49 211-775502
Fax: +49 211-775611

Heiner Leiska
Mühlenkamp 6c
D-22303 Hamburg
Germany
Tel: +49 40-27 09 54-10
Fax: +49 40-27 09 54-49

Mitsuo Matsuoka Architectural
Photo Office
2-40-13-1004 Hongo 2-chome
Bunkyo-ku
Tokyo 113-0033
Japan
Tel: +81 3 3818 9217
Fax: +81 3 3818 9256

Michael Moran Photography Inc
371 Broadway, 2nd floor
New York
NY 10013
USA
Tel: +1 212 334 4543
Fax: +1 212 334 3854

# PHOTOGRAPHER CONTACTS

Robert Peebles
111 Hightrees House
Nightingale Lane
London SW12 8AH
Tel: +44 07939 023634.

Matthew Priestman Architects
20-22 Rosebery Avenue
London EC1R 4SX
t +44 207 833 8843
f +44 207 833 8418
www.matthewpriestmanarchitects.com

Christian Richters
Dettenstrasse 1
D-48147 Münster
Germany
Tel: +49 251 27 74 47
Fax: +49 251 27 43 88

Margherita Spiluttini
Schönlaterngasse 8
A-1010 Vienna
Austria
Tel/fax: +43 1 512 59 08
email: spiluttini@eunet.at

Kim Zwarts
Hertogsingel 29a
6211 NC Maastricht
The Netherlands
Tel: +31 43 3250761
Fax: +31 43 3219138

This bibliography lists articles from the international technical press from the years 1990 to 2000. The subject matter of these articles covers general issues about the nature and the future of building construction, from a materials-based standpoint.

AA FILES
no. 31, Summer 1996.
'Truth to material' vs 'the principle of cladding': the language of materials in architecture.

A+T
no. 14, 1999.
Special issue. Materiales sensibles [Sensitive materials].

ARCA
no. 129, September 1998.
Special issue. Superfici [Surfaces].

ARCHITECT (THE HAGUE)
vol. 30, no. 5, May 1999.
Grotere rijkheid met eenvoudiger middelen. Vlies- en metselwekgevels van Rudy Uytenhaak [Greater richness with a choice of materials. Prefabrication and preshaping of materials are the primary methods used].

ARQUITECTURA VIVA
no. 54, May/June 1997.
Berlin de piedra. Revestimientos de fachada: ?variedad en la unidad? [Berlin in stone. Recladding facades: variety in unity?]

ARCHITECTURE (NEW YORK)
vol. 88, no. 4, April 1999.
Wood vs steel: two industries scuffle in a public relations battle for green bragging rights.

ARCHITECTURE NEW ZEALAND
May/June, 1999.
Skin game.

ARCHITECTURE INTERIEURE CREE
no. 289, 1999.
Special issue. Friches / renovation / reconversion [Renovation and conversion of disused buildings].

ARCHITECTURE MOUVEMENT CONTINUITE
no. 20, April 1988.
Facades industrielles [Industrial facades].

ARCHITECTURE TODAY
no. 98, May 1999.
Innovation: a vision of the construction industry twenty years from now predicts far-reaching changes.

ARCHITECTURAL RECORD
October 1995.
The Intelligent Exterior.

ARCHITECTURAL RECORD
vol. 186, no. 4, April 1998.
The nature of green architecture.

ARCHITECTURAL RECORD
vol. 187, no. 4, April 1999.
Mentors. Getting involved in 'green' design.

ARCHITECTURAL RECORD
vol. 187, no. 8, August 1999.
What it means to be green.

ARCHITECTURAL REVIEW
vol. 194, no. 1167, May 1994.
Special issue. Materiality.

ARCHITECTURAL REVIEW
January 1995.
Light spirited.

ARCHITECTURAL REVIEW
vol. 202, no. 1208, October 1997.
Special issue. Nature of materials.

ARCHITECTURAL REVIEW
vol. 207, no. 1239, May 2000.
Special issue. Materiality.

ARCHITEKT
no. 1, January 1997.
Okologische Bewertung von Baustoffen [The ecological assessment of construction materials].

ARCHITEKT
no. 1, January 1998.
Neue Dammstoffe - (k)eine Alternative? [Natural insulation materials - true alternatives (or not) ?]

ARCHITEKT
no. 5, May 1998.
Special issue. Fassade - Gesicht, Haut oder Hulle? [A facade - the face, the skin or the cladding?]

ARCHITEKT
no. 11, November 1999.
Special issue. Im Reich der Erfindung [In the realm of invention].

ARCHITEKT
no. 3, March 2000.
Planung und Ausfuhrung: Glasfassaden [Design and implementation: glass facades].

ARCHITEKTUR (BERLIN)
vol. 40, no. 8, August 1991.
Vorgehangte Fassaden [Facades].

ARCHITHESE
vol. 22, no. 3, May/June 1992.
Special issue. Nur Fassade [Facades].

ARCHITHESE
vol. 26, no. 5, September / October 1996.
Special issue. Masse - Korper - Gewicht [Mass - body - weight].

BAUMEISTER
vol. 93, no. 8, August 1996.
Neue Baustoffe [New materials].

BAUMEISTER
vol. 95, no. 7, July 1998.
Special issue. Neue Oberflachen - Material als architektonisches Programm [New surfaces - materials as architectural programme].

BAUMEISTER
vol. 97, no. 9, September 2000.
Special issue. Baumaterial - der Stoff, aus dem die Raume sind [Building materials].

BAUWELT
vol. 87, no. 16, April 26, 1996.
Aufs Ganze gehen Glasfassaden [Go all the way with glass facades].

BAUWELT
vol. 87, no. 43/44, November 22, 1996.
Auf dem Prufstand [From the test bed].

BAUWELT
vol. 88, no. 7, February 1997.
Naturstein [Natural stone].

BAUWELT
vol. 91, no. 3, January 21, 2000
Von den Materialien [On materials].

CASABELLA
June 1996
Le pareti ventilate [Ventilated walls].

DETAIL
vol. 30, no. 4, August/September 1990.
Special issue. Fassaden-Konstruktionen [Facade systems].

DETAIL
vol. 33, no. 3, June/July 1993.
Special issue. Metallfassaden [Metal facade constructions].

DETAIL
vol. 35, no. 4, August/September 1995.
Special issue. Dachtragwerke [Roof structures].

DETAIL
vol. 36, no. 4, June 1996.
Special issue. Fassade, Fenster [Facades and fenestration].

DETAIL
vol. 38, no. 1, January/February 1998.
Special issue. Einfaches Bauen [Simple forms of building].

DETAIL
vol. 38, no. 7, October/November 1998.
Special issue. Fassaden [Facades].

DETAIL
vol. 40, no. 1, January/February 2000.
Special issue. Bauen mit Holz [Timber construction].

DETAIL
vol. 40, no. 5, July/August 2000.
Special issue. Flache Dacher [Flat roof construction].

DEUTSCHE BAUZEITSCHRIFT
vol. 38, no. 4, April 1990.
Fassaden mit Stahlbauteilen [Facades with steel elements].

DEUTSCHE BAUZEITSCHRIFT
vol. 40, no. 8, August 1992.
Die Dreidimensionalitat der Fassaden-Verschraubung [The three dimensionality of screwing together facades].

DEUTSCHE BAUZEITSCHRIFT
vol. 45, no. 5, May 1997.
Auswahl von Baumaterialien. Gegenwartige trends und zukunftiges Potential [Choice of building materials. Contemporary trends and future potential].

DEUTSCHE BAUZEITSCHRIFT
vol. 46, no. 3, March 1998.
Dacher und Fassaden in Titanzink: Wirtschaftliche und gestalterische Alternativen mit Metall [Roofs and facades in titanium zinc: research and design alternatives with metal].

DEUTSCHE BAUZEITUNG
vol. 130, no. 1, January 1996.
Materialien [Materials].

DEUTSCHE BAUZEITUNG
vol. 131, no. 10, October 1997.
Chemie im Schafspelz? Dammstoffe aus Altpapier oder Naturfasern - (k)eine Alternative? [Insulating material from old paper or natural fibre - an alternative/no alternative?]

DEUTSCHE BAUZEITUNG
vol. 133, no. 8, August 1999.
Fluch und Segen [PVC recycling in building materials].

DOMUS
no. 756, January 1994.
Materiali e progetto [Materials and design].

DOMUS
no. 760, May 1994.
La vita dei prodotti: note sul progetto ecologico [Product life: notes on ecological design].

DOMUS
no. 789, January 1997.
Leapfrog - progettare la sostenibilita [Leapfrog - designing sustainability].

DOMUS
no. 801, February 1998.
Materialita [Materiality].

DOMUS
no. 818, September 1999.
Special issue. Impara dalla natura [Learning from nature].

ECO
vol. 38, no. 7, October/November 1998.
For green, try blue.

ECO DESIGN
vol. 6, no. 2, 1998.
Special issue. Eco design around the world.

GLASFORUM
vol. 39, no. 3, June 1989.
Tendenzen der Glasarchitektur: Glasfassadenkonzepte aus England [Trends in glass architecture: glass facades from England].

MONITEUR ARCHITECTURE AMC
no. 22, June 1991.
Details: les facades metalliques [Details: metal facades].

MONITEUR ARCHITECTURE AMC
no. 42/43, June/July 1993.
Facades legeres (2) [Lightweight facades (2)].

MONITEUR ARCHITECTURE AMC
no. 49, March 1994.
Ingenierie des facades [Facade engineering].

MONITEUR ARCHITECTURE AMC
no. 70, April 1996.
Facades: les bardages metalliques [Facades: metal cladding].

MONITEUR ARCHITECTURE AMC
no. 75, November 1996.
Facade: panneaux de bois [Facades: timber panels].

MONITEUR ARCHITECTURE AMC
no. 83, October 1997.
Les facades.

PROGRESSIVE ARCHITECTURE,
February 1994.
What makes a good curtain wall?

PROGRESSIVE ARCHITECTURE
March 1994.
The ends of finishing.

PROGRESSIVE ARCHITECTURE
June 1994.
Amazing glazing.

PROGRESSIVE ARCHITECTURE
December 1995.
Cladding"s ticking time-bombs.

QUADERNS
No.202, 1994.
La flexibilidad come dispositivo [Flexibility as a device].

RECUPERARE
vol. 10, no. 9, November/December 1991.
La valutazione delle facciate ventilate [The evaluation of ventilated facades].

RECUPERARE EDILIZIA DESIGN IMPIANTI
vol. 2, no. 8, November/December 1983.
Ventilated facades in building rehabilitation.

SOLAR ENERGY
June 1996.
Numerical study of a ventilated facade panel.

TECHNIQUES & ARCHITECTURE
December-January 1994.
Facade Legere et menuiserie metallique [Lightweight facades and metal joinery].

TECHNIQUES & ARCHITECTURE
no. 413, April/May 1994.
La dimension ecologique [The ecological dimension].

TECHNIQUES & ARCHITECTURE
no. 422, October/November 1995.
Revetements de facade [Covering facades].

TECHNIQUES & ARCHITECTURE
no. 430, February/March 1997.
Beton en parement [Concrete as adornment].

TECHNIQUES & ARCHITECTURE
no. 447, February/March 2000.
CFAO [Conception fabrication assistee par ordinateur] et chantier naval: Le Georges, restaurant, Centre Pompidou, Paris [Computer aided manufacture and the naval yard: Georges Restaurant at the Pompidou Centre].

TECHNIQUES & ARCHITECTURE
no. 448, April/May 2000.
De la matiere [Material matters].

WORLD ARCHITECTURE
no. 33, 1995.
Smart cars versus smart facades.

WORLD ARCHITECTURE
no. 47, June 1996.
The art of glass.

WORLD ARCHITECTURE
no. 48, July/August 1996.
Cladding and roofing.

WORLD ARCHITECTURE
no. 84, March 2000.
Race to the finish.

F



G



H



I



L

ANDREW WATTS conceived the book, wrote the text, hand-drew the illustrations and set out the pages. Andrew Watts has nearly 20 years' experience working as an architect specialising in facade detailing on international projects with a wide range of construction technologies. He was a project architect for Jean Nouvel in Paris, working on some of his most notable buildings. Andrew Watts has a Masters Degree from the University of Cambridge in Interdisciplinary Design. More recently as a facade specialist, he has worked on some well-known projects around the world including Federation Square, Melbourne and the Millennium Bridge, London. He presented a paper on passive and low energy design to the PLEA Conference 2000. Andrew Watts is currently working on a companion volume in the Modern Construction series. Andrew can be contacted at: awatts@newtecnic.com.

The text was technically-edited by STEVEN WRIGHT in New York, who was a project manager/architect with both Santiago Calatrava and Rafael Viñoly and was also an Editor of Perspecta magazine at Yale University. The authors maintain links with the universities of Cambridge, Virginia and Yale.

YASMIN WATTS designed the book. She undertook both the illustrations on CAD and the graphic design of the layouts with Andrew Watts. Yasmin Watts was an architect at the Renzo Piano Building Workshop in Paris, where she worked on the Cultural Centre in New Caledonia, and the Cité Internationale in Lyon, France. Yasmin can be contacted at: ywatts@newtecnic.com.

The Environment Chapter was written in collaboration with PIERS HEATH, who is a Principal Mechanical Engineer at Battle McCarthy in London, specialising in the design of low energy environmentally responsive buildings.

The essay on embodied energy was written in collaboration with NAZAR GEORGIS, who is a Structural Engineer leading structural design at Battle McCarthy in London. He specialises in light structures with a low environmental impact.

The German language edition of this book has been translated by NORMA KESSLER, who has adapted the book specifically for use in the German-speaking countries, adding DIN standards and giving it a greater focus for use both in those countries and around the world.

DAVID MAROLD is Editor for Architecture and Philosophy at Springer-Verlag in Vienna. He has driven this book from a set of basic layouts to a completed book. He has a passion for books and their design, ranging from their wider content to the quality of print paper.

# AUTHOR'S THANKS

I would like to thank Cormac Deavy of Ove Arup & Partners for his assistance with the Structure chapter, and my mother, Mrs Helena Watts, for proof reading in the final stages. I would also like to thank the following people who have been involved with this project : Craig Anders, Madeleine Brown of the History of Art and Architecture Library at the University of Cambridge, Ed Ford, Jon Linton, Peter MacKeith, Tim Quick, Catherine Redgwell, Neville Surti, Patricia Westerburg, Nick Wood.

I would like to thank all the following architects providing images of their work: Hiroshi Araki at Tadao Ando Architect & Associates, Simon Gardiner at Allgood plc, Askim Lantto Architects, Yoko Watanabe at Shigeru Ban Architects, Peter Wilson at Architekturbüro Bolles Wilson, Nicole Woodman at David Chipperfield Architects, Joaquín Pérez-Goicoechea at Cruz y Ortiz arquitectos, Brent Richards at Design Antenna, Brenda Kamphuis at Erick van Egeraat Associated Architects, Jörn Hustedt at Architekten von Gerkan, Marg und Partner, Hayley Franklin at Sean Godsell Architects, Franziska v. Wedel at Thomas Herzog + Partner, Hideki Yamaguchi at Y. Ikehara Architect & Associates, Dr. Jan Esche at Ingenhoven, Overdiek und Partner, Yoko Sugasawa at Arata Isozaki & Associates, Mariko Nishimura at Toyo Ito & Associates, Peter Janesch and Tamas Karacsony, Jim Dunster and Robert Peebles of Kohn Pedersen Fox International, Dirk Jan Postel at Kraaijvanger-Urbis Architects, Kengo Kuma & Associates, Labfac Architectes, Louise Ashcroft at Daniel Libeskind Architects, Hanneke Hollander at Mecanoo Architekten, Lisa Green at Richard Meier & Partners Architects, Anna Moça at Morphosis, Rafael Moneo Arquitectos, Kazunori Hiruta, Naito Architect & Associates, Lars Spuybroek at NOX Architects, Jan Knikker at the Office for Metropolitan Architecture, Andre Houdart at Oosterhuis, Andres Perea Ortega, Paredes / Pedrosa Arquitectos, Linda Coeuret at Gilles Perraudin Architecte, Gaëlle Lauriot-Prévost at Dominique Perrault Architecte, Chiara Casazza at Renzo Piano Building Workshop, Nicky Walker at Porphyrios Associates Architects, Matthew Priestman at Matthew Priestman Architects, Philipp Tschofen at propeller z, João Álvaro Rocha and José Manuel Gigante, Tina Wilson at Richard Rogers Partnership, Liz Kim, Bernard Tschumi Architects, Dr Nicola Pearsall at the University of Northumbria, Koen van Velsen Architects, Lauren Crahan at Weiss Manfredi Architects, Todd Williams, Billie Tsien and Associates, Yoshiharu Tsukamoto, Minako Ueda at Riken Yamamoto & Field Shop.

I would like to thank all the following photographers for providing images of their work: Christian Richters, Luís Ferreira Alves, Javier Azurmendi, Jim Dunster, Esto Photographics Inc, Dennis Gilbert / VIEW, David Grandorge, Hiroyuki Hirai, Holger Knauf, Heiner Leiska, Mitsuo Matsuoka, Michael Moran, Robert Peebles, Margherita Spiluttini, Kim Zwarts.

Information on the strength properties of stone has been provided by CERAM Building Technology, as a result of DETR Partners in Innovation Project (Department of the Environment Transport and the Regions in the United Kingdom).

Information on Solarchis Solar Houses, by the Solar Architecture Studio and Maeta Concrete Industry Ltd in Japan, has been provided by the CADDET Centre for Renewable Energy, in England.